T0338048

Statistical Thermodynamics

Statistical Thermodynamics

Basics and Applications to Chemical Systems

Iwao Teraoka
Tandon School of Engineering
New York University
Brooklyn, NY, US

This edition first published 2019
© 2019 John Wiley & Sons Inc.

Registered Office
John Wiley & Sons, Inc., 111 River Street, Hoboken, NJ 07030, USA

Editorial Office
111 River Street, Hoboken, NJ 07030, USA

For details of our global editorial offices, customer services, and more information about Wiley products visit us at www.wiley.com.

Wiley also publishes its books in a variety of electronic formats and by print-on-demand. Some content that appears in standard print versions of this book may not be available in other formats.

Library of Congress Cataloging-in-Publication Data

Names: Teraoka, Iwao, author.
Title: Statistical thermodynamics : basics and applications to chemical systems / Iwao Teraoka.
Description: First edition. | Hoboken, NJ : John Wiley & Sons, 2019. | Identifiers: LCCN 2018036497 (print) | LCCN 2019002835 (ebook) | ISBN 9781119375258 (Adobe PDF) | ISBN 9781119375289 (Epub) | ISBN 9781118305119 (hardcover)
Subjects: LCSH: Statistical thermodynamics.
Classification: LCC QC311.5 (ebook) | LCC QC311.5 .T47 2019 (print) | DDC 536/.70727–dc23
LC record available at https://lccn.loc.gov/2018036497

Cover Design: Wiley
Cover Image: Courtesy of Erika Teraoka

Set in 10/12pt WarnockPro by SPi Global, Chennai, India

Printed in the United States of America

V10008041_020519

This book is dedicated to my wife, Sadae Teraoka. Without her encouragement, I would not have been able to finish it.

Contents

Preface

This book was born out of my long-time wish that chemistry and chemical engineering students should learn statistical thermodynamics using a book written for them. Many of the books on that subject were written for physics students, and the contents are not appropriate for those who deal with molecules. There are a few good books that nonphysics students can rely on, but they look old-fashioned.

Many research papers in chemistry, chemical engineering, materials science, biochemistry, and biophysics are written using the concepts of statistical thermodynamics, whether the authors of the papers are aware of that or not. The papers are mostly about molecules, and that is why molecular-level description of quantities observed in experiments has a big presence in these areas of science. Statistical mechanics offers a right tool for that purpose.

Admittedly, the concepts of statistical mechanics are not easy to grasp. The fundamental hypotheses are philosophical, and they are translated into equations. By applying the tools of statistical mechanics to different thermodynamic systems and solving practice problems, you will be able to get a hang of the fundamental concepts. All the chapters, except for Chapter 1, have practice problems at the end. It is essential to solve them. You can find the answers at Wiley's web site: www.wiley.com/go/Teraoka_StatsThermodynamics.

One of the axioms in Confucius' famous book, the Analects, is shown below. Its translation is as follows: If you just learn and do not think, you will be left in the dark. If you just think and do not learn, you are on a precarious footing. My translation may not be authentic, but those who study statistical mechanics should be always reminded of this axiom.

學而不思則罔思而不學則殆

There are different ways to introduce the concepts to early-stage learners. I like the method adopted in Atlee Jackson's textbook titled Equilibrium Statistical Mechanics [1]. The partition function is derived elegantly in a simple method. Chapters 2–7 of this book follow mostly the Atlee Jackson's book.

Some of the practice problems are borrowed from a Japanese book, Statistical Mechanics [2]. Its English translation is available [3].

The prerequisites for this book are the first semester of undergraduate physical chemistry and some math (calculus and linear algebra). In calculus, series and derivatives are far more important for statistical mechanics than the integrals are. In linear algebra, 2×2 matrices are all that are required. Familiarity with probability theory will be a big help. If you are not strong in math, it will be a good idea to go through Appendix A, first. As the thermodynamics is the best place to practice partial differentiation, statistical mechanics is the best place to use Taylor expansion.

The construction of chapters in this book is shown here. The arrows indicate relationships of prerequisites. Chapters 1–7 cover all the concepts and tools of statistical mechanics. The remaining chapters are applications to chemical systems.

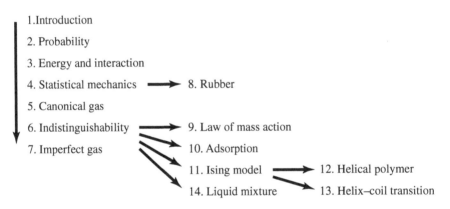

1. Introduction

2. Probability

3. Energy and interaction

4. Statistical mechanics ⟶ 8. Rubber

5. Canonical gas

6. Indistinguishability ⟶ 9. Law of mass action

7. Imperfect gas   10. Adsorption

        11. Ising model ⟶ 12. Helical polymer

        14. Liquid mixture   13. Helix–coil transition

An appendix is a collection math formulas necessary for deriving the equations in this book. Physical constants are listed in Symbols and Constants.

This book does not include pairing of single-strand DNA, liquid crystals, surface phenomena (other than adsorption), and liquid–vapor equilibria. I want to cover them in future revisions.

This book does not cover quantum statistical mechanics. The latter is most prominently applied to a system of electrons in metals and semiconductors. If you need to learn quantum statistical mechanics, you should read one of the good books written for physics students [4].

References

1 Jackson, E.A. (2000). *Equilibrium Statistical Mechanics*. Mineola, NY: Dover Publications. Originally, Prentice Hall, Englewood Cliffs, NJ (1968).

2 Kubo, R., Ichimura, H., Usui, T., and Hashitsume, N. (1961). *Daigaku Enshu: Netsugaku, Tokei Rikigaku*. Tokyo, Japan: Shokabo.

3 Kubo, R. (1965). *Statistical Mechanics: An Advanced Course with Problems and Solutions*. Amsterdam, Netherlands: Elsevier.

4 For example, Reichel, L.E. (1980). *A Modern Course in Statistical Physics*. Austin, TX: University of Texas Press.

Acknowledgments

The photograph on the cover of this book was taken by Erika Teraoka in the Wiener Zentralfriedhof (Vienna Central Cemetery).

About the Companion Website

This book is accompanied by a companion website:
www.wiley.com/go/Teraoka_StatsThermodynamics

The website includes:
- Answers are available to the questions in the main text book.

Symbols and Constants

The numbers are from Physicists' Desk Reference [1] and other sources.

Constant	Symbol	Value
Avogadro's number	N_A	$6.022\,137 \times 10^{23}$ mol^{-1}
Boltzmann constant	k_B	$1.380\,658 \times 10^{-23}$ J K^{-1}
Vacuum permittivity	ε_0	$8.854\,188 \times 10^{-12}$ F m^{-1}
Gas constant	R	$8.314\,510$ J (K·mol)$^{-1}$
Planck constant	h	$6.626\,070 \times 10^{-34}$ J s
Speed of light in vacuum	c	$2.997\,925 \times 10^8$ m s^{-1}
Elementary charge	e	$1.602\,177 \times 10^{-19}$ C

Reference

1 Anderson, H.L. and Cohen, E.R. (1989). General section. In: *A Physicist's Desk Reference* (ed. H.L. Anderson). New York, NY: American Institute of Physics.

1

Introduction

Section 1.1 looks at the similarities and differences between classical thermodynamics and statistical thermodynamics. Then, in Section 1.2, we see several examples of phenomena that are beautifully described by statistical mechanics. Section 1.3 lists practices of notation adopted by this book.

1.1 Classical Thermodynamics and Statistical Thermodynamics

Classical thermodynamics, when applied to a closed system, starts with two fundamental laws. The first law of thermodynamics accounts for a balance of energy:

$$\mathrm{d}'Q + \mathrm{d}'W = \mathrm{d}U \tag{1.1}$$

where the system receives heat $\mathrm{d}'Q$ and work $\mathrm{d}'W$ to change its internal energy by $\mathrm{d}U$ (see Figure 1.1). The prime in "d'" indicates that the quantity may not be a thermodynamic variable, i.e. not expressed as a total derivative. When the volume of the system changes from V to $V + \mathrm{d}V$, $\mathrm{d}'W = -p\,\mathrm{d}V$, where p is the pressure.

The second law of thermodynamics expresses $\mathrm{d}'Q$ by a thermodynamic variable, but only when the change is reversible:

$$\mathrm{d}'Q = T\,\mathrm{d}S \tag{1.2}$$

where T is the temperature. The second law introduces the entropy S.

In classical thermodynamics, we try to find relationships between macroscopic variables, S, T, U, V, and p. The equation of state is one of the relationships. We also learned different types of energy, specifically, enthalpy H, Helmholtz free energy F, and Gibbs free energy G. These measures of energy are convenient when we consider equilibria under different constraints. For example, at constant T and V, it is F that minimizes when the system is at equilibrium. Certainly, we can always maximize S of the universe (system + the

Statistical Thermodynamics: Basics and Applications to Chemical Systems, First Edition. Iwao Teraoka.
© 2019 John Wiley & Sons, Inc. Published 2019 by John Wiley & Sons, Inc.
Companion website: www.wiley.com/go/Teraoka_StatsThermodynamics

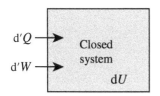

Figure 1.1 A closed system received heat d$'Q$ and work d$'W$ from the surroundings to change its internal energy by dU.

surroundings), but knowing the details of the surroundings is not feasible or of our concern. Rather, we want to focus on the system, although it is the maximization of the entropy of the universe that dictates the equilibrium of the system. People have devised F for that purpose. If we minimize F of the system under given T and V, we are equivalently maximizing S of the universe. Likewise, G minimizes when the system's temperature and pressure are specified.

As you may recall, classical thermodynamics does not need to assume anything about the composition of the system – whether it is a gas or liquid, what molecules constitute the system, and so on. The system is a continuous medium; and it is uniform at all length scales, if it consists of a single phase. In other words, there are no molecules in this view.

Statistical thermodynamics, in contrast, starts with a molecule-level description of the system – what types of molecules make up the system, whether interactions are present between molecules, and, if they are, how the interaction depends on the distance between molecules, and so on. Furthermore, statistical thermodynamics specifies microscopic states of the molecules, for example, their positions and velocities. If the molecules are monatomic, specifying the positions and velocities may be sufficient for our purposes. When the molecules are diatomic, however, we need to specify the states of rotation and vibration as well. If the molecule is polyatomic, specifying these states becomes more complicated. Even for a system of monatomic molecules, specifying the positions and velocities requires an astronomical number of variables. Typically, the number is close to Avogadro's number. Listing and evaluating all the variables is a daunting task. Fortunately, evaluating thermodynamics variables such as U, F, and G does not require all the details. It is rather the averages of the microscopic variables that count in evaluating the thermodynamic variables, and that is where statistical thermodynamics comes in.

1.2 Examples of Results Obtained from Statistical Thermodynamics

Here, we take a quick look at some of the results of applying statistical thermodynamics to different systems.

1.2.1 Heat Capacity of Gas of Diatomic Molecules

Figure 1.2 shows how the molar heat capacity C_V/n of a gas consisting of diatomic molecules changes with temperature T. There are two characteristic

Figure 1.2 Molar heat capacity C_V/n of a gas consisting of diatomic molecules, plotted as a function of temperature T. At T around Θ_{rot}, the characteristic temperature of rotation, C_V/n increases from $\frac{3}{2}R$ to $\frac{5}{2}R$; and at around Θ_{vib}, the characteristic temperature of vibration, further increases to $\frac{7}{2}R$.

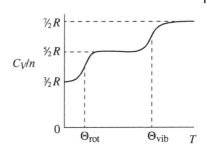

Table 1.1 Characteristic temperature of rotation, Θ_{rot}, and characteristic temperature of vibration, Θ_{vib}, for some diatomic molecules.

Molecule	Θ_{rot} (K)	Θ_{vib} (K)
H_2	87.6	6331
N_2	2.88	3393
O_2	2.08	2239

temperatures Θ_{rot} and Θ_{vib} (rotation and vibration). Each diatomic molecule has its own Θ_{rot} and Θ_{vib}, and some of them are listed in Table 1.1.

The molar heat capacity is $\frac{3}{2}R$ at $T \ll \Theta_{rot}$, where R is a gas constant. We see this range only for H_2; for other gases, the boiling point is above Θ_{rot}. As T increases and surpasses Θ_{rot}, C_V/n increases to $\frac{5}{2}R$. There is a broad range of temperature that gives a nearly constant value of C_V/n before it increases to $\frac{7}{2}R$ as T exceeds Θ_{vib}. For most diatomic molecules that are gas at room temperature (RT), $\Theta_{rot} \ll RT \ll \Theta_{vib}$, and that is why a gas of diatomic molecules has $C_V/n = \frac{5}{2}R$.

1.2.2 Heat Capacity of a Solid

Figure 1.3 depicts the molar heat capacity C_V/n of a molecular solid (nonionic), plotted as a function of temperature T. At low temperatures, $C_V/n \sim T^3$, and increases to a plateau value of $3R$ as T increases. Vibration in a lattice (crystal) accounts for this heat capacity. Einstein attempted to explain the heat capacity in his 1905 paper [1]. His statistical model correctly predicted $3R$, but not T^3. It is Debye who explained the $\sim T^3$ dependence by improving the Einstein model [2].

1.2.3 Blackbody Radiation

Anything with a temperature $T > 0$ radiates. A blackbody is a perfect emitter of the radiation (light) and also a perfect absorber. The radiation has different

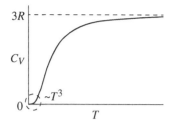

Figure 1.3 Heat capacity C_V of a molecular solid, plotted as a function of temperature T. At close to $T = 0$, $C_V \sim T^3$. With an increasing T, C_V approaches a plateau value of $3R$.

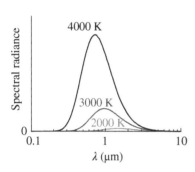

Figure 1.4 Irradiance of a blackbody at different temperatures, per wavelength, is plotted as a function of wavelength λ. The temperature is indicated adjacent to the curve.

wavelength components and is not visible unless the wavelength falls in the visible range, 450–750 nm. When the radiation intensity is plotted as a function of T, the curve peaks at some wavelength λ_{peak} (see Figure 1.4). With an increasing T, λ_{peak} moves to a shorter wavelength, and the peak intensity increases. Stars exhibit different colors, and it is due to temperature differences. The radiation from the sun peaks at around 500 nm (blue–green), since its surface temperature is around 5800 K. A red star has a lower temperature, and a white star ($\lambda_{peak} \cong 300$ nm) has a higher temperature.

The λ_{peak} decreases as $\sim T^{-1}$ as T increases, which is called Wien's displacement law, discovered in 1893. The profile of the spectrum has tails at both ends. The long-wavelength tail follows $\sim \lambda^{-4}$, and short-wavelength tail $\sim e^{-const./\lambda}$. The long-wavelength tail was explained using classical electromagnetism, but it could not explain the short-wavelength tail, or the Wien's law. Max Planck proposed a photon hypothesis – light consists of energy particles called photons, each carrying energy reciprocally proportional to λ – in 1900 [3]. He succeeded in explaining the whole radiation spectrum.

1.2.4 Adsorption

This example is more chemical than are the preceding examples. When a clean surface (glass, graphite, etc.) is exposed to a vapor, some molecules adsorb onto the surface (Figure 1.5a). The surface coverage θ (fraction of surface covered with the molecules) increases with an increasing partial pressure p of the vapor

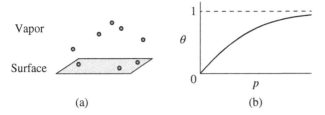

(a)   (b)

Figure 1.5 (a) Surface is in contact with a vapor, and some of the molecules adsorb onto the surface. (b) Surface coverage θ, plotted as a function of the partial pressure p of the vapor.

(see Figure 1.5b). The plot is called an adsorption isotherm, as it is taken at a constant temperature.

The adsorption phenomenon can be explained by reaction kinetics, but statistical thermodynamics provides a molecular-level description of the isotherm. For example, we can estimate, from the experimentally obtained isotherm, the cohesive energy of adsorption per molecule.

1.2.5 Helix–Coil Transition

A polypeptide is a polymer of identical amino acid residues. For example, poly(L-lysine) is a polymer of L-lysine. The polypeptide adopts a helix conformation, a random-coil conformation, or a mixture of them (a part of the polymer chain is in helix conformation), see Figure 1.6(a). Which conformation the polymer takes depends on the environment such as temperature and solvent. Figure 1.6 (b) is a sketch for a plot of percent helix as a function of temperature. The polypeptide is benzyl ester of poly(glutamic acid), and therefore soluble in a polar organic solvent. At low temperatures, nearly all of the polymer is in a coil conformation, and gradually changes to an all-helix conformation as the temperature rises. It may appear counterintuitive that the

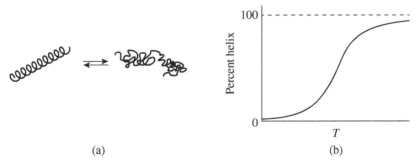

(a)   (b)

Figure 1.6 (a) Polypeptide chain in helix conformation, in equilibrium with a chain in coil conformation. (b) Percent helix in a polypeptide chain, plotted as a function of temperature, in a polar organic solvent.

ordered state of the polypeptide is seen at high temperatures, rather than at low temperatures. The helix conformation is made possible by hydrogen bonds between a donor (H atom in an amide bond) and an acceptor (O atom in the amide bond) several residues away along the chain. If there is a mechanism that supersedes this intrachain hydrogen bond, then the chain may adopt a coil conformation. The latter can occur if the solvent molecules provide a stronger hydrogen bond to the H atoms and O atoms of the amide bonds. It is a competition between the two types of hydrogen bonds that gives rise to the inverted temperature dependence of the percent helix.

1.2.6 Boltzmann Factor

You would have learned about the Boltzmann factor, $e^{-\Delta E/(RT)}$, where ΔE is the energy difference per mole, without being shown its derivation. The Boltzmann factor appears in many different situations. It appears, for example, in the barometric formula, $p(h) = p(0)e^{-Mgh/(RT)}$, where $p(h)$ is the pressure at altitude h, M is the molar mass of the gas, and g is the gravitational constant. The Debye–Hückel theory for electrolyte solutions and the Gouy–Chapman theory for colloidal suspensions also use the Boltzmann factor. In nuclear magnetic resonance (NMR), the population of an up spin $(+\frac{1}{2})$ of a magnetic dipole μ in magnetic field B with respect to the down spin $(-\frac{1}{2})$ is $\exp(N_A\mu B/RT)$, where N_A is the Avogadro's number.

Statistical thermodynamics derives the Boltzmann factor from fundamental hypotheses. We learn the hypotheses and the derivation in Chapter 4.

1.3 Practices of Notation

This section lists some practices used in this book.

(1) *Symbols*: This book uses italic symbols for variables; roman-typefaced symbols are not variables. For example, the base of the natural logarithm is e, not e. Likewise, the circumference ratio is π, not π.

(2) $O(x^n)$ *represents a quantity proportional to* x^n: It may include higher or lower order terms. For example, the Taylor expansion of e^x is $1 + x + \frac{1}{2}x^2$, up to $O(x^2)$.

(3) *Limit and asymptote*: We strictly distinguish them. We often consider behaviors of thermodynamic functions at low temperatures and at high temperatures. We evaluate the limiting value of a function $f(T)$ as $T \to 0$, and this is the low-temperature limit. We also evaluate the dominant term of the function when T is low and when T is high. Such a term retains the temperature dependence and is called an asymptote.

Figure 1.7 Plot of $y = f(x) = \ln(1 + \cosh x)$ (solid line) and its large-x asymptote (dashed line).

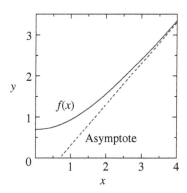

For example, the molar heat capacity C_V/n of a molecular solid has the low-temperature limit of zero, but its low-temperature asymptote is $C_V/n \sim T^3$. The high-temperature limit of C_V/n is $3R$.

Let us consider the following function $f(x)$ for a practical example.

$$f(x) = \ln(1 + \cosh x), \quad x \geq 0 \tag{1.3}$$

The small-x limit is $\ln 2$. There is no large-x limit, as $f(x) \to \infty$ as $x \to \infty$. However, we can get the large-x asymptote:

$$f(x) \cong x - \ln 2 \tag{1.4}$$

Figure 1.7 shows a plot of $y = f(x)$ and its large-x asymptote. The plot of $y = f(x)$ runs close to the asymptote that is a straight line.

The leading term of the large-x asymptote is x, but including the constant makes the asymptote a better approximation. The asymptote, Eq. (1.4), neglects $O(e^{-x})$.

2

Review of Probability Theory

In this chapter, we first review the concepts of probability, typical probability distributions relevant to statistical mechanics, and a few quantities that characterize each probability distribution. Then, we learn the concept of uncertainty, since it is related to the entropy discussed in Chapter 4.

2.1 Probability

As is the case with most probability theory books, we start with considering the tossing of a coin. Its outcome is either "head" or "tail." We do not know the result before tossing the coin. Tossing a coin is a trial, since we know the result *a posteriori*.

We call a specific outcome of the trial an "event." There are two or more events for a trial, and we index them by i. In tossing a coin, $i =$ h (head) and t (tail). When we roll a die, $i = 1, 2, 3, 4, 5$, and 6.

We can repeat the trials for a total N times. Event i occurs n_i times. In tossing a coin N times, $n_h + n_t = N$. The ratios, n_h/N and n_t/N, are called **relative frequencies**. The sum of all possible frequencies is equal to 1. In rolling a die N times, the relative frequencies are $n_1/N, n_2/N, \ldots$, and n_6/N, and their sum is 1.

Consider starting a series of trials and updating the running sums, n_h and n_t. At an early stage, the relative frequencies, n_h/N and n_t/N, vary a lot as the result of a new trial being added. With an increasing N, the relative frequencies approach their respective constant numbers. If the coin is a uniform disc, then n_h/N will approach ½ as $N \to \infty$. Likewise, if the die is not rigged, then all of $n_1/N, n_2/N, \ldots$, and n_6/N will approach ⅙. The limiting value is called a probability. We denote by P_i the probability for the event i. The P_i satisfies

$$\sum_{i \in \text{events}} P_i = 1 \qquad (2.1)$$

Statistical Thermodynamics: Basics and Applications to Chemical Systems, First Edition. Iwao Teraoka.
© 2019 John Wiley & Sons, Inc. Published 2019 by John Wiley & Sons, Inc.
Companion website: www.wiley.com/go/Teraoka_StatsThermodynamics

and

$$P_i \geq 0 \qquad (2.2)$$

just as the relative frequencies do.

The fundamental requirement for P_i (Eqs. (2.1) and (2.2)) alone is sufficient for developing a whole probability theory. What is required of P_i? The answer is "exclusive" and "complete." In the example of coin tossing, there are only two outcomes. "Head" and "tail" constitute a complete set of events. Furthermore, there is no overlap between them (exclusive, aka σ-additive). Therefore, $P_h + P_t = 1$. In the example of rolling a die, the six events $-\{i = 1\}, \{i = 2\},\ldots,$ and $\{i = 6\}$ – are mutually exclusive and complete: $P_1 + P_2 + \cdots + P_6 = 1$.

In rolling a die, $\{i = 1, 2, 3\}$ and $\{i = 1, 5, 6\}$ are not mutually exclusive. $\{i = 1, 2, 3\}$ and $\{i = 5, 6\}$ are exclusive, but they are not complete. $\{i = 1, 2, 3\}, \{i = 4, 5\},$ and $\{i = 6\}$ are exclusive and complete.

Consider two events:

$$\text{Event A} = \{i \text{ is odd}\} = \{i = 1, 3, \text{or } 5\} \qquad (2.3)$$

$$\text{Event B} = \{i \leq 3\} = \{i = 1, 2, \text{or } 3\} \qquad (2.4)$$

The probability of getting one of the outcomes in a trial is $\frac{1}{6}$. After many trials, we can assign the probability of $\frac{1}{6}$ to each outcome a priori. The probabilities of events A and B defined are

$$P(\text{A}) = \sum_{i \in \text{A}} P_i = \frac{1}{6} \times 3 = \frac{1}{2} \qquad (2.5)$$

$$P(\text{B}) = \sum_{i \in \text{B}} P_i = \frac{1}{6} \times 3 = \frac{1}{2} \qquad (2.6)$$

Figure 2.1 shows a diagram (Venn diagram in set theory) of the event space. The two events share the outcomes of $\{1\}$ and $\{3\}$. We denote by AB (or A $\cap$ B) the shared event:

$$\text{event AB} = \{i \text{ is odd and } \leq 3\} = \{i = 1 \text{ or } 3\} \qquad (2.7)$$

Its probability is

$$P(\text{AB}) = \frac{1}{6} \times 2 = \frac{1}{3} \qquad (2.8)$$

and is called a **joint probability**.

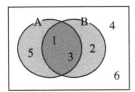

Figure 2.1 Venn diagram of the event space. The numbers indicate the outcome of rolling a die.

A compound event of A and B, denoted as A ∪ B, combines the two events:

event A ∪ B = {i is odd or ≤ 3}

$$= \{i = 1, 2, 3, \text{ or } 5\} \tag{2.9}$$

As seen in Figure 2.1,

$$P(A \cup B) = P(A) + P(B) - P(AB) \tag{2.10}$$

Subtracting $P(AB)$ takes care of double counting.

Consider the following situation: Someone rolls a die and that person knows the outcome, but you do not. The person tells you that event B has occurred, meaning that the outcome was either 1, 2, or 3. Given this piece of information, what is the probability that event A has also occurred, i.e. the outcome was either 1 or 3?

This probability is called a **conditional probability**, and we denote it by $P(A|B)$. The event after "|" defines the condition. $P(A|B)$ is the probability of A when we know that B has occurred. For the present example,

$$P(A \mid B) = \frac{2}{3} \tag{2.11}$$

Note that $\frac{2}{3}$ is equal to $(\frac{1}{3})/(\frac{1}{2})$, that is $P(AB)/P(B)$. In general,

$$P(A \mid B) = \frac{P(AB)}{P(B)} \tag{2.12}$$

Division by $P(B)$ is due to narrowing the space for event A.

In our example, $\frac{2}{3}$ is different from $P(A) = \frac{1}{2}$. However, a different condition, say, C, can make $P(A|C)$ equal to $P(A)$. The latter happens if limiting the space for A to the one simultaneously satisfying C does not change the probability. We call the two events mutually independent or say that event A is independent of event C. Recapping, If $P(A \mid C) = P(A)$ then A and C are independent of each other. Apparently, our events A and B are not independent.

If event C = {$i \leq 4$}, then $P(C) = \frac{2}{3}$ and $P(A|C) = \frac{1}{2}$, and therefore $P(A|C) = P(A)$. Events A and C are independent of each other. As you may have guessed by now, the requirement for the independence is that event C contains an equal number of odd integers and even integers (A and not A).

In the next two sections, we look at typical distributions of probability in two types of events. One is "discrete" and the other is "continuous."

2.2 Discrete Distributions

In the example of tossing a coin n times, the number of trials that get a "head" varies from 0 (all tails) to n (all heads). The number i can be any integer between 0 and n. In the example of rolling a die n times, the number i of trials in which you get "2" varies from 0 to n. The value of i cannot be known prior to completing all the trials. These numbers are called **random variables**. In these two examples, $i = 0, 1, 2, \ldots, n$ and can take only discrete values (integers).

Let us denote by $P(i)$ the probability that the random number is i. Since the events $\{i = 0\}$, $\{i = 1\}$, ..., and $\{i = n\}$ are exclusive and complete, $P(0)$, $P(1)$, ..., and $P(n)$ satisfy

$$\sum_{i=0}^{n} P(i) = 1 \tag{2.13}$$

and

$$P(i) \geq 0 \tag{2.14}$$

Here, we look at three typical discrete distributions.

2.2.1 Binomial Distribution

An outcome for n times coin tossing is expressed as a sequence of "h" and "t," for example, htthhthhthhthhtt. A sequence of binary values (h and t) is called a **Bernoulli sequence**. By now, we know that the probability of getting "h" in each trial is $1/2$ (a priori probability). So is the probability of getting "t." For $n = 2$, the outcomes are hh, ht, th, and tt, and each of them has a probability of $(1/2)^2 = 1/4$ (joint probability). For $n = 3$, the outcomes are hhh, hht, hth, htt, thh, tht, tth, and ttt, and each of them has a probability of $(1/2)^3 = 1/8$. In general, the probability of a specific sequence in n trials is $(1/2)^n$.

We choose a random variable i as the number of trials that have resulted in "h." The i times occurrence of "h" can be anywhere within the sequence. There are

$$\binom{n}{i} = {}_nC_i = \frac{n!}{i!(n-i)!} \tag{2.15}$$

ways for placing "h" i times in a sequence of n trials. Therefore, the probability for i is

$$P(i) = \left(\frac{1}{2}\right)^n \binom{n}{i} = \left(\frac{1}{2}\right)^n {}_nC_i = \left(\frac{1}{2}\right)^n \frac{n!}{i!(n-i)!} \tag{2.16}$$

This distribution is called a **binomial distribution**. The coefficient ${}_nC_i$ is called a **binomial coefficient**.

So far, we have assumed that the coin is not rigged and the probability of seeing "h" in each trial is $1/2$. In general, however, the single-trial probability can be any number between 0 and 1. Let us denote the a priori probability of getting "h" by p. Then, the probability of getting "t" is $q = 1 - p$. A specific sequence of n trials that has i occurrences of "h" and $n - i$ occurrences of "t" has a probability of $p^i q^{n-i} = p^i (1-p)^{n-i}$. Therefore,

$$P(i) = \binom{n}{i} p^i (1-p)^{n-i} \tag{2.17}$$

Figure 2.2 Binomial distribution for $n = 100$ and $p = 0.3, 0.5$, and 0.7.

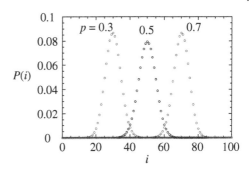

Figure 2.2 shows a plot of $P(i)$ for $n = 100$ and $p = 0.3, 0.5$, and 0.7.
The sum of $P(i)$ is equal to 1:

$$\sum_{i=0}^{n} \binom{n}{i} p^i (1-p)^{n-i} = 1 \qquad (2.18)$$

The abovementioned equality is a part of the **binomial theorem** (also Eq. (A.25)) in which we lift the restriction that q be equal to $1 - p$:

$$\sum_{i=0}^{n} \binom{n}{i} p^i q^{n-i} = (p+q)^n \qquad (2.19)$$

The left-hand side is an expansion of $(p + q)^n$ by a polynomial, and the expansion coefficient is $_nC_i$.

2.2.2 Poisson Distribution

Another important discrete distribution is a **Poisson distribution**. Suppose customers arrive at random at an automated teller machine (ATM) of a bank. We count the number of customers who arrive every five minutes, for instance. The counts follow a Poisson distribution. In another example, let us stand at the roadside of a freeway to count the number of cars that pass us every minute or every 10 seconds. The counts follow another Poisson distribution. The count i is a random variable.

A Poisson distribution is given as

$$P(i) = e^{-\lambda} \frac{\lambda^i}{i!} \qquad (2.20)$$

where λ is a parameter. Since i is a count, $i = 0, 1, 2, \ldots$ There is no upper limit to i. With the coefficient $e^{-\lambda}$, the sum of $P(i)$ for all possible values of i is equal to 1. Figure 2.3 shows three examples of the Poisson distribution.

We can derive the Poisson distribution from a binomial distribution. See Problem 2.1.

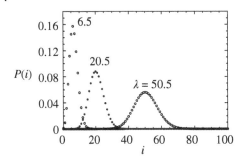

Figure 2.3 Poisson distribution with $\lambda = 6.5, 20.5$, and 50.5.

Other examples of the Poisson distribution are listed here.

A. The number of numbers that fall in a given section of a number line, when they are generated at random in a much broader section.

B. Number of points that fall on a given area when they are thrown at a planar target that is much broader than the area.

C. (colloid chemistry or light scattering) Number of particles contained in a small part of the volume of a suspension.

D. (biology) Number of mutations that occur in a given section of DNA when it is exposed to radiation.

E. (biology) Number of colonies per area on a culture in a Petri dish.

F. (polymer chemistry) Degree of polymerization in living anionic polymerization. The polymerization is initiated simultaneously. Adding repeat units of the polymer occurs randomly.

G. Number of trains that arrive at a platform per unit time in a New York City (NYC) subway.

H. Score in a basketball game.

Think what conditions are required in the last two examples.

2.2.3 Multinomial Distribution

The outcome of each trial in the binomial distribution is either one of the two possibilities. There are trials whose outcome has three or more possibilities. For example, in rolling a die, each trial has six possible outcomes. Suppose we roll the die n times, and count the total number of trials for each of the six possibilities. Let the count be i_j for the outcome of j ($j = 1, 2, \ldots, 6$). The random variable is now a set of $i_1, i_2, \ldots,$ and i_6 with a restriction of $i_1 + i_2 + \cdots + i_6 = n$. Since each sequence of the outcomes has a probability of $(1/6)^n$, the probability $P(i_1, i_2, \ldots, i_6)$ is expressed as

$$P(i_1, i_2, \ldots, i_6) = \frac{n!}{i_1! i_2! \ldots i_6!} \left(\frac{1}{6}\right)^n \tag{2.21}$$

The **multinomial coefficient** $n!/(i_1!i_2!\ldots i_6!)$ is obtained as the number of ways to get "1" i_1 times out of n, multiplied by the number of ways to get "2" i_2 times out of the remaining $n - i_1$, and so on:

$$\frac{n!}{i_1!i_2!\ldots i_6!} = \binom{n}{i_1}\binom{n-i_1}{i_2}\binom{n-i_1-i_2}{i_3}\binom{n-i_1-i_2-i_3}{i_4}$$
$$\times \binom{n-i_1-i_2-i_3-i_4}{i_5} \tag{2.22}$$

Once i_1, i_2, i_3, i_4, and i_5 are specified, the rest is i_6.

The same distribution occurs when you randomly distribute n beads (each has its own face) into M cans. The probability of having i_1 beads in the first can, i_2 beads in the second can,..., and i_M beads in the Mth can is given as

$$P(i_1, i_2, \ldots, i_M) = \frac{n!}{i_1!i_2!\ldots i_M!}p_1^{i_1}p_2^{i_2}\ldots p_M^{i_M} \tag{2.23}$$

where $n!/(i_1!i_2!\ldots i_M!)$ is the number of ways to divide n into M parts with $i_1, i_2,\ldots$, and i_M being the numbers in the M parts, and p_i is the probability that a bead is placed into the ith can ($i = 1, 2,\ldots, M$). It is required that

$$i_1 + i_2 + \cdots + i_M = n \tag{2.24}$$

and the sum of the probabilities is equal to 1:

$$\sum_{i_1, i_2, \cdots, i_M} P(i_1, i_2, \ldots, i_M) = 1 \tag{2.25}$$

from the multinomial theorem (Eq. (A.27)), where the series is calculated for nonnegative integers i_1, i_2,..., and i_M that satisfy Eq. (2.24).

Tossing a thick coin with a smooth edge can land the coin on the edge. Its probability is small, but not zero. The tossing results in three outcomes, head, tail, and edge with probabilities p_h, p_t, and p_e ($p_h + p_t + p_e = 1$). The probability of having i_h times of head, i_t times of tail, and i_e times of edge in a total n times of tossing is

$$P(i_h, i_t, i_e) = \frac{n!}{i_h!i_t!i_e!}p_h^{i_h}p_t^{i_t}p_e^{i_e} \tag{2.26}$$

Equation (2.25) is now

$$\sum_{i_h=0}^{n}\sum_{i_t=0}^{n-i_h}\sum_{i_e=0}^{n-i_h-i_t} P(i_h, i_t, i_e) = 1 \tag{2.27}$$

2.3 Continuous Distributions

Now we move to probability distributions with continuous random variables. In the first example, we draw a line on a horizontal table and drop a pin

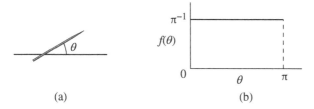

(a) (b)

Figure 2.4 (a) A pin is dropped onto a horizontal plane, and the angle it forms with a straight line on the plane is measured as θ. (b) The θ is a continuous random variable between 0 and π. Its probability distribution is uniform.

(see Figure 2.4a). We cannot know the angle θ the pin forms with the line before it lands. The θ is anywhere between 0 and π, and is a typical example of a continuous random variable. We discuss the distribution of θ in Problem 2.5.

Another example involves movement of a particle in a liquid. An ink is a suspension of pigmented particles in water or another solvent. When we drop a tiny amount of the ink into the same solvent, the ink will be diluted to the whole volume of the solvent. The dilution is a result of movement of ink particles. It is impossible to predict where the particle will be in a second or 10 minutes. The location of the particle at a given time is another example of a continuous random variable.

We use an example of effusion of gas to bridge a discrete distribution and a continuous distribution. Imagine a box filled with gas molecules (Figure 2.5a). The box has a tiny hole on one of its faces, and there is a flat screen parallel to the face. The space between the face and the screen is evacuated. The molecules in the box will fly out of the box through the hole one by one to hit the screen. The screen has equally spaced vertical lines to form strips, and we count the number of molecules that hit each strip.

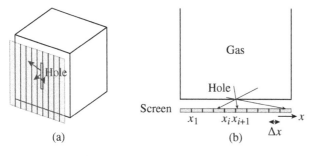

(a) (b)

Figure 2.5 (a) A box with a hole is filled with gas molecules. They fly out of the box to hit a screen placed in front of the hole. The screen is divided into vertical strips. (b) Top view. The width of each strip is Δx.

Figure 2.6 Histogram for the relative frequencies of molecules hitting strips at x_i.

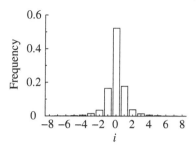

Here, the relative frequency is the count of molecules divided by the total count, and is a function of the position x_i of the ith strip. When we plot the resultant discrete distribution, it is a histogram that looks like the bar graph in Figure 2.6.

When the total count is low, the histogram is not symmetric, and there are ups and downs in the bar heights. When the count is sufficiently high, these irregularities are smoothed out, and the relative frequencies approach the probabilities. We denote by $P(x, \Delta x)$ the relative frequency for a flying molecule to land on a strip at x of width Δx. We can narrow the strips, and the distribution will look like the one shown in Figure 2.7a. A further decrease in Δx will bring the plot to the one in Figure 2.7b.

If the strips are sufficiently narrow (Δx is small) and the tally is taken for a sufficiently long time, the bar graph will approach a continuous curve (see Figure 2.8). We denote by $f(x)$ the limiting form of $P(x, \Delta x)$:

$$\lim_{\Delta x \to 0} P(x, \Delta x) = f(x) \qquad (2.28)$$

Now, x is the random variable of the continuous distribution. The function $f(x)$ is called a **probability density function** or a **probability distribution function**.

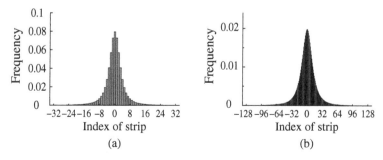

Figure 2.7 (a) Relative frequency of the count is plotted as a function of the position of the strip. (b) When the strips narrow, the plot will look like this.

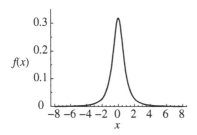

Figure 2.8 Limiting form of the relative frequency is a continuous distribution $f(x)$.

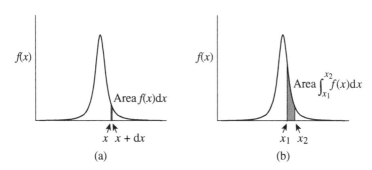

(a) (b)

Figure 2.9 (a) Probability that the random variable falls between x and $x + dx$ is equal to the area $f(x)dx$ of the shaded narrow strip. (b) The area rule can be extended to a broad section.

In the continuous distribution, it is the area under the curve that represents the probability (see Figure 2.9a). The shaded area is equal to $f(x)dx$, if dx is sufficiently small. The area is equal to the probability for a molecule to hit the screen between x and $x + dx$. The area rule applies to a large section of x; see Figure 2.9b. The probability P to locate the molecule between x_1 and x_2 is

$$P = \int_{x_1}^{x_2} f(x)dx \tag{2.29}$$

The probability of finding x between $-\infty$ and $+\infty$ is one:

$$\int_{-\infty}^{\infty} f(x)dx = 1 \tag{2.30}$$

This integral is called **normalization**, and the distribution $f(x)$ that integrates to one is called normalized.

Often, we obtain the distribution $f(x)$ without being normalized. For example, $\exp(-x^2)$. Then, we write $f(x) = a \exp(-x^2)$, where a is a constant, and use Eq. (2.30) to obtain a. For this example, $a = \pi^{-1/2}$.

Here, we look at typical continuous distributions.

2.3.1 Uniform Distribution

The example of a pin dropping in the beginning of this section shows a **uniform distribution**. There is no preference to the random variable θ, and therefore $f(\theta)$ is constant at π^{-1}. Figure 2.4b shows a plot of $f(\theta)$.

Another example of the uniform distribution is the position of a molecule in vapor phase in a box. Its x coordinate can be any real number as long as it is within the box. Again, there is no preference.

Likewise, when we place a random number x generated on [0, 1], its distribution is $f(x) = 1$. If x is generated on [0, a], $f(x) = a^{-1}$.

2.3.2 Exponential Distribution

Suppose that we generate many random numbers on [0, 1] (see Figure 2.10a). They are represented by the ticks on the number line. Note that the ticks are not equally spaced, and that is the nature of the uniform distribution. Now, we pay attention to the spacings between adjacent pairs of the ticks. Obviously, the spacings are not equal. There are more narrow spacings than are broad spacings. The distribution of the spacing looks like the plot in Figure 2.10b.

The spacing x between an adjacent pair of random numbers follows an **exponential distribution**:

$$f(x) = ae^{-ax}, \quad x \geq 0 \tag{2.31}$$

where a is a constant. This distribution is normalized. See Figure 2.11a for $f(x)$. Part b of the figure compares the exponential distributions with large and small values of a.

We can derive the exponential distribution from a uniform distribution of n points on [0, 1]. For a spacing between adjacent points to be x, the section of length x must contain no other points. We divide the section into N ($N \gg 1$) equally spaced subsections (each is x/N long). Since the probability of a subsection to have a point is nx/N, the probability for all of the N subsections not to have any points is $(1 - nx/N)^N$. In the limit of $N \rightarrow \infty$, the probability is equal

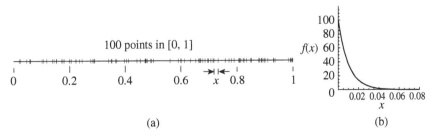

(a) (b)

Figure 2.10 (a) Hundred random numbers were generated in [0, 1]. (b) The spacing x between an adjacent pair of numbers is distributed with an exponential distribution.

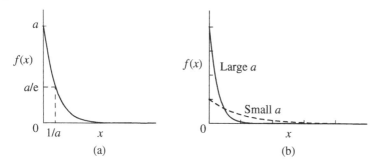

Figure 2.11 (a) Exponential distribution, $f(x) = ae^{-ax}$. (b) Solid and dashed lines are for large and small values of a.

to e^{-nx} (that is, how the exponential function is defined). Thus, we find a in Eq. (2.31) represents the number of points in $[0, 1]$ in this example.

The probability of the exponential distribution peaks at $x = 0$, regardless of the value of a, and decreases toward 0 with an increasing x. Therefore, in placing many points on the number line, the most probable position of the adjacent point is immediately next. This situation creates an interesting misunderstanding.

NYC's subway system had been chaotic until some time ago. Train arrivals were next to random, and the headquarter of the subway system was not aware where their trains were. Arrivals of trains followed a uniform distribution, and therefore the time t between a pair of consecutive train arrivals should have followed an exponential distribution. A direct consequence of it is that the probability is the highest at $t = 0$ and decreases with an increasing t. Chances are that the next train will arrive on the heels of the departing train. The next train will arrive in no time. Your experience tells another story. Rather, you wait for a long time for the next train. This discrepancy is called a **waiting time paradox**.

Here is a trick. In the time line that records train arrivals, place also the event of your arrival at the platform. It is not likely that your arrival will get into the smallest spacing between an adjacent pair. Rather, it is likely that your arrival will be in one of the largest spacings. The greater the spacing, the more likely it will contain your arrival. Therefore, the spacing of train arrivals in which you arrive at the platform is distributed with

$$f(x) = a^2 x e^{-ax} \tag{2.32}$$

which is normalized in $[0, \infty)$. Figure 2.12 compares Eqs. (2.31) and (2.32). Now you understand why you do not benefit from the exponential distribution.

Once you enter a train car and the train departs, you do not know when the next train arrives. It is highly likely that the next train is immediately behind. Therefore, if you miss a train, you do not need to worry.

Figure 2.12 Exponential distribution ae^{-ax} and a distribution weighted by x.

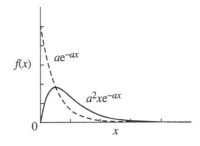

2.3.3 Normal Distribution

The most important distribution among continuous distributions is a **normal distribution**. Its probability density function is given as

$$f(x) = \frac{1}{\sqrt{2\pi\sigma^2}} \exp\left(-\frac{(x-x_0)^2}{2\sigma^2}\right) \tag{2.33}$$

with two parameters x_0 and σ. As seen in Figure 2.13, the bell-shaped distribution function is symmetric around its center at $x = x_0$. The typical width of the curve is σ. Between $x_0 - \sigma$ and $x_0 + \sigma$, the curve is upward convex.

The normal distribution appears here and there in natural and social phenomena. Given here is a list of examples.

A. Position of an ink particle at a given time after an aliquot of the aqueous ink is dropped into water. Here, we take the x coordinate of the position.
B. Velocity of a molecule in a gas phase. The velocity has x, y, and z components. The x component v_x is distributed with a normal distribution centered at $v_x = 0$. We examine this distribution in detail in Section 5.1.

2.3.4 Distribution of a Dihedral Angle

Now we look at a more chemical example of continuous distribution. It is a dihedral angle in disubstituted ethane. Figure 2.14a defines the dihedral angle ϕ in 1,2-dichloroethane. The trans conformation (t) of the figure gives $\phi = 0$.

Figure 2.13 Probability density function of a normal distribution. At $x = x_0 \pm \sigma$, $f(x)$ is 0.601 times the peak value.

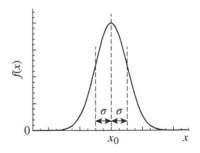

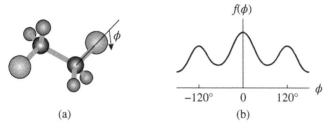

$f(\phi)$

$-120°$ 0 $120°$ ϕ

(a)   (b)

Figure 2.14 (a) Ball-stick model of 1,2-dichloroethane, dihedral angle ϕ is defined in the figure that shows a trans conformation. (b) Distribution of ϕ.

Figure 2.14b shows a sketch for the distribution $f(\phi)$. The trans conformation is the most likely, and $f(\phi)$ maximizes at $\phi = 0$. There are two other local maxima that represent gauche and gauche prime conformations (g, g'). Their respective likelihood is less compared with t.

2.4 Means and Variances

So far, we have avoided calculating or referring to the **mean** and **variance** of a distribution. In this section, we learn their definitions and how to calculate them when the distribution is given. The mean, also known as the **expectation** or simply the **average**, and the variance are the most important quantities that characterize each probability distribution.

2.4.1 Discrete Distributions

For a discrete distribution $P(i)$, the mean of the random variable i is defined as

$$\langle i \rangle = \sum_i iP(i) \tag{2.34}$$

It is a sum of i weighted by $P(i)$. Old books use $\bar{i}$ for $\langle i \rangle$. Likewise, we can calculate the mean of i^2:

$$\langle i^2 \rangle = \sum_i i^2 P(i) \tag{2.35}$$

which is also called the **second moment** ($\langle i \rangle$ is called the first moment; first, second, etc. refer to the power of the random variable). We call $\langle i^2 \rangle^{1/2}$ the **root-mean-square** of i. The name combines three operations – taking a square root, calculating the mean, and squaring – arranged in the reversed order.

The variance is the second moment around the mean:

$$\text{var}(i) = \langle (i - \langle i \rangle)^2 \rangle = \sum_i (i - \langle i \rangle)^2 P(i) \tag{2.36}$$

where $i - \langle i \rangle$ is called a deviation. Since $(i - \langle i \rangle)^2 = i^2 - 2i\langle i \rangle + \langle i \rangle^2$,

$$\langle (i - \langle i \rangle)^2 \rangle = \sum_i i^2 P(i) - 2\langle i \rangle \sum_i i P(i) + \langle i \rangle^2 = \langle i^2 \rangle - \langle i \rangle^2 \tag{2.37}$$

The variance is equal to the mean square minus the square of the mean. Since the variance is nonnegative, the root-mean-square is greater than or equal to the absolute value of the mean. The square root of the variance is called the **standard deviation.**

Calculating the averages is not limited to i and i^2. The average can be calculated for any function $g(i)$ of i:

$$\langle g(i) \rangle = \sum_i g(i)P(i) \tag{2.38}$$

For example, $g(i) = i^{-1}$, $i^{1/2}$, $|i|$, and so on.

So far, we have assumed that the probability distribution is normalized. If not, we need to divide the sum of $g(i)P(i)$ by the sum of $P(i)$:

$$\langle g(i) \rangle = \frac{\sum_i g(i)P(i)}{\sum_i P(i)} \tag{2.39}$$

Let us calculate the mean and variance of a binomial distribution. First, the mean is calculated as

$$\langle i \rangle = \sum_{i=0}^{n} i \binom{n}{i} p^i (1-p)^{n-i} \tag{2.40}$$

The difference of this series from the one in Eq. (2.18) is an extra factor of i in each term. We replace $1 - p$ by q for now:

$$\langle i \rangle = \sum_{i=0}^{n} i \binom{n}{i} p^i q^{n-i} \tag{2.41}$$

Note that ip^i is obtained by differentiating p^i by p and then multiplying p. Therefore,

$$\langle i \rangle = \sum_{i=0}^{n} \binom{n}{i} p \frac{\partial}{\partial p} p^i q^{n-i} \tag{2.42}$$

Since taking the sum and operating $p\partial/\partial p$ can be interchanged,

$$\langle i \rangle = p\frac{\partial}{\partial p} \sum_{i=0}^{n} \binom{n}{i} p^i q^{n-i} = p\frac{\partial}{\partial p}(p+q)^n = np(p+q)^{n-1} \tag{2.43}$$

where the binomial theorem, Eq. (2.19) (also Eq. (A.25)), was used. Now, we bring back q to $1 - p$ to obtain

$$\langle i \rangle = np \tag{2.44}$$

Likewise, $\langle i^2 \rangle$ is calculated by operating $p\partial/\partial p$ twice on $(p+q)^n$; each operation ends up with multiplying i:

$$\langle i^2 \rangle = p\frac{\partial}{\partial p}p\frac{\partial}{\partial p} \sum_{i=0}^{n} \binom{n}{i} p^i q^{n-i} = p\frac{\partial}{\partial p}p\frac{\partial}{\partial p}(p+q)^n = p\frac{\partial}{\partial p}np(p+q)^{n-1}$$

$$= np(p+q)^{n-1} + n(n-1)p^2(p+q)^{n-2} \tag{2.45}$$

Now, we set $p+q = 1$ to obtain

$$\langle i^2 \rangle = np + n(n-1)p^2 = (np)^2 + np(1-p) \tag{2.46}$$

Therefore, the variance is

$$\text{var}(i) = \langle i^2 \rangle - \langle i \rangle^2 = np(1-p) \tag{2.47}$$

We can write this result as npq as well.

How about the Poisson distribution? The mean of the random variable i is calculated as

$$\langle i \rangle = \sum_{i=0}^{\infty} ie^{-\lambda}\frac{\lambda^i}{i!} = \sum_{i=1}^{\infty} ie^{-\lambda}\frac{\lambda^i}{i!} = \lambda \sum_{i=1}^{\infty} e^{-\lambda}\frac{\lambda^{i-1}}{(i-1)!} = \lambda \tag{2.48}$$

We can also use the same method as the one used for the binomial distribution:

$$\langle i \rangle = e^{-\lambda} \sum_{i=0}^{\infty} i\frac{\lambda^i}{i!} = e^{-\lambda}\lambda\frac{\partial}{\partial \lambda} \sum_{i=1}^{\infty} \frac{\lambda^i}{i!} \tag{2.49}$$

Since the sum of the series is equal to e^λ,

$$\langle i \rangle = e^{-\lambda}\lambda\frac{\partial}{\partial \lambda}e^\lambda = \lambda \tag{2.50}$$

To calculate the variance, we first calculate $\langle i(i-1) \rangle$:

$$\langle i(i-1) \rangle = \sum_{i=0}^{\infty} i(i-1)e^{-\lambda}\frac{\lambda^i}{i!} = \lambda^2 \sum_{i=2}^{\infty} e^{-\lambda}\frac{\lambda^{i-2}}{(i-2)!} = \lambda^2 \tag{2.51}$$

Therefore,

$$\langle i^2 \rangle = \langle i(i-1) \rangle + \langle i \rangle = \lambda^2 + \lambda \tag{2.52}$$

and

$$\text{var}(i) = \langle i^2 \rangle - \langle i \rangle^2 = \lambda \tag{2.53}$$

Table 2.1 summarizes the mean and variance for the binomial and Poisson distributions.

The means and variances of the multinomial distribution, Eq. (2.23), can be calculated in the same way as we did for the binomial distribution. The mean

Table 2.1 Mean and variance of a binomial distribution and a Poisson distribution.

Distribution	$P(i)$	Mean	Variance
Binomial	$\binom{n}{i} p^i (1-p)^{n-i}$	np	$np(1-p)$
Poisson	$e^{-\lambda} \dfrac{\lambda^i}{i!}$	λ	λ

of i_1 is calculated according to

$$\langle i_1 \rangle = \sum_{i_1, i_2, \ldots, i_M} i_1 P(i_1, i_2, \ldots, i_M) = \sum_{i_1, i_2, \cdots, i_M} p_1 \frac{\partial}{\partial p_1} \frac{n!}{i_1! i_2! \ldots i_M!} p_1^{j_1} p_2^{i_2} \cdots p_M^{i_M}$$

$$= p_1 \frac{\partial}{\partial p_1} (p_1 + p_2 + \cdots + p_M)^n = p_1 n (p_1 + p_2 + \cdots + p_M)^{n-1} \quad (2.54)$$

Since $p_1 + p_2 + \cdots + p_M = 1$, we get

$$\langle i_1 \rangle = np_1 \quad (2.55)$$

Likewise, the mean of i_1^2 is calculated as

$$\langle i_1^2 \rangle = \sum_{i_1, i_2, \ldots, i_M} i_1^2 P(i_1, i_2, \ldots, i_M) = p_1 \frac{\partial}{\partial p_1} p_1 \frac{\partial}{\partial p_1} (p_1 + p_2 + \cdots + p_M)^n$$

$$= p_1 \frac{\partial}{\partial p_1} p_1 n (p_1 + p_2 + \cdots + p_M)^{n-1}$$

$$= np_1 (p_1 + p_2 + \cdots + p_M)^{n-1} + p_1^2 n(n-1)(p_1 + p_2 + \cdots + p_M)^{n-2} \quad (2.56)$$

which leads to

$$\langle i_1^2 \rangle = np_1 + n(n-1)p_1^2 \quad (2.57)$$

Therefore, the variance is

$$\langle (i_1 - \langle i_1 \rangle)^2 \rangle = np_1(1 - p_1) \quad (2.58)$$

These results are identical to those for the binomial distribution.

Unlike the binomial counterpart, the multinomial distribution has a covariance, $\langle (i_1 - \langle i_1 \rangle)(i_2 - \langle i_2 \rangle) \rangle$. We first calculate $\langle i_1 i_2 \rangle$:

$$\langle i_1 i_2 \rangle = p_1 \frac{\partial}{\partial p_1} p_2 \frac{\partial}{\partial p_2} (p_1 + p_2 + \cdots + p_M)^n$$

$$= n(n-1) p_1 p_2 (p_1 + p_2 + \cdots + p_M)^{n-2} \quad (2.59)$$

which leads to $\langle i_1 i_2 \rangle = n(n-1) p_1 p_2$. Then,

$$\langle (i_1 - \langle i_1 \rangle)(i_2 - \langle i_2 \rangle) \rangle = \langle i_1 i_2 \rangle - \langle i_1 \rangle \langle i_2 \rangle = -np_1 p_2 \quad (2.60)$$

The negative sign is reasonable, since a positive deviation of i_1 causes i_2 to deviate negatively.

Related useful formulae are listed here:

$$\langle (i_1 - i_2)^2 \rangle = n(p_1 + p_2) + n(n-1)(p_1 - p_2)^2 \tag{2.61}$$

$$\langle (i_1 - i_2 - \langle i_1 - i_2 \rangle)^2 \rangle = n[p_1 + p_2 - (p_1 - p_2)^2] \tag{2.62}$$

2.4.2 Continuous Distributions

For a continuous distribution, the mean of the random variable x is calculated as

$$\langle x \rangle = \int_{-\infty}^{\infty} x f(x) \mathrm{d}x \tag{2.63}$$

Likewise, the mean square is

$$\langle x^2 \rangle = \int_{-\infty}^{\infty} x^2 f(x) \mathrm{d}x \tag{2.64}$$

Then, the variance is

$$\mathrm{var}(x) = \int_{-\infty}^{\infty} (x - \langle x \rangle)^2 f(x)\,\mathrm{d}x = \langle x^2 \rangle - \langle x \rangle^2 \tag{2.65}$$

As it was for the discrete distribution, the average can be calculated for any function $g(x)$ of the random variable x:

$$\langle g(x) \rangle = \int_{-\infty}^{\infty} g(x) f(x)\,\mathrm{d}x \tag{2.66}$$

If $f(x)$ is not normalized,

$$\langle g(x) \rangle = \frac{\int_{-\infty}^{\infty} g(x) f(x) \mathrm{d}x}{\int_{-\infty}^{\infty} f(x) \mathrm{d}x} \tag{2.67}$$

Now we calculate the mean and variance for the exponential distribution given by Eq. (2.31). The mean is calculated as

$$\langle x \rangle = \int_{-\infty}^{\infty} x a e^{-ax} \mathrm{d}x = \frac{1}{a} \tag{2.68}$$

The mean square is

$$\langle x^2 \rangle = \int_{-\infty}^{\infty} x^2 a e^{-ax} \mathrm{d}x = \frac{2}{a^2} \tag{2.69}$$

Therefore, the variance is

$$\mathrm{var}(x) = \langle x^2 \rangle - \langle x \rangle^2 = \frac{1}{a^2} \tag{2.70}$$

Next, we calculate the mean and variance for the normal distribution. Since $f(x)$ given by Eq. (2.33) is an even function of $x - x_0$, $(x - x_0)f(x)$ is an odd function of $x - x_0$. Therefore, its integral over $(-\infty, \infty)$ vanishes, and we obtain

$$\langle x \rangle = \langle x - x_0 \rangle + x_0 = x_0 \tag{2.71}$$

The variance, expressed as

$$\langle (x - \langle x \rangle)^2 \rangle = \langle (x - x_0)^2 \rangle = \frac{1}{\sqrt{2\pi\sigma^2}} \int_{-\infty}^{\infty} (x - x_0)^2 \exp\left(-\frac{(x - x_0)^2}{2\sigma^2}\right) dx \tag{2.72}$$

is calculated using Eq. (A.31), with $n = 1$. The result is

$$\langle (x - \langle x \rangle)^2 \rangle = \sigma^2 \tag{2.73}$$

Now we know that σ^2 is the variance and σ is the standard deviation. The whole Eq. (2.33), complete with the normalization constant, is worth memorizing. Whenever a type of integral as the one in Eq. (2.72) appears, we do not need to look at the integral formula.

Table 2.2 summarizes the mean and variance of the two continuous distributions.

2.4.3 Central Limit Theorem

The two discrete distributions shown in Figures 2.2 and 2.3 look close to a bell-shaped curve of the normal distribution. When certain conditions are satisfied, these two discrete distributions can be approximated by normal distributions. In fact, if the variance ($np(1-p)$ or λ) is sufficiently large compared with one, either distribution can be approximated by a normal distribution that has the same mean and variance as those of the discrete distribution. Its proof is given in Problems 2.9 and 2.10. The reason why the approximation is possible is that, with an increasing n or λ, the difference from the normal distribution diminishes, as the ratio of the standard deviation to the mean (aka **coefficient of variation**) decreases with the increasing n or λ. The high-order moments around the mean decrease even faster.

Table 2.2 Mean and variance of an exponential distribution and a normal distribution.

Distribution	$f(x)$	Range	Mean	Variance
Exponential	ae^{-ax}	$x \geq 0$	a^{-1}	a^{-2}
Normal	$\dfrac{1}{\sqrt{2\pi\sigma^2}} \exp\left(-\dfrac{(x - x_0)^2}{2\sigma^2}\right)$	$-\infty < x < +\infty$	x_0	σ^2

This approximation rule applies to any distribution, discrete or continuous. It is a part of the **central limit theorem** or the **law of large numbers**. The approximation will be useful in many situations.

2.5   Uncertainty

In this section, we learn to quantify the concept of uncertainty using $P(i)$ or $f(x)$. We follow the definition of H by Tolman [4]. Boltzmann introduced the same concept for the first time many decades earlier. Boltzmann's definition was for the probability distribution of the kinetic energy of a molecule in gas phase. Here, we find a mathematical expression of the uncertainty for discrete and continuous distributions of probability.

Which do you bet on, getting a head in the coin tossing or getting 3 in die rolling, if the reward is the same? Of course, you will bet on getting a head, since its probability is higher. We can say that getting a head is more certain than getting 3. The greater the probability, the smaller the uncertainty.

Considering successive trials allows us to choose a right expression for the uncertainty. First, the uncertainty should add up as more trials are added. For example, in tossing a coin twice, there are four possibilities (head + head, head + tail, tail + head, and tail + tail). The uncertainty in the twice tossing should be the sum of the uncertainty of the first tossing and the uncertainty of the second tossing. Each of the four events in the twice tossing has a probability of $\frac{1}{4}$, which is the product of $\frac{1}{2}$ and $\frac{1}{2}$. Conversion of the product to the sum is made possible by employing $-\ln P$ as a part of the expression of the uncertainty. The minus sign is included, because we consider uncertainty, not certainty. When the coin is tossed three times, the uncertainty should be the sum of the three individual uncertainties, and the probability of each of the eight events is the product of probabilities of the events in the three trials.

Second, we want to extend the concept of uncertainty from the one for the event to the one for the trial. Each trial has multiple events that are mutually exclusive and complete. The ith event has a probability of P_i. The uncertainty of the trial, H, is defined as the sum of $-\ln P_i$, weighted by the probability:

$$H \equiv - \sum_i P_i \ln P_i \tag{2.74}$$

In other words, H is the mean of $-\ln P_i$. Since $P_i < 1$, H is always positive.

We can show that the uncertainty of two independent trials is equal to the sum of two uncertainties, each for one trial. Let P_{ij} be the probability to have the outcome of i and j in the compound trial. From the independence, $P_{ij} = P_i P_j$, where P_i and P_j are the probabilities of the individual trials. Then,

$$- \sum_{i,j=1}^{n} P_{ij} \ln P_{ij} = - \sum_{i,j=1}^{n} P_i P_j \ln P_i - \sum_{i,j=1}^{n} P_i P_j \ln P_j$$

$$= - \sum_{i=1}^{n} P_i \ln P_i \sum_{j=1}^{n} P_j - \sum_{i=1}^{n} P_i \sum_{j=1}^{n} P_j \ln P_j = - \sum_{i=1}^{n} P_i \ln P_i - \sum_{j=1}^{n} P_j \ln P_j$$

$$(2.75)$$

Calculation of H is as follows. In tossing a coin,

$$H = - \left(\frac{1}{2} \ln \frac{1}{2} \right) \times 2 = \ln 2 \tag{2.76}$$

In rolling a die once, H is calculated as

$$H = - \left(\frac{1}{6} \ln \frac{1}{6} \right) \times 6 = \ln 6 \tag{2.77}$$

Rolling a die has a greater uncertainty compared with tossing a coin.

If a die is rolled twice, and each outcome (getting i in the first rolling and getting j in the second rolling; $i, j = 1, 2, \ldots, 6$) is regarded as an individual event, then

$$H = - \left(\frac{1}{36} \ln \frac{1}{36} \right) \times 36 = \ln 36 = 2 \ln 6 \cong 3.58 \tag{2.78}$$

The uncertainty is twice as large the one for one-time rolling.

If your interest is not in the outcomes in the two rollings, but in the sum of the two numbers, the uncertainty is different. Let us denote the sum by i ($i = 2, 3, \ldots, 12$). Since the probability of getting i is $(6 - |i - 7|)/36$, the uncertainty is calculated as

$$H = - \sum_{i=2}^{12} \frac{6 - |i - 7|}{36} \ln \frac{6 - |i - 7|}{36} \cong 2.27 \tag{2.79}$$

This value of H is less than the one in Eq. (2.78), because of more inclusive event criterion in tallying the sum compared with considering each outcome separately in twice rolling.

Consider a binary event which occurs with probabilities of p and $1 - p$. You can imagine tossing a rigged coin. The uncertainty is

$$H = -p \ln p - (1 - p) \ln(1 - p) \tag{2.80}$$

Figure 2.15 shows a plot of H as a function of p. $H = 0$ at $p = 0$ and 1, a reasonable result, since there is no uncertainty. The uncertainty maximizes at $p = \frac{1}{2}$.

The definition of H was first proposed by Ludwig Boltzmann in 1872. He rather defined H for a probability in continuous space:

$$H \equiv - \int f(x) \ln f(x) dx = -\langle \ln f(x) \rangle \tag{2.81}$$

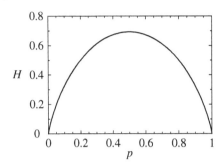

Figure 2.15 Uncertainty H for one event of tossing a rigged coin, plotted as a function of p, the probability of having the head.

This definition of H can be extended to a probability distribution $f(x, y)$ that takes two random variables x and y. It can be easily shown that, if $f(x, y)$ is equal to the product of individual distributions, $g(x)h(y)$, the uncertainty of $f(x, y)$ is equal to the sum of two uncertainties:

$$-\iint f(x, y) \ln f(x, y) \, dx \, dy = -\int g(x) \ln g(x) dx - \int h(y) \ln h(y) dy$$

$$(2.82)$$

Examples of calculating the uncertainty are shown here for uniform distributions. The examples are for the position (x coordinate) of a particle in a box. If $0 \le x \le 2, f(x) = \frac{1}{2}$, and

$$H = -\int_0^2 \frac{1}{2} \ln \frac{1}{2} dx = \ln 2 \tag{2.83}$$

If x extends from 0 to 5, $H = \ln 5$. The broader the range, the greater the H.

Unlike discrete distributions, H for continuous distributions can be negative. For example, in a uniform distribution with $f(x) = 2$ for $0 \le x \le 1/2$,

$$H = -\int_0^{1/2} 2 \ln 2 \, dx = -\ln 2 \tag{2.84}$$

What matters is not the absolute value or the sign of H, but the difference in H. Among uniform distributions, H decreases with a decreasing width of the distribution.

It is interesting to calculate H for a normal distribution. From Eq. (2.33),

$$H = -\int_{-\infty}^{\infty} f(x) \left[-\frac{1}{2} \ln(2\pi\sigma^2) - \frac{(x - x_0)^2}{2\sigma^2} \right] dx \tag{2.85}$$

Since the variance of the distribution is σ^2, it is calculated as

$$H = \frac{1}{2} \ln(2\pi e) + \ln \sigma \tag{2.86}$$

Figure 2.16 Uncertainty H, plotted as a function of σ for a normal distribution with variance σ^2.

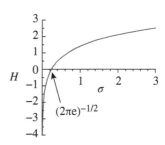

The mean x_0 does not show in the expression of H, as the uncertainty is about a deviation from the mean. Figure 2.16 shows a plot of H as a function of σ. With a decreasing σ, H decreases. The dependence is logarithmic.

Problems

2.1 *Deriving a Poisson distribution from a binomial distribution.* This problem derives the Poisson distribution from a binomial distribution using an example of dropping points along a line. We distribute points randomly along a number line at a rate of λ per unit length. We pick up a section of $[0, 1]$. The number of points the section gets, i, is supposed to be distributed with a Poisson distribution with mean $= \lambda$. We will prove it here, starting with a binomial distribution.

For that purpose, we divide the $[0, 1]$ section into N subsections of equal length, where $N \gg 1$. If $N \gg \lambda$, each section will contain at most one point, and most subsections will be free of a point.

(1) Note that the probability of a subsection to have a point is λ/N ($\ll 1$). What is the probability $P(i)$ of the random variable i?

(2) Calculate $\ln P(i)$ using Stirling's approximation. Since $i/N \ll 1$ and $\lambda/N \ll 1$, the resultant approximation can be further simplified. Simplify the expression up to $O(i)$ and $O(\lambda)$.

2.2 *Peak in the Poisson distribution.* The Poisson distribution has a peak. When i is increased, $P(i)$ first increases and then decreases.

(1) Find j that satisfies $P(j) = P(j+1)$. Disregard whether the j thus found is an integer or not.

(2) You will find that, if λ is an integer, these two points will share the maximum. For $i \leq j$, $P(i)$ increases; for $j + 1 \leq i$, $P(i)$ decreases. If λ is not an integer, the answer in (1) is not an integer. Then, either j or

$j + 1$ will maximize $P(i)$. Find which value of i maximizes $P(i)$ when λ is not an integer.

2.3 *Convolution.* The number n of cars and sport-utility vehicles (SUVs) that pass a point on a freeway per unit time follows a Poisson distribution $P_{pois}(n; \lambda)$ with parameter λ. The fraction of SUVs is p, and therefore, the number i of SUVs out of n vehicles follows a binomial distribution P_{bin} $(i; n, p)$. Use convolution to show that the number of SUVs that pass the point per unit time follows a Poisson distribution. What is the parameter of the Poisson distribution?

2.4 *Weighted exponential distribution.* Calculate the mean and variance of the random variable x that follows Eq. (2.32) derived for the waiting-time paradox. Compare them with those for the exponential distribution.

2.5 *Pin drop.* In the example of pin drop, the angle θ the pin forms with a given straight line is uniformly distributed in $[0, \pi]$. Suppose we repeat the pin drop 100 times and record the angle each time. After the trials, we arrange the angles in increasing order. What is the distribution between an adjacent pair of the angles?

2.6 *Two orthogonal normal distributions.* Two random variables x and y follow independently

$$f(x) = (2\pi\sigma_x^2)^{-1/2} \exp\left(-\frac{(x - x_0)^2}{2\sigma_x^2}\right)$$

$$f(y) = (2\pi\sigma_y^2)^{-1/2} \exp\left(-\frac{(y - y_0)^2}{2\sigma_y^2}\right)$$

(1) What is the mean of $x - y$? What is its variance?
(2) If x and y are not independent, can the variance of $x - y$ be zero?

2.7 *Normal distribution, mean of the absolute value.* $\langle x \rangle = 0$ in the normal distribution of x with zero mean. Calculate $\langle |x| \rangle$ and compare it with $\langle x^2 \rangle^{1/2}$. Which is greater, and why is it so?

2.8 *Waiting-time paradox in two dimensions.* The effect we saw in the waiting-time paradox is more serious in two or higher dimensions. Let us consider a simple problem here. The figure below shows vertical and horizontal lines drawn randomly on a square. The lines tessellate the square into many rectangles. Suppose there are n vertical lines and

n horizontal lines, where $n \gg 1$. The figure shows a square with 10 horizontal lines and 10 vertical lines.

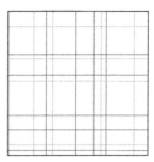

(1) What probability distribution does the area of the rectangle follow?
(2) Calculate the mean and variance of the area.
(3) You throw in a point randomly onto the square. The point will find itself in one of the rectangles. What probability distribution does the area of the rectangle follow?
(4) Calculate the mean and variance of the area for the distribution in (3).

2.9 *Continuous approximation of the binomial distribution.* The binomial distribution $P_{\text{bin}}(i; n, p)$, Eq. (2.17), has a bell-shaped curve, especially when n is large. We can prove that the binomial distribution approaches a normal distribution in the limit of $n \to \infty$ and the step size becomes negligible relative to n and therefore i/n can be approximated as continuous. Let $x = i - np$, the deviation of i from the mean.

(1) Use Stirling's approximation, Eq. (A.29), to show that $\ln P_{\text{bin}}(i; n, p)$ is approximated as

$$\ln P_{\text{bin}}(i; n, p) \cong (np + x) \ln \frac{np}{np + x} + [n(1-p) - x] \ln \frac{n(1-p)}{n(1-p) - x}$$
$$+ \frac{1}{2} \ln \frac{n}{2\pi[n(1-p) - x](np + x)}$$

(2) Use the Taylor expansion of the logarithmic function to evaluate the first two terms on the right-hand side of the equation in (1) up to $O((x/np)^2)$ or $O((x/[n(1-p)])^2)$.

(3) Then, approximate the third term up to $O(x^0)$. This part is just the normalization constant. The final result will be

$$\ln P_{\text{bin}}(i; n, p) \cong -\frac{1}{2\sigma^2}x^2 + \frac{1}{2} \ln \frac{1}{2\pi\sigma^2}$$

with $np(1-p) = \sigma^2$.

2.10 *Continuous approximation of the Poisson distribution.* The Poisson distribution $P_{\text{pois}}(i; \lambda)$, Eq. (2.20), has a bell-shaped curve, especially if λ is large. Show that $P_{\text{pois}}(i; \lambda)$ is approximated by a normal distribution with the same mean and variance in the limit of $\lambda \to \infty$.

2.11 *Uncertainty of an exponential distribution.* Calculate the uncertainty for the exponential distribution given by Eq. (2.31). Draw a sketch for the plot of H as a function of a.

3

Energy and Interactions

For a given chemical system, applying the statistical mechanics starts with writing a function called the partition function. The writing requires that we know the energy of each molecule in the system. We learn in this chapter how to describe and specify the energy of each molecule.

In the absence of an applied external field and in the absence of interaction between molecules, the energy of a molecule consists of the energy due to the movement of the molecule as a whole and the energy for the movement of atoms within the molecule. We first use classical mechanics to describe the movements and the kinetic energy. Toward the end of the chapter, we learn quantum-mechanical methods to describe the kinetic energy. "Classical" means that the movement of a molecule follows Newton's equation of motion. It does not only describe the center-of-mass translation but also the rotation and the vibration.

In Sections 3.1 and 3.2, we first consider kinetic energy of atoms and ions and kinetic energy of diatomic molecules, respectively, that do not interact with each other. In Section 3.3, we look at the energy of polyatomic molecules. Then, in Section 3.4, we learn different types of interaction between molecules. In these four sections, we rely on classical mechanics. After examining briefly an extensive nature of the interaction in Section 3.5, we turn to quantum mechanics to describe all the components of the energy in Section 3.6.

3.1   Kinetic Energy and Potential Energy of Atoms and Ions

3.1.1   Kinetic Energy

Consider an atom i isolated from the other atoms. There is no field applied. The energy of the atom consists of the **kinetic energy** $\varepsilon_{i,\mathrm{kin}} = \tfrac{1}{2} m_i \mathbf{v}_i^2$ and the energy of the electron $\varepsilon_{i,\mathrm{elec}}$:

$$\varepsilon_i = \frac{1}{2} m_i \mathbf{v}_i^2 + \varepsilon_{i,\mathrm{elec}} \tag{3.1}$$

Statistical Thermodynamics: Basics and Applications to Chemical Systems, First Edition. Iwao Teraoka.
© 2019 John Wiley & Sons, Inc. Published 2019 by John Wiley & Sons, Inc.
Companion website: www.wiley.com/go/Teraoka_StatsThermodynamics

where m_i is the mass of the atom that moves at velocity $\mathbf{v}_i$. Electrons are not excited to higher energy levels unless the temperature is as high as the one on the surface of the sun. We examine this situation in Section 3.6.4. In practice, $\varepsilon_{i,\text{elec}}$ is a constant, and therefore we do not consider $\varepsilon_{i,\text{elec}}$ in this book, unless otherwise mentioned. The atom is isotropic (no specific orientation), and therefore there is no rotation. Equation (3.1) applies also to an isolated ion in the absence of applied field.

Classically, the kinetic energy of an atom of mass m moving at velocity $\mathbf{v}$ is $\frac{1}{2}m\mathbf{v}^2$. Let us estimate a typical value. At room temperature, the speed of a helium molecule is around $1400\,\text{m s}^{-1}$. Its mass is $(4 \times 10^{-3}\,\text{kg mol}^{-1})/(6.02 \times 10^{23}\,\text{mol}^{-1}) = 6.6 \times 10^{-27}\,\text{kg}$. Therefore, $\frac{1}{2}m\mathbf{v}^2$ is around $6.5 \times 10^{-21}\,\text{J}$. This value is close to thermal energy, $k_B T = 4.1 \times 10^{-21}\,\text{J}$, for a reason.

In the presence of a field, the atom and the ion may acquire additional energy. This energy is called a **potential**. Here, we consider two potentials.

3.1.2 Gravitational Potential

An atom of mass m at height h from a reference level has the potential ϕ given as

$$\phi = mgh \tag{3.2}$$

where g is the acceleration by gravity. The position of the atom determines ϕ. Since the difference in mgh among molecules in a typical box we see in the laboratory is small, this potential is not important.

3.1.3 Ion in an Electric Field

When an electric field $\mathbf{E}$ is applied, it does not give an additional potential to the neutral atom, but a charged particle such as an ion gains an extra potential. Suppose that an electrostatic potential φ generates $\mathbf{E}$ ($\mathbf{E} = -\nabla\varphi$). Then, the potential of the ion (of charge q_i) is $\phi = q_i\varphi$. A uniform $\mathbf{E}$, say vertically up, may be generated by a pair of parallel electrodes carrying charges of the opposite signs. The potential for a positive ion decreases as it gets closer to the negative electrode (thus, φ decreases); see Figure 3.1. Since $\varphi = \varphi(\mathbf{r})$, $\phi = \phi(\mathbf{r})$, where $\mathbf{r}$

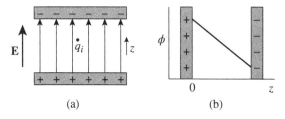

Figure 3.1 (a) An ion of charge q_i in a uniform electric field $\mathbf{E}$ vertically upward (+z direction). The field may be generated by a pair of parallel electrodes carrying the opposite charges. (b) The potential ϕ of the ion (assume $q_i > 0$), plotted as a function of the distance of the charge from the positive electrode.

is the position of the ion. For a negative ion, $q_i < 0$, and therefore ϕ increases with an increasing distance from the positive electrode. As is the case with the gravitational potential, the potential by the electric field is determined by the position of the ion.

3.1.4 Total Energy of Atoms and Ions

Combining the energy ε_i of the isolated atom at $\mathbf{r}_i$ and the potential energy $\phi(\mathbf{r}_i)$, the total energy E of a system consisting of N atoms is expressed as

$$E = \sum_{i=1}^{N} \varepsilon_i + \sum_{i=1}^{N} \phi(\mathbf{r}_i) \tag{3.3}$$

Note that E is expressed microscopically: E depends on the position and the velocity of all the atoms within the system.

3.2 Kinetic Energy and Potential Energy of Diatomic Molecules

3.2.1 Kinetic Energy (Translation, Rotation, Vibration)

A diatomic molecule can move its center of mass in x, y, and z directions (translation), rotate around its center of mass, and change its bond length (vibration); see Figure 3.2.

The degree of freedom is listed in Table 3.1 for each of the three modes of motion. We learn subsequently that the rotation of a diatomic molecule has two degrees of freedom. The total number is 6, equal to the degree of freedom for each atom times the number of atoms in the molecule. The covalent bond

Figure 3.2 Three modes of motion of a diatomic molecule: (a) center-of-mass translation, (b) rotation, and (c) vibration.

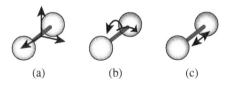

(a) (b) (c)

Table 3.1 Degrees of freedom in three modes of motion of a diatomic molecule.

Mode of motion	Degree of freedom
Translation	3
Rotation	2
Vibration	1
Total	6

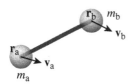

Figure 3.3 Two atoms A and B (mass m_a and m_b) that constitute a diatomic molecule are at $\mathbf{r}_a$ and $\mathbf{r}_b$, respectively, and moving at velocities $\mathbf{v}_a$ and $\mathbf{v}_b$.

between the two nuclei just adds a restriction on the movement, but does not change the total degrees of freedom.

Now we consider movement of the two atoms in the diatomic molecule. The molecule consists of two atoms with mass m_a and m_b at $\mathbf{r}_a$ and $\mathbf{r}_b$, respectively, moving at velocities $\mathbf{v}_a$ and $\mathbf{v}_b$, connected by a covalent bond (see Figure 3.3). The covalent bond places a restriction on the internuclear distance $r_{ab} = |\mathbf{r}_{ab}|$, where $\mathbf{r}_{ab} \equiv \mathbf{r}_a - \mathbf{r}_b$. The restriction can be expressed by a potential $\Phi(r_{ab})$. This potential is not due to an external field, and therefore is not a part of ϕ we considered in Section 3.1. The kinetic energy of the molecule is expressed as

$$\varepsilon_{kin} = \frac{1}{2} m_a \mathbf{v}_a{}^2 + \frac{1}{2} m_b \mathbf{v}_b{}^2 + \Phi(r_{ab}) \tag{3.4}$$

Here we convert Eq. (3.4) into a three-part, molecular-level expression of the kinetic energy:

$$\varepsilon_{kin} = \varepsilon_{trans} + \varepsilon_{rot} + \varepsilon_{vib} \tag{3.5}$$

We start with conversion of $m_a \mathbf{v}_a{}^2 + m_b \mathbf{v}_b{}^2$ by introducing the center of mass $\mathbf{R}$:

$$\mathbf{R} \equiv \frac{m_a \mathbf{r}_a + m_b \mathbf{r}_b}{m_a + m_b} \tag{3.6}$$

Then, the velocity $\mathbf{V}$ of the center of mass is

$$\mathbf{V} \equiv \frac{d\mathbf{R}}{dt} = \frac{m_a \mathbf{v}_a + m_b \mathbf{v}_b}{m_a + m_b} \tag{3.7}$$

We can show that $m_a \mathbf{v}_a{}^2 + m_b \mathbf{v}_b{}^2$ is rewritten as

$$m_a \mathbf{v}_a^2 + m \mathbf{v}_b^2 = M \mathbf{V}^2 + \mu (\mathbf{v}_a - \mathbf{v}_b)^2 \tag{3.8}$$

where

$$M = m_a + m_b \tag{3.9}$$

is the mass of the molecule, and

$$\mu \equiv \frac{m_a m_b}{m_a + m_b} \tag{3.10}$$

is called the **reduced mass** of the molecule. It is the harmonic mean of m_a and m_b:

$$\frac{1}{\mu} = \frac{1}{m_a} + \frac{1}{m_b} \tag{3.11}$$

Figure 3.4 Potential Φ (solid line) for the covalent bond in a diatomic molecule minimizes to $-\varepsilon_0$ when the internuclear distance r is equal to its equilibrium length r_0. The potential at around $r = r_0$ can be approximated by a harmonic potential (dashed line).

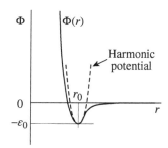

If $m_a = m_b$ (homonuclear diatomic molecule), $\mu = m_a/2$. If $m_a \ll m_b$ as in HCl (a is H), $\mu \cong m_a$; the reduced mass is close to the mass of the lighter atom.

In Eq. (3.8), $\mathbf{v}_a - \mathbf{v}_b = d\mathbf{r}_{ab}/dt$. Therefore,

$$\varepsilon_{kin} = \frac{1}{2}M\mathbf{V}^2 + \frac{1}{2}\mu\left(\frac{d\mathbf{r}_{ab}}{dt}\right)^2 + \Phi(r_{ab}) \tag{3.12}$$

Next, we consider $\Phi(r_{ab})$. Figure 3.4 shows a sketch of the potential Φ, plotted as a function of the internuclear distance $r = r_{ab}$. There is a minimum in Φ, and the minimum $-\varepsilon_0$ is reached when r is equal to its equilibrium length r_0. Either stretching the bond from r_0 or compressing the bond requires an extra energy. The zero level of the potential is arbitrarily set to the energy when r is infinite, that is, when the two atoms are dissociated. We may call ε_0 the bond energy. If r tries to become a lot smaller compared with r_0, the potential sharply rises, as the nonbonding electrons on the two atoms start to overlap.

The Taylor expansion of $\Phi(r)$ around $r = r_0$ is written as

$$\Phi(r) = \Phi(r_0) + (r - r_0)\left(\frac{d\Phi}{dr}\right)_{r_0} + \frac{1}{2}(r - r_0)^2\left(\frac{d^2\Phi}{dr^2}\right)_{r_0} + \cdots \tag{3.13}$$

where $d\Phi/dr$ at $r = r_0$ is zero. If we neglect the higher order terms and approximate $\Phi(r)$ by a constant plus the second-order term only (harmonic approximation), then

$$\Phi(r) = -\varepsilon_0 + \frac{1}{2}\kappa(r - r_0)^2 \tag{3.14}$$

where $\Phi(r_0) = -\varepsilon_0$ and

$$\kappa \equiv \left(\frac{d^2\Phi}{dr^2}\right)_{r_0} \tag{3.15}$$

is the force constant of the spring that is a model for the bond length change around its equilibrium length. As a rule of thumb, the force constant is around $500\,\text{N m}^{-1}$ for a single bond, and a double bond has a spring twice as strong. The force constant of a triple bond is three times as large.

The potential in Eq. (3.14) is called a **harmonic potential**, and the particle that moves in this potential is called a **harmonic oscillator**. From Eqs. (3.12) and (3.14), we obtain

$$\varepsilon_{kin} = \frac{1}{2}MV^2 + \frac{1}{2}\mu\left(\frac{dr_{ab}}{dt}\right)^2 + \frac{1}{2}\kappa(r_{ab} - r_0)^2 - \varepsilon_0 \tag{3.16}$$

We now examine dr_{ab}/dt. The $\mathbf{r}_{ab}$ changes when its length r_{ab} changes or its direction changes. We decompose $d\mathbf{r}_{ab}$, the change of $\mathbf{r}_{ab}$ in time dt, into these two parts. The length change is written as $dr_{ab} = (dr_{ab}/dt)dt$. The change in the direction causes $\mathbf{r}_{ab}$ to change by $\omega r_{ab}\,dt$ in the direction perpendicular to $\mathbf{r}_{ab}$, where ω is the angular velocity due to the rotation of the molecule. The molecule can rotate in two mutually perpendicular directions, and we express the two angular velocities as ω_1 and ω_2 (see Figure 3.5).

The three components are orthogonal. Therefore,

$$\left(\frac{d\mathbf{r}_{ab}}{dt}\right)^2 = \left(\frac{dr_{ab}}{dt}\right)^2 + r_{ab}^2(\omega_1^2 + \omega_2^2) \tag{3.17}$$

Since r_{ab} is close to its equilibrium value, r_0,

$$\left(\frac{d\mathbf{r}_{ab}}{dt}\right)^2 = \left(\frac{dr_{ab}}{dt}\right)^2 + r_0^2(\omega_1^2 + \omega_2^2) \tag{3.18}$$

Combining Eqs. (3.16) and (3.18), we obtain

$$\varepsilon_{kin} = \frac{1}{2}MV^2 + \frac{1}{2}\mu r_0^2(\omega_1^2 + \omega_2^2) + \frac{1}{2}\mu\left(\frac{dr_{ab}}{dt}\right)^2 + \frac{1}{2}\kappa(r_{ab} - r_0)^2 - \varepsilon_0 \tag{3.19}$$

It is obvious which terms in the equation belong to the three components of ε_{kin}:

$$\varepsilon_{trans} = \frac{1}{2}MV^2 \tag{3.20}$$

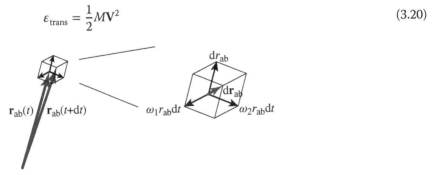

Figure 3.5 The bond vector changes from $\mathbf{r}_{ab}$ at time t to $\mathbf{r}_{ab} + d\mathbf{r}_{ab}$ in time dt by rotation and bond length change. The change $d\mathbf{r}_{ab}$ can be decomposed into a component dr_{ab} parallel to $\mathbf{r}_{ab}$ and two other components ($\omega_1 r_{ab}\,dt$, $\omega_2 r_{ab}\,dt$) perpendicular to $\mathbf{r}_{ab}$ (zoomed view).

$$\varepsilon_{\text{rot}} = \frac{1}{2}\mu r_0^2(\omega_1^2 + \omega_2^2) \tag{3.21}$$

$$\varepsilon_{\text{vib}} = \frac{1}{2}\mu\left(\frac{dr_{\text{ab}}}{dt}\right)^2 + \frac{1}{2}\kappa(r_{\text{ab}} - r_0)^2 \tag{3.22}$$

The constant $-\varepsilon_0$ simply shifts the energy level. Since the setting of the energy level is arbitrary, we can neglect the constant.

Here is a comment on the rotation. The rotational movement of a diatomic molecule can be recast into the movement of a point on a sphere. Imagine a sphere that has its center at the center of mass of the molecule and the most distant atom on the surface. As the molecule rotates, the atom moves on the sphere (Figure 3.6). Just as a person on the earth can travel in the north–south direction and in the east–west direction, the linear molecule can rotate in two independent directions.

In Eq. (3.21), μr_0^2 is the **moment of inertia**, and is usually expressed by the symbol I:

$$I = \mu r_0^2 \tag{3.23}$$

Table 3.2 lists I for some diatomic molecules. The moment of inertia represents the difficulty of changing the angular velocity, in the same way as the

Figure 3.6 A linear molecule such as a diatomic molecule can rotate in two directions (a), just as we can move north–south and east–west on the earth (b).

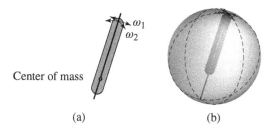

Center of mass

(a)   (b)

Table 3.2 Moment of inertia I for some diatomic molecules.

Molecule	$10^{47} I$ (kg m^2)
H_2	0.046 001
N_2	1.400 87
O_2	1.936 42
Cl_2	11.473
HCl	0.264 247
CO	1.449 44

Each element is the most abundant species.
Source: Data from Anderson 1989 [5].

mass expresses the difficulty of changing the linear velocity. The heavier the molecule, the greater the I.

3.2.2 Dipolar Potential

3.2.2.1 Potential of a Permanent Dipole

Homonuclear diatomic molecules such as H_2 and O_2 are nonpolar. Unless all the parts of the molecule are nonpolar, every molecule or a part of the molecule has an electric **dipole moment**. For example, each of the three bonds in a BF_3 molecule has an electric dipole, since the F atom is more electronegative than the B atom is. The three dipoles add up to zero; the molecule as a whole is nonpolar.

Here we consider a heteronuclear diatomic molecule, HCl. The difference in electronegativity of the two atoms leads to a partial positive charge on the H atom and a partial negative charge on the Cl atom. The HCl molecule may be modeled as a charge $+q$ (>0) and the opposite charge $-q$, separated by the bond length r. We define $\mathbf{r}$ as a bond vector that originates from the negative center within the Cl nucleus and ends at the positive center within the H nucleus. The dipole moment $\boldsymbol{\mu}$ of the HCl molecule is defined as

$$\boldsymbol{\mu} = q\mathbf{r} \tag{3.24}$$

See Figure 3.7.

Note that we use the same symbol μ to denote the dipole and the reduced mass. They are used in different situations, and there is no confusion.

You may have learned in organic chemistry that the dipole is in the direction from the positive center to the negative center. However, in all the other areas of chemistry and in the other disciplines of science, the dipole is defined as it is being introduced here.

The SI unit for $\mu = |\boldsymbol{\mu}|$ is C m. If a pair of unit charges ($q = e$, where e is the proton charge) are separated by $1\,\text{Å} = 10^{-10}$ m, the dipole moment is

$$1.602 \times 10^{-19}\,\text{C} \times 10^{-10}\,\text{m} = 1.602 \times 10^{-29}\,\text{C m} \tag{3.25}$$

Another unit, debye, is widely used, after a Dutch electrical engineer, Peter Debye (1884–1966), who made important contributions to chemical physics. One debye is equal to $10^{-21}/299\,742\,458$ C m. Then,

$$1.602 \times 10^{-29}\,\text{C m} = 4.8\,\text{debye} \tag{3.26}$$

See Figure 3.8.

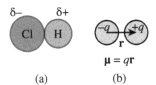

$$\boldsymbol{\mu} = q\mathbf{r}$$

(a) (b)

Figure 3.7 (a) Two atoms in a polar molecule HCl have partial charges. (b) Dipole moment μ is defined as the product of charge q and charge separation $\mathbf{r}$.

Figure 3.8 A pair of elementary charges with the opposite signs separated by 1 Å is equal to a dipole moment of 4.8 debye.

$|\boldsymbol{\mu}| = 4.8$ debye

Table 3.3 Dipole moments of diatomic and polyatomic molecules.

Molecule	μ (debye)
HF	1.82
HCl	1.08
CO	0.11
NO	0.15
HC≡N	2.98
H_2O	1.85
NH_3	1.47
HCHO	2.33

NIST Computational Chemistry Comparison and Benchmark Database, http://cccbdb.nist.gov/.

Table 3.3 lists values of the dipole moment for some polar molecules in the vapor phase.

In a polyatomic molecule, partial charges are distributed to all the atoms of the molecule. We can locate the mean position of all positive charges, $\mathbf{r}_{pos}$, weighted by the charges. Likewise, we can locate the mean position of all negative charges, $\mathbf{r}_{neg}$. The sum of the positive charges, q_{pos}, and the sum of the negative charges add up to zero. The dipole moment of the molecule is defined as $q_{pos}(\mathbf{r}_{pos} - \mathbf{r}_{neg})$.

In Section 3.1.3, we learned that a charged particle has a potential in the electric field. Since the dipole is equivalent to a charge q at $\mathbf{R} + \mathbf{r}$ and a charge $-q$ at $\mathbf{R}$, the potential of the dipole is

$$\phi_{dipole} = q\varphi(\mathbf{R} + \mathbf{r}) - q\varphi(\mathbf{R}) \tag{3.27}$$

where φ is the electrostatic potential. This equation is rewritten to

$$\phi_{dipole} = q[\varphi(\mathbf{R} + \mathbf{r}) - \varphi(\mathbf{R})] = q\mathbf{r} \cdot \nabla\varphi = \boldsymbol{\mu} \cdot \nabla\varphi = -\boldsymbol{\mu} \cdot \mathbf{E} \tag{3.28}$$

The dipole moment $\boldsymbol{\mu}$ changes its orientation, as the molecule rotates or tumbles. In the presence of external electric field $\mathbf{E}$, the dipole has a potential energy given as

$$\phi_{dipole} = -\mu E \cos\theta \tag{3.29}$$

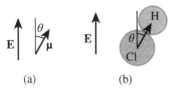

Figure 3.9 (a) The dipole μ is oriented in the direction at θ from the electric field **E**. (b) The potential energy of the dipole in an HCl molecule depends on θ.

where $E = |\mathbf{E}|$, and θ is the angle between **E** and μ (see Figure 3.9). The energy is the lowest when μ is parallel to **E**, and the highest when μ is antiparallel to **E**.

As opposed to the other type of electric dipole discussed subsequently, μ in the polar molecule or on a polar bond is called a **permanent dipole moment**.

3.2.2.2 Potential of an Induced Dipole

There is a dipole moment that is zero in the absence of the applied electric field **E**, but nonzero in its presence. The dipole moment is induced within a nonpolar molecule. Even a helium atom can have an **induced dipole**.

In the He atom, two electrons are moving around the positively charged nucleus. In the absence of **E**, the center of the two electrons coincides with the nucleus on the average. There is no dipole on the average.

Now we apply **E**. The latter can be realized by a pair of parallel electrodes of the opposite polarities. The upward **E** is generated by a positive electrode at the bottom and a negative electrode at the top (see Figure 3.10).

We place the helium atom somewhere between the two electrodes. Suppose the nucleus fixed. Then, the electrons are attracted toward the positive electrode. The center of the negative charges is closer to the positive electrode than the nucleus is. As a result, the helium atom gains a dipole moment μ_{ind}, oriented upward. When **E** is inverted, μ_{ind} is in the other direction; μ_{ind} is proportional to **E**. The proportionality constant is called a **polarizability** (molecular polarizability) and is usually denoted by α

$$\mu_{\text{ind}} = \alpha \mathbf{E} \tag{3.30}$$

The potential energy of the atom polarized in this way is expressed as

$$\phi = -\frac{1}{2}\alpha E^2 \tag{3.31}$$

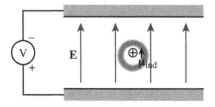

Figure 3.10 A nonpolar helium atom in an electric field **E** generated by a pair of parallel electrodes. Electrons are attracted toward the positive electrode, resulting in an induced dipole, μ_{ind}.

The coefficient " $\frac{1}{2}$ " is due to the same reason as the one in the elastic energy of a spring, $\frac{1}{2}\kappa x^2$. The minus sign indicates that the dipole lowers the energy. As a rule of thumb, α is proportional to the number of electrons in the molecule. Therefore, α is roughly proportional to the molar mass of the molecule. Variations come from the mobility of electrons: delocalized electrons increase α. Tight binding of orbital electrons by strong Coulomb attraction of the nucleus decreases α. The latter is most clearly seen in fluorine atoms or in covalent bonds with F atoms.

A polar molecule has also an induced dipole. The latter is generated on top of the permanent dipole.

It is customary to list the **polarizability volume** α' defined as

$$\alpha' \equiv \frac{\alpha}{4\pi\varepsilon_0} \qquad (3.32)$$

rather than α, where $\varepsilon_0 = 8.854 \times 10^{-12}$ F m^{-1} is the vacuum permittivity. The α' has the dimension of (length)^3. Table 3.4 lists α' for some monatomic, diatomic, and polyatomic molecules.

There are several simple physical chemistry applications that calculate μ and α' when you construct a molecular structure on a drawing window. These applications include ChemAxon, J-Chem, and ChemDraw.

Table 3.4 Polarizability volumes of some monatomic, diatomic, and polyatomic molecules.

Molecule	α' (Å^3)
He	0.208
Ne	0.381
Ar	1.664
H_2	0.787
N_2	1.710
O_2	1.562
F_2	0.997
Cl_2	4.671
HCl	2.515
CO	1.953
CO_2	2.507
HC≡CH	3.487

Calculated with Marvin Sketch from ChemAxon.

3.3 Kinetic Energy of Polyatomic Molecules

A molecule consisting of three or more atoms has more degrees of freedom than the diatomic molecule does. Then, the expression of the kinetic energy is more complicated than the one given by Eq. (3.19). Both ε_{rot} and ε_{vib} can be different from those in that equation. Table 3.5 lists the degrees of freedom in translation, rotation, and vibration for diatomic, linear polyatomic, and nonlinear polyatomic molecules. In this section, we learn what we should expect for the kinetic energy of a polyatomic molecule without going into details.

Polyatomic molecules are divided into linear molecules and nonlinear molecules. A linear molecule is straight. CO_2, H—C≡N, and H—C≡C—H are typical examples. In a nonlinear molecule, not all of its atoms are on a straight line. Some of the nonlinear molecules are planar (H_2O, HCHO, etc.) and others are nonplanar (NH_3, CH_4, etc.).

3.3.1 Linear Polyatomic Molecule

The parts of ε_{trans} and ε_{rot} in a linear polyatomic molecule are identical to those in the diatomic molecule, but the vibrational part is different. The polyatomic molecule has two or more bonds. Each of the bonds can change its length, and the angle between adjacent bonds can also change. As we did for the diatomic molecule, we isolate the vibrational modes of motion from the center-of-mass translation and rotation.

Let us consider a CO_2 molecule (see Figure 3.11a). Suppose we stretch one of the bonds while holding the other bond and the bond angle unchanged. This mode of change causes a translation of the center of mass. To prevent the center-of-mass translation, stretching one of the two C=O bonds must be compensated for by stretching the other C=O bond by the same amount. Under the condition that the molecule as a whole will not change its center of mass or rotate, stretching one of the bonds necessitates stretching or contraction of the other bonds, in general. Consequently, all the bond length changes and

Table 3.5 Degrees of freedom in three modes of motion for diatomic, linear polyatomic, and nonlinear polyatomic molecules. A polyatomic molecule has n atoms.

Mode of motion	Diatomic	Linear polyatomic	Nonlinear polyatomic
Translation	3	3	3
Rotation	2	2	3
Vibration	1	$3n - 5$	$3n - 6$
Total	6	$3n$	$3n$

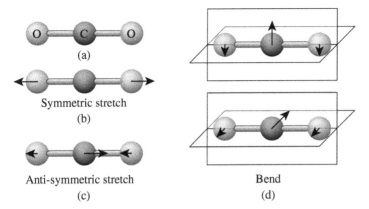

Figure 3.11 Normal modes in a CO_2 molecule (a). Movements of atoms are shown for (b) symmetric stretch, (c) antisymmetric stretch, and (d) bend modes. The bending can occur in two directions (a degeneracy of two).

bond angle changes in the molecule are related to each other. This situation is greatly simplified by adopting normal coordinates. The latter reorganize the mutually related movements of atoms into mutually independent modes, and each of them is a specific combination of atom movements or changes in the bond lengths and angles. These modes are called **normal modes**.

In the CO_2 molecule, for example, the two C=O bonds can stretch in phase or stretch out of phase (see Figures 3.11b,c). The in-phase stretch is called a symmetric stretch, and the out-of-phase stretch is called an antisymmetric stretch. The symmetric stretch does not move the center of mass of the molecule. Since the antisymmetric stretch moves the two O atoms in the same direction relative to the C atom, the C atom must move in the other direction to keep the center of mass at the same position. The molecule has two more normal modes – a bond angle change or a bend. The bend can occur in two directions (see Figure 3.11d). We say that the bend mode has a degeneracy of two.

Each of the n normal modes (ith mode; $i = 1, 2, ..., n$) has its own reduced mass μ_i and force constant κ_i. Thus, the vibrational part of the energy is expressed as

$$\varepsilon_{\text{vib}} = \frac{1}{2} \sum_{i=1}^{n} \left[\mu_i \left(\frac{dX_i}{dt} \right)^2 + \kappa_i X_i^2 \right] \tag{3.33}$$

where X_i is the ith normal coordinate that represents the collective deviation of the atoms' positions from those at the equilibrium, specific to the mode. For example, X_i for the symmetric stretch of $O_a C_b O_c$ (a, b, and c are labels) is $2^{-1/2}[(r_{ab} - r_{ab0}) + (r_{bc} - r_{bc0})]$, where r_{ab} and r_{bc} are the lengths of the two CO bonds, and r_{ab0} and r_{bc0} are their equilibrium lengths. The kinetic energy of the linear polyatomic molecule is expressed as

$$\varepsilon_{\text{kin}} = \frac{1}{2}MV^2 + \frac{1}{2}\mu r_0^{\ 2}(\omega_1^2 + \omega_2^2) + \frac{1}{2}\sum_{i=1}^{n}\left[\mu_i\left(\frac{dX_i}{dt}\right)^2 + \kappa_i X_i^2\right] \qquad (3.34)$$

3.3.2 Nonlinear Polyatomic Molecule

For a diatomic molecule and a linear polyatomic molecule, we did not include the rotation around its axis, since such a movement does not lead to any change. When the molecule is not straight, however, we need to take into account the rotation around the major axis of the molecule.

We consider a water molecule here. The molecule is usually placed on a y–z plane with the z axis passing the oxygen atom and the midpoint of the two hydrogen atoms (Figure 3.12).

The origin of the x, y, z coordinates is at the center of mass. Rotation around the z axis is easy, since it moves only the light hydrogen atoms. In contrast, the rotation around the y axis and rotation around the x axis are not as easy, since they move the heavy oxygen atom. The difficulty of rotation is expressed by the moment of inertia, and it is defined as

$$I = \sum_j m_j r_j^2 \qquad (3.35)$$

where j runs through all the atoms of the molecule, m_j is the mass of the jth atom and r_j is the distance of the atom from the axis of rotation (Figure 3.13).

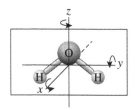

Figure 3.12 Rotation of a water molecule. The molecule is placed on the yz plane, and the z axis passes the oxygen atom and the midpoint of the two hydrogen atoms. The origin is at the center of mass of the molecule. There are three axes of rotation, each with a different moment of inertia.

Axis of rotation

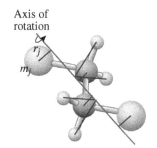

Figure 3.13 The moment of inertia around the axis of rotation is equal to the sum of the product of mass m_j of atom j and the square of its distance r_j to the axis, and the sum is taken with respect to all the atoms in the molecule. The drawing is shown for dichloroethane.

The moment of inertia is different for each of the three rotations, and the energy of rotation is expressed as

$$\varepsilon_{\text{rot}} = \frac{1}{2} \sum_{i=1}^{3} I_i \omega_i^2 \tag{3.36}$$

where the three values of i are for the rotations around x, y, and z directions, and ω_i is the angular velocity for the ith rotational mode. The same expression applies to other nonlinear molecules, but I_i are different for each molecule and for each mode of rotation, in general. The bulkier the molecule, the greater the I_i, since r_j is greater. The heavier the atoms are, the larger the I_i. We note here that, in some molecules, two or three of I_i are identical.

The vibrational part of the energy in the nonlinear molecule is similar to the one in the linear molecule. We just need to list all the normal modes. Figure 3.14 shows the three normal modes in a water molecule that is planar.

The situation is a bit more complicated for a nonplanar molecule. Here we look at an ammonium molecule. In Table 3.5, $n = 4$, and therefore the molecule has six normal modes, as illustrated in Figure 3.15. Here, we pay attention to asymmetric stretch and asymmetric scissoring. The asymmetric stretch has two independent modes, and we say that the mode has a degeneracy of two. You can envision two of the three N—H bonds stretching while the other bond is contracting. There are three ways to choose two out of three bonds for the stretch,

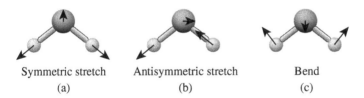

Symmetric stretch	Antisymmetric stretch	Bend
(a)	(b)	(c)

Figure 3.14 Normal modes in a H_2O molecule. Movements of atoms are shown for symmetric stretch, antisymmetric stretch, and bend modes. All the modes are nondegenerate.

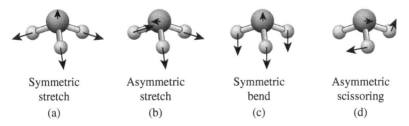

Symmetric stretch	Asymmetric stretch	Symmetric bend	Asymmetric scissoring
(a)	(b)	(c)	(d)

Figure 3.15 Normal modes in a NH_3 molecule. Movements of atoms are shown for symmetric stretch, asymmetric stretch, symmetric bend (umbrella), and asymmetric scissoring modes. Each of the two asymmetric modes has a degeneracy of two.

but only two of the three asymmetric stretch modes are independent, since the third can be constructed from the first two asymmetric stretch modes. The same statement applies to asymmetric scissoring modes.

Combining the three elements of kinetic energy, we have the expression of ε_{kin} for a nonlinear molecule as

$$\varepsilon_{kin} = \frac{1}{2}M\mathbf{V}^2 + \frac{1}{2}\sum_{i=1}^{3} I_i\omega_i^2 + \frac{1}{2}\sum_{i=1}^{n}\left[\mu_i\left(\frac{dX_i}{dt}\right)^2 + \kappa_i X_i^2\right] \tag{3.37}$$

where n is the number of normal modes for vibration. We need to keep in mind that for degenerate modes, κ_i of the modes are identical and so are μ_i.

3.4 Interactions Between Molecules

In many systems, molecules interact with each other, and gases are not an exception. The interaction is an extra component of the energy that comes on top of the energy that each molecule has, independent of the other molecules. The latter consists of the kinetic energy and the potential energy, as we learned in the preceding sections.

The interaction is primarily between a pair of molecules (**binary interaction**). When the pressure is low, the interaction is negligible, as the closest other molecule is far away. With an increasing pressure, the distance between the nearest pair of molecules decreases; and, concomitantly, the binary interaction becomes stronger. In the kinetic theory of gas, chances of a molecule colliding with another molecule increase. Only when the binary interaction is sufficiently strong, higher-order interactions such as those among three molecules (tertiary interactions) come into scope. Here, we focus on the binary interactions.

We consider a system of N molecules. There are $_NC_2 = \frac{1}{2}N(N-1)$ pairs of molecules in the system, and that is the number of possible interactions. Figure 3.16 shows a system of five molecules, and there are 10 pairs.

In the system of N molecules, the total interaction E_{int} is expressed as a series of pairwise interactions Φ_{ij} between molecules i and j:

$$E_{int} = \frac{1}{2}\sum_{i\neq j}^{N} \Phi_{ij} \tag{3.38}$$

where the series in the Eq. (3.38) is a short-hand notation of the double series:

$$\frac{1}{2}\sum_{i\neq j}^{N} = \sum_{i>j}^{N} = \sum_{i=2}^{N}\sum_{j=1}^{i-1} \tag{3.39}$$

Figure 3.17 depicts two molecules i and j close to each other that may interact. For simplicity, we assume that the interaction depends on the distance r_{ij}

Figure 3.16 Five interacting particles can have a total of 10 interactions.

Figure 3.17 Repulsive interaction between two molecules *i* and *j* at $\mathbf{r}_i$ and $\mathbf{r}_j$. The force $\mathbf{F}_i$ on molecule *i* by the interaction is just the opposite of the force $\mathbf{F}_j$ on molecule *j*.

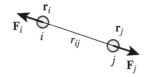

between the two molecules and not on the orientation of the two molecules or the direction of molecule *i* relative to molecule *j*. Let us denote by $\mathbf{r}_i$ and $\mathbf{r}_j$ the positions of the two molecules. Then, $r_{ij} = |\mathbf{r}_i - \mathbf{r}_j|$, and the interaction Φ is expressed as $\Phi(r_{ij})$. The Φ depends on $\mathbf{r}_i$ and $\mathbf{r}_j$ through r_{ij}. Since Φ is a potential, the force $\mathbf{F}_i$ on molecule *i* by the interaction is obtained by differentiating Φ by $\mathbf{r}_i$:

$$\mathbf{F}_i = -\frac{\partial \Phi}{\partial \mathbf{r}_i} = -\Phi'(r_{ij})\frac{\partial r_{ij}}{\partial \mathbf{r}_i} \tag{3.40}$$

where the prime denotes the derivative by the argument, i.e. $\Phi'(r_{ij}) = d\Phi/dr_{ij}$, and $\partial/\partial \mathbf{r}_i$ is a short-hand notation for $[\partial/\partial x_i, \partial/\partial y_i, \partial/\partial z_i]^{\mathrm{T}}$ (T indicates a transpose). Likewise, the force $\mathbf{F}_j$ on molecule *j* by the interaction is expressed as

$$\mathbf{F}_j = -\frac{\partial \Phi}{\partial \mathbf{r}_j} = -\Phi'(r_{ij})\frac{\partial r_{ij}}{\partial \mathbf{r}_j} \tag{3.41}$$

Since $\partial r_{ij}/\partial \mathbf{r}_i = -\partial r_{ij}/\partial \mathbf{r}_j$, we find that $\mathbf{F}_i = -\mathbf{F}_j$. It is an example of the law of action and reaction.

Note that $\mathbf{r}_i$ and $\mathbf{r}_j$ being used here denote the positions of the molecules, whereas the same symbols denoted the positions of nuclei in the diatomic molecule in Section 3.2. Likewise, Φ in this section is for the interaction, and not the potential for the bond length.

Molecules interact with each other through a variety of mechanisms. Commonly observed interactions are listed in Table 3.6 for repulsive interactions and attractive interactions, separately. The table also classifies the interactions by their ubiquity. The prevalent interactions are **excluded-volume interaction** (repulsive) and **van der Waals interaction**, aka, **London dispersion force** (attractive). We learn later why these two interactions exist for any pair of molecules. Less common compared with these two interactions, but still quite often observed interactions are **dipole–dipole interaction** (usually attractive) and **electron donor–acceptor interaction** (attractive). The latter interaction can be strong. Hydrogen bonding belongs to the donor–acceptor interaction.

Table 3.6 Common molecular interactions.

Present in	Repulsive interaction	Attractive interaction
Any system	Excluded volume	van der Waals interaction (London dispersion force)
Most systems		Dipole–dipole interaction
Some systems	Coulomb repulsion	Donor–acceptor interaction (including hydrogen bonding)
		Coulomb attraction

For charged molecules and ions, Coulomb interactions between the same charges and Coulomb interactions between the opposite charges are also important.

Here, we look at these interactions, paying attention to how the interactions depend on the distance between the two molecules. The hydrogen bond is somewhere between a covalent bond and the dipolar interaction, and it is not discussed here.

3.4.1 Excluded-Volume Interaction

If a molecule is modeled by a hard, impenetrable sphere of radius R_s, the distance r between the centers of the two spheres cannot be less than $2R_s$. The potential Φ by the excluded volume is $+\infty$ for $r < 2R_s$ and 0 for $r > 2R_s$:

$$\Phi(r) = \begin{cases} +\infty & (r < 2R_s) \\ 0 & (r > 2R_s) \end{cases} \tag{3.42}$$

See Figure 3.18a. Sometimes, this hard-core potential is approximated by

$$\Phi(r) = \frac{\text{constant}}{r^{12}} \tag{3.43}$$

The large exponent on r makes the two functions close to each other (see Figure 3.18b).

3.4.2 Coulomb Interaction

The Coulomb interaction Φ between a pair of charges q_1 and q_2 separated by r is simple in a gas phase:

$$\Phi(r) = \frac{q_1 q_2}{4\pi\varepsilon_0 r} \tag{3.44}$$

The Φ is positive (repulsive) for the same charges, and negative (attractive) for the opposite charges. Unlike any other interactions, the Coulomb interaction

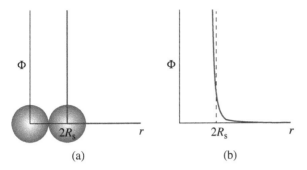

Figure 3.18 Excluded-volume interaction. (a) Hard-core potential Φ between a pair of hard spheres of radius R_s, plotted as a function of the center-to-center distance r. $\Phi = +\infty$ for $r < 2R_s$; 0 for $r > 2R_s$. (b) The hard-core potential can be approximated by a r^{-12} potential.

is long-ranged, since the decay of $\Phi(r)$ with an increasing r is slow. When the distance is 10 times as long, the interaction is still just one-tenth.

3.4.3 Dipole–Dipole Interaction

Although this interaction is not as universal as the excluded volume and the van der Waals interaction, and is not as strong as the Coulomb interaction and the acid–base interaction, the dipolar interaction exists between a pair of polar molecules and between polar bonds of molecules. In principle, the dipole–dipole interaction is just the sum of interactions between the partial charges that constitute the dipoles. Since the separation between the two partial charges of the opposite signs within a dipole is small, it can be shown that the interaction decays more rapidly with an increasing distance between the dipoles compared with the decay for the Coulomb interaction. The exact formula for the interaction Φ between dipoles μ_i and μ_j, separated by r, is as follows:

$$\Phi(r) = \frac{\mu_i \mu_j}{4\pi\varepsilon_0 r^3}[3\cos\theta_i \cos\theta_j - \cos(\theta_i + \theta_j)] \tag{3.45}$$

where angles θ_i and θ_j specify the orientations of the two dipoles relative to the vector that connects the two dipoles (Figure 3.19a). The dipolar interaction decays as r^{-3}, as opposed to the Coulomb interaction that decays as r^{-1}.

The interaction depends also on θ_i and θ_j. When the two dipoles are side by side and in the same direction as shown in Figure 3.19b, $\theta_i = \theta_j = \pi/2$, and Eq. (3.45) gives repulsive interaction: $\Phi(r) = \mu_i \mu_j/(4\pi\varepsilon_0 r^3)$. In the head-to-tail alignment (Figure 3.19c), $\theta_i = 0$ and $\theta_j = \pi$, and Eq. (3.45) gives attractive interaction: $\Phi(r) = -2\mu_i \mu_j/(4\pi\varepsilon_0 r^3)$. If the two dipoles are in opposite directions, the signs of $\Phi(r)$ will be reversed.

The interaction between the two dipoles results from the electric field at dipole j generated by the dipole i. Since the interaction on the dipole by the

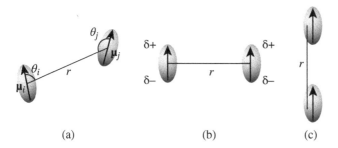

Figure 3.19 Interaction between dipole moments. (a) The interaction depends on r as r^{-3}, and also on the orientation of the dipoles relative to the direction of the two dipoles. (b) Side-by-side alignment of two dipoles oriented in the same direction leads to repulsive interaction. (c) Head-to-tail alignment of two dipoles oriented in the same direction leads to attractive interaction.

field is proportional to the magnitude of the field, we find that the dipolar field decays as r^{-3} with an increasing distance r from the dipole.

3.4.4   van der Waals Interaction

In Section 3.2.2.2, we learned that a nonpolar molecule does not have a dipole in the absence of the external field. The statement is true for the average (either long-time or ensemble) only. Let us consider a He atom. Certainly, the center (midpoint) of its two electrons overlaps with the nucleus on the average. However, at a given instance, the two electrons do not have their center at the nucleus position, in general. As electrons move, their center position changes, and it is usually away from the nucleus. As a result, there is an instantaneous dipole moment, which changes its magnitude and orientation all the time. The long-time average is zero, and so is the ensemble average. We can say that a nonpolar molecule has a **fluctuating dipole**. The origin of the fluctuating dipole is identical to the one for the induced dipole.

If a nonpolar molecule has a temporary dipole moment, there should be an interaction between a dipole in one molecule and a dipole in another molecule. Let us consider the ramification of the fluctuating dipole in the He atom. Figure 3.20 depicts two He atoms i and j displaced horizontally. In Figure 3.20a, the atom i temporarily acquires a dipole moment $\mu_{ind}(i)$ vertically up, which spawns an electric field. Its force line that passes the other He atom j is indicated by a dashed line with arrows. The electric field induces a dipole moment $\mu_{ind}(j)$ oriented vertically down in atom j. The two dipoles are in the opposite directions, leading to attractive interaction. If $\mu_{ind}(i)$ flips, $\mu_{ind}(j)$ also flips, and the same attractive interaction will result. In Figure 3.20b, $\mu_{ind}(i)$ is horizontal and directed toward atom j. The electric field at atom j spawned by

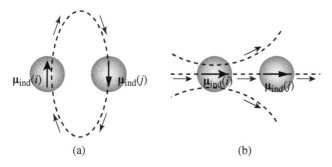

(a)   (b)

Figure 3.20 Origin of the van der Waals interaction between two He atoms *i* and *j* displaced horizontally. (a) A fluctuating dipole in atom *i*, $\mu_{ind}(i)$, is vertically up for the moment and induces a dipole $\mu_{ind}(j)$ in atom *j* vertically down. (b) A fluctuating dipole in atom *i* toward atom *j* induces $\mu_{ind}(j)$ in the same direction. In both situations, the dipole–dipole interaction is attractive.

$\mu_{ind}(i)$ induces $\mu_{ind}(j)$ in the same direction. The interaction between the two dipoles is attractive. Thus, we find that the fluctuating dipole leads to attractive interaction, regardless of the direction of $\mu_{ind}(i)$.

Now we consider how the magnitude of the interaction varies with the distance *r* between the two atoms. The electric field at atom *j* generated by $\mu_{ind}(i)$ is proportional to $\mu_{ind}(i)/r^3$. Therefore, $\mu_{ind}(j)$ is proportional to $\mu_{ind}(i)/r^3$, and the interaction between the two dipoles is proportional to $\mu_{ind}(i)\cdot\mu_{ind}(j)/r^3$, which is proportional to $[\mu_{ind}(i)]^2/r^6$. Since the interaction is attractive, we can write the van der Waals interaction as

$$\Phi(r) = -\frac{\text{constant}}{r^6} \qquad (3.46)$$

where the constant is positive. Recall that $\mu_{ind}(i)$ represents an induced dipole, and is therefore proportional to the polarizability of the molecule, which in turn is roughly proportional to the molar mass of the atom or molecule. Then, we find that the constant in Eq. (3.46) is proportional to the square of the molar mass.

3.4.5 Lennard-Jones Potential

Now we combine the two universal interactions – the repulsion by the excluded volume and the attraction by the fluctuating dipoles. For the excluded volume, we adopt the r^{-12} potential, since it is the square of the potential for the van der Waals interaction. The following expression is often used for the potential Φ of the combined interaction:

$$\Phi(r) = \varepsilon_0 \left[\left(\frac{r_0}{r}\right)^{12} - 2\left(\frac{r_0}{r}\right)^6 \right] \qquad (3.47)$$

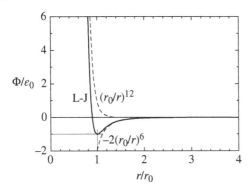

Figure 3.21 Lennard-Jones potential Φ, Eq. (3.47) (solid line). The potential Φ reduced by ε_0 minimizes to -1 when the center-to-center distance r is equal to r_0. The potential has a short-range repulsion and a long-range attraction in a continuous curve. The dashed lines represent the repulsion and attraction parts of the potential.

where r_0 is a length close to the linear dimension of the molecule. Note that ε_0 in this equation is not the vacuum permittivity. This potential is usually referred to as a **Lennard-Jones potential**.

The potential minimizes to $-\varepsilon_0$ at $r = r_0$. Figure 3.21 shows a plot of the Lennard-Jones potential Φ reduced by ε_0, together with its repulsive component $(r_0/r)^{12}$ and attractive component $-2(r_0/r)^6$. The Lennard-Jones potential is repulsive at short distances and attractive at long distances. At $r = 2r_0$, $\Phi/\varepsilon_0 \cong -0.03$; we can assume that the Lennard-Jones potential reaches the distance of $\sim 2r_0$.

Table 3.7 lists the values of ε_0 and r_0 for some nonpolar molecules. The values are different from reference to reference. There are still efforts to obtain the Lennard-Jones parameters, empirically and by quantum-chemical calculations.

Table 3.7 Lennard-Jones parameters. ε_0 is divided by the Boltzmann constant k_B.

Molecule	ε_0/k_B (K)	r_0 (Å)
Argon (Ar)	112	4.07
Krypton (Kr)	155	4.37
Xenon (Xe)	214	4.78
Nitrogen (N_2)	92	4.40
Oxygen (O_2)	113	4.10
Fluorine (F_2)	104	4.01
Clorine (Cl_2)	296	5.03
Methane (CH_4)	140	4.51
Ethane (C_2H_6)	216	5.37
Propane (C_3H_8)	255	6.14
Ethylene (C_2H_4)	201	5.15
Benzene (C_6H_6)	377	6.93
Carbon dioxide (CO_2)	202	4.99

Source: Data from Cuadros et al. 1996 [6].

As seen in Table 3.7, $2r_0 \cong 1$ nm for these small molecules. The magnitude of ε_0 depends on the molar mass. For O_2, for example, $\varepsilon_0/k_B = 113$ K. Therefore, at room temperature, the thermal energy is sufficient for an oxygen molecule to get out of the attraction of other oxygen molecules. At low temperatures, the molecules tend to get close to each other.

3.5 Energy as an Extensive Property

We have learned that the energy of a system of N molecules is expressed as

$$E = \sum_{i=1}^{N} \varepsilon_i + \sum_{i=1}^{N} \phi(\mathbf{r}_i) + \frac{1}{2} \sum_{i \neq j}^{N} \Phi(r_{ij}) \tag{3.48}$$

where ε_i is the energy of isolated molecule i in the absence of an external field, $\phi(\mathbf{r}_i)$ is the potential of the molecule by external fields, and $\Phi(r_{ij})$ represents the interaction between molecules i and j separated by r_{ij}.

Each of the first two terms in Eq. (3.48) scales with the size of the system, typically the number N of molecules if the system consists of the same type of molecules. We call it an **extensive** property.

In contrast, the interaction term in Eq. (3.48) consists of $\frac{1}{2}N(N-1)$ combinations of molecules i and j. Therefore, strictly speaking, E is not an extensive variable. However, as proved subsequently, E is can be regarded as extensive. To prove it, let us divide the whole system into two parts, A and B, as shown in Figure 3.22. There is no wall between A and B; just imagine a boundary without any physical barrier.

We learned that the interaction is short-ranged and only the pairs of molecules in close proximity count in the series in $\Sigma_{i \neq j}{}^N$. The pair is either between two molecules in part A, between two molecules in part B, or between a molecule in A and a molecule in B. We separate the series into three parts: a partial series Σ_{AA} for the pairs within A, a partial series Σ_{BB} for the pairs within B, and a partial series Σ_{AB} for the pairs across the A–B boundary. For the latter, the molecule in part A must be close to the boundary, typically within 1 nm, and the molecule in part B must be also close to the boundary, on top of the condition that A and B be close to each other. The extra requirement for the

Figure 3.22 A system divided into two parts A and B. The number of pairs of molecules in close proximity across the boundary is negligible compared with the number of pairs within A and the number of pairs within B.

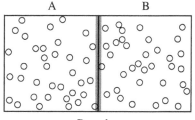

Boundary

interaction across the boundary leads to the conclusion that the number of such pairs is orders of magnitude as small as the number of nearby pairs in Σ_{AA} and the number of nearby pairs in Σ_{BB}, as long as the linear dimension of the system is a lot larger than 1 nm: $\Sigma_{AA} \gg \Sigma_{AB}$ and $\Sigma_{BB} \gg \Sigma_{AB}$. The division into A and B is arbitrary, i.e. we can place the boundary anywhere within the parent system. Therefore, the interaction also qualifies as an extensive variable. And we can write the energy for the system of A + B as consisting of two parts:

$$E_{A+B} = E_A + E_B \qquad (3.49)$$

3.6 Kinetic Energy of a Gas Molecule in Quantum Mechanics

In the preceding sections of this chapter, we used classical mechanics to derive the molecular level expressions for the kinetic energy of a molecule in vapor phase. However, classical mechanics fails to describe the microscopic world of molecules, in general. Rather, quantum mechanics rules the world. The quantum mechanics is built upon just a handful of assumptions. Since there are no experimental results that go against the predictions of the quantum mechanics, we consider that the quantum mechanics correctly describes the movement of a molecule and movement of electrons and nuclei within the molecule. We also consider that the formulas derived from the quantum mechanics are correct and exact.

Quantum mechanics applied to chemistry is called quantum chemistry. It requires one or two semesters dedicated to the subject just to scratch the surface. Here, we quickly run through the ideas of the quantum mechanics applied to the center-of-mass translation, rotation, and vibration, and the results obtained from the application. We do not consider electrons here, and pay attention to the states of the molecule as a whole and their energy levels. In Chapter 5, we find that the classical expression of ε_{kin} we derived in Section 3.2 is the high-temperature approximation of the exact quantum mechanical expression.

Here, we look at the quantum mechanical versions of ε_{trans}, ε_{rot}, and ε_{vib} separately. The essential part is the **quantization** (aka discretization) of the energy.

3.6.1 Quantization of Translational Energy

In classical mechanics, x, y, z components of the velocity can be any real number. Consequently, ε_{trans} can be any nonnegative number. We say that the energy is continuous. In quantum mechanics, ε_{trans} is discrete.

Imagine a molecule (or a particle) of mass m in a box of L_x, L_y, and L_z (Figure 3.23). The particle is represented by a function called a **wave function,**

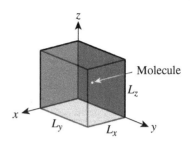

Figure 3.23 A molecule in a rectangular box of L_x, L_y, and L_z.

$\psi(\mathbf{r})$. Presence of walls forces $\psi(\mathbf{r})$ to be zero when $\mathbf{r}$ is on one of the walls; but otherwise, the box does not pose any restriction to the position or movement of the particle. This requirement leads to limiting $\psi(\mathbf{r})$ to $\psi_x(x)\psi_y(y)\psi_z(z)$, where

$$\psi_x(x) = \left(\frac{2}{L_x}\right)^{1/2} \sin\frac{\pi n_x x}{L_x} \tag{3.50}$$

where $n_x = 1, 2, 3, \ldots$.

The other two functions, $\psi_y(y)$ and $\psi_z(z)$, are similar to Eq. (3.50) except that x is now replaced with y and z, respectively. The three positive integers, n_x, n_y, and n_z, are the **quantum numbers** for the translation. The prefactor $(2/L_x)^{1/2}$ normalizes $\psi_x(x)$, i.e.

$$\int_0^{L_x} [\psi_x(x)]^2 dx = 1 \tag{3.51}$$

Therefore, the overall wave function $\psi(\mathbf{r}) = \psi_x(x)\psi_y(y)\psi_z(z)$ satisfies

$$\int_{\text{box}} [\psi(\mathbf{r})]^2 d\mathbf{r} = 1 \tag{3.52}$$

Equation (3.52) indicates that $[\psi(\mathbf{r})]^2$ may qualify as a probability density function. In fact, $[\psi(\mathbf{r})]^2 d\mathbf{r}$ represents the probability of finding the particle in a small volume $d\mathbf{r}$ around $\mathbf{r}$. Figure 3.24 shows a plot of $[\psi_x(x)]^2$ for $n_x = 1, 2, 3,$ 4, and 5. With an increasing n_x, the number of waves in the box increases. You can expect that, when n_x is sufficiently large, the waves will be almost invisible, and thus $[\psi_x(x)]^2$ is almost constant at $1/L_x$. It means that the particle is uniformly distributed along the x direction. Likewise, when n_x, n_y, and n_z are large, $[\psi(\mathbf{r})]^2$ is almost constant at $1/(L_x L_y L_z)$, and the particle is uniformly distributed within the box. In classical mechanics, the particle should be uniformly distributed, since there is no potential to preferentially locate the particle around one spot or toward a wall. Thus, we find that the large n_x, n_y, and n_z reproduces the situation that we know in classical mechanics.

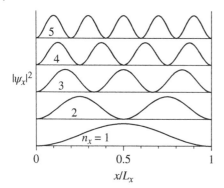

Figure 3.24 The square of the x component of the wave function, ψ_x, for a particle contained in a box of length L_x, plotted as a function of the position in the x direction. The square is proportional to the probability of finding the particle at x. The plot is shown for the five lowest-energy states. For clarity, the curves are shifted vertically. The numbers adjacent to the curves indicate the quantum numbers n_x.

The three integers, n_x, n_y, and n_z specify the **state** of the particle in the box. Each state has its own energy, and that is

$$\varepsilon_{\text{trans}} = \frac{h^2}{8m}\left(\frac{n_x^2}{L_x^2} + \frac{n_y^2}{L_y^2} + \frac{n_z^2}{L_z^2}\right) \tag{3.53}$$

where $h = 6.626 \times 10^{-34}$ J s is the **Planck's constant**.

The state of $n_x = n_y = n_z = 1$ gives the lowest $\varepsilon_{\text{trans}}$, and therefore called the **ground state**. If we choose $L_x \geq L_y \geq L_z$, the state with $n_x = 2$ and $n_y = n_z = 1$ gives the second lowest $\varepsilon_{\text{trans}}$, and so on. With an increasing n_x, n_y, and n_z, $\varepsilon_{\text{trans}}$ increases. Since only integral values of n_x, n_y, and n_z are allowed, the values of $\varepsilon_{\text{trans}}$ are discrete. However, if the spacing between adjacent levels is sufficiently small, the energy may look nearly continuous, as shown subsequently.

First, we find how large typical values of n_x, n_y, and n_z are. Consider a gas in a cube ($L_x = L_y = L_z = L$). We equate $\varepsilon_{\text{trans}}$ in Eq. (3.53) with its classical expression:

$$\frac{h^2}{8mL^2}(n_x^2 + n_y^2 + n_z^2) = \frac{1}{2}mv^2 \tag{3.54}$$

which leads to

$$(n_x^2 + n_y^2 + n_z^2)^{1/2} = \frac{2mvL}{h} \tag{3.55}$$

At room temperature, the speed v of a hydrogen molecule is around $2000\,\text{m s}^{-1}$. Its mass is $(2 \times 10^{-3})/(6.02 \times 10^{23}) \cong 3.3 \times 10^{-27}$ kg. In a cube of 10 cm, $2mvL/h = (n_x^2 + n_y^2 + n_z^2)^{1/2} \cong 1.9 \times 10^9$. Therefore, a typical value for n_x, n_y, and n_z is around 1.1×10^9. A smaller cube will have smaller n_x, n_y, and n_z.

Then, we find how small the spacing is between adjacent energy levels. For simplicity, we increase n_x by 1 while holding n_y, and n_z unchanged. The change in $\varepsilon_{\text{trans}}$ is

$$\Delta\varepsilon_{\text{trans}} = \frac{h^2(n_x+1)^2}{8mL_x^2} - \frac{h^2 n_x^2}{8mL_x^2} = \frac{h^2(2n_x+1)}{8mL_x^2} \cong \frac{h^2 n_x}{4mL_x^2} \tag{3.56}$$

For a hydrogen molecule in a box of $L_x = 10\,\text{cm}$, $\Delta\varepsilon_{\text{trans}}$ is 3.7×10^{-30} J. The fractional change is

$$\frac{\Delta\varepsilon_{\text{trans}}}{\varepsilon_{\text{trans}}} = \frac{2}{n_x} \tag{3.57}$$

For the hydrogen molecule, it is just ~2 ppb. Since the difference is tiny, we can regard that the energy is continuous in practice and safely apply the classical model to describe the center-of-mass translation of a molecule.

3.6.2 Quantization of Rotational Energy

The classical expression of the energy of rotation is $\tfrac{1}{2}I(\omega_1^2 + \omega_2^2)$ (Eq. (3.21)). The energy is continuous, as ω_1 and ω_2 are any real numbers. In quantum mechanics, only discrete values are allowed for the energy. The rotational energy level is specified by a nonnegative integer J and given as

$$\varepsilon_{\text{rot}}(J) = \frac{h^2}{8\pi^2 I}J(J+1) \tag{3.58}$$

This J is called the **rotational quantum number**. Figure 3.25 shows the levels of $\varepsilon_{\text{rot}}(J)$ for $J = 0, 1, 2, 3, 4, 5$, and 6. The spacing between adjacent levels increases with an increasing J:

$$\varepsilon_{\text{rot}}(J+1) - \varepsilon_{\text{rot}}(J) = \frac{h^2}{8\pi^2 I}2(J+1) \tag{3.59}$$

For instance, the first two levels ($J = 0$ and 1) differ by

$$\varepsilon_{\text{rot}}(1) - \varepsilon_{\text{rot}}(0) = \frac{h^2}{4\pi^2 I} \tag{3.60}$$

Figure 3.25 Energy levels for rotation of a diatomic molecule. The spacing between an adjacent pair of levels widens with an increasing energy level.

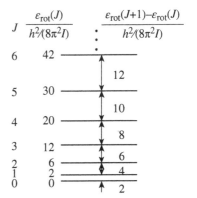

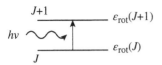

Figure 3.26 A molecule at the Jth rotational level may absorb radiation to transition to the $(J + 1)$th energy level, if the energy of the radiation, $h\nu$, matches the energy level difference, $\varepsilon_{rot}(J + 1) - \varepsilon_{rot}(J)$.

For an HCl molecule with $I = 2.64 \times 10^{-48}$ kg m^2, the gap is 4.2×10^{-21} J. This gap is orders of magnitude as large as the one in ε_{trans}, and is comparable to $k_B T$ at room temperature, 4.1×10^{-21} J. Therefore, the **thermal energy** is sufficient to excite the molecule to higher energy levels ($J = 1, 2, \ldots$), but not sufficiently large for the molecule to have J on the order of 1000. Typical values of J are around 10 or 20 for most small molecules. We learn this situation in detail in Section 5.4.

The discreteness of the energy levels can be most obviously seen in the **absorption spectrum**. A molecule at energy level J absorbs **electromagnetic radiation** to be excited to the level $J + 1$ (see Figure 3.26).

Only those transitions that increase J by one are allowed. The energy of the radiation required for the transition must match the energy difference. That is why the absorption spectrum consists of narrow lines. The radiation of frequency ν has the energy of $h\nu$. Therefore, the frequency of the radiation for the rotational transition is given as

$$\nu = \frac{\varepsilon_{rot}(J + 1) - \varepsilon_{rot}(J)}{h} = \frac{h}{8\pi^2 I} 2(J + 1) \tag{3.61}$$

For $J = 10$ and $I = 1.449 \times 10^{-47}$ kg m^2 (CO), the frequency is $\nu = 1.27 \times 10^{13}$ Hz, which falls in the microwave to far-infrared (IR) region. Since molecules are distributed to different values of J prior to the absorption of radiation, we observe equally spaced absorption lines. Spectrometers to observe these transitions are not commercially available. Fortunately, however, similar rotational transition lines can also be observed in the IR region of radiation as rotational transitions are added to the main transition between adjacent energy levels of vibration.

There is an added complexity to the rotational states: The integer J alone is not sufficient to specify the state of the diatomic molecule. We need another integer M_J to completely specifying the state. The states with the same J but different values of M_J have the same energy. We then say that these M_J states are **degenerate**. The second index ranges from $-J$ to J, and therefore the Jth energy level consists of $2J + 1$ states. We say that the **degeneracy** of the Jth energy level is $2J + 1$. The diagram of the energy levels looks like the one in Figure 3.27. The number of seats of an identical level increases with J. The situation is similar to the orbitals of an electron in a hydrogen atom: J is like the orbital angular moment quantum number, and M_J is the magnetic quantum number.

Each rotational state has its own wave function. Here we look at the wave functions for $J = 0$ and 1 (see Figure 3.28). The ground state ($J = 0$) represents a uniform distribution for the orientation of the molecule. The state is isotropic,

Figure 3.27 Energy levels for rotation of a diatomic molecule are shown together with the degeneracy. Each bar represents a specific state of rotation.

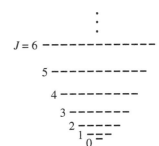

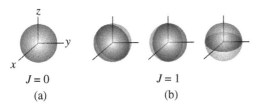

$J = 0$ $J = 1$

(a)   (b)

Figure 3.28 Wave functions for rotation represented by a sphere surface. (a) The function for $J = 0$ is isotropic. (b) The three functions for $J = 1$ are polarized in x, y, and z directions. The sign of the function is the opposite for $x > 0$ and $x < 0$, for example.

as there is no preferred direction, analogous to the s orbital of the electron. The next energy level ($J = 1$) has three states ($M_J = -1$, 0, and 1), as there are three p orbitals for the electron. The two wave functions for $M_J = -1$ and 1 can construct spherical functions polarized in the x and y directions in the xyz Cartesian coordinates, as $2p_x$ and $2p_y$ orbitals are constructed for the electron. The function with $M_J = 0$ is polarized in the z direction, and is similar to the $2p_z$ orbital. The wave functions for $J = 2$ or greater can be drawn in a similar way.

A nonlinear polyatomic molecule has three moments of inertia. The expression of the rotational energy is complicated. If you are interested in this matter, refer to molecular spectroscopy books.

3.6.3 Quantization of Vibrational Energy

In Eq. (3.22), the vibrational part ε_{vib} was expressed as a sum of the kinetic energy of a particle with mass μ and the elastic energy of a spring with force constant κ. We substitute $r_{ab} - r_0$ with $r_{amp}\cos \omega t$, where r_{amp} is the amplitude of the vibration, and ω is the angular frequency of vibration. Then, we have

$$\varepsilon_{vib} = \frac{1}{2}r_{amp}^2(\mu\omega^2\sin^2\omega t + \kappa\cos^2\omega t) \tag{3.62}$$

If $\omega = (\kappa/\mu)^{1/2}$, then ε_{vib} is a constant of time: $\varepsilon_{vib} = \frac{1}{2}\kappa r_{amp}^2$. The frequency

$$\nu = \frac{\omega}{2\pi} = \frac{1}{2\pi}\sqrt{\frac{\kappa}{\mu}} \tag{3.63}$$

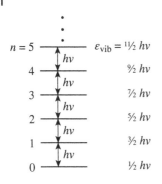

Figure 3.29 Energy levels of a harmonic oscillator. The spacing between an adjacent pair of levels is hv for all levels.

is called the **characteristic frequency of vibration**. Since r_{amp} can be any positive number, ε_{vib} is continuous in classical mechanics.

We can quantize this classical treatment of the harmonic oscillator. The vibrational state of the diatomic molecule of characteristic frequency v is specified by a nonnegative integer n. The nth energy level is given as

$$\varepsilon_{vib} = \left(n + \frac{1}{2}\right)hv \tag{3.64}$$

The energy levels are equally spaced by hv, starting at $hv/2$ for $n = 0$ (see Figure 3.29). Unlike the rotation, the vibration has a positive energy in its ground state. This $\frac{1}{2}hv$ is called the **zero-point energy**. Even at absolute zero temperature, molecules vibrate. Another difference from the rotational states is that all the energy levels are **nondegenerate**, i.e. their degeneracy is one.

Table 3.8 lists characteristic frequency of vibration v for some diatomic molecules. The frequency is also listed in wave number (cm^{-1}).

Table 3.8 Characteristic frequency of vibration, v, and the associated wave number $\bar{v}$, for diatomic molecules. The atoms are most abundant species.

Molecule	$10^{-13}v$ (Hz)	$\bar{v}$ (cm^{-1})
H_2	13.195	4401.2
N_2	7.0711	2358.6
O_2	4.7374	1580.2
F_2	2.748	916.6
Cl_2	1.678	559.7
HCl	8.9667	2990.9
CO	6.5051	2169.8
NO	5.7088	1904.2

Source: Data from Anderson 1989 [5].

Let us estimate the energy difference between the ground state and the first excited state. In an H_2 molecule, for example, $h\nu = 8.743 \times 10^{-20}$ J, much greater than $k_B T$ at room temperature, 4.1×10^{-21} J. As a result, nearly all of H_2 molecules are at the ground state. This situation is different from the distribution of rotational states.

A molecule with a smaller ν has the smaller gap between the adjacent energy levels. In $^{35}Cl_2$, $h\nu$ is only 1.11×10^{-20} J. It is still larger than $k_B T$ at room temperature, but this gap is a lot smaller compared with the gap in the H_2 molecule. The chlorine gas is sufficiently heavy to exhibit properties of vibrationally excited states at room temperature. We examine the difference between different diatomic molecules in Section 5.3 when we consider the heat capacity.

A polyatomic molecule has three or more vibrational modes. Quantization is performed for each mode independently. The vibration part of the energy is then expressed as

$$\varepsilon_{\text{vib}} = \sum_i \left(n_i + \frac{1}{2} \right) h\nu_i \tag{3.65}$$

where n_i is the quantum number for the ith normal mode that has a characteristic frequency ν_i.

Vibrational modes can be observed in IR absorption spectroscopy, if the mode changes the dipole moment of the molecule. That is a prerequisite for us to observe the mode in the IR spectroscopy (**IR-active**). Here, we briefly consider the IR activity.

A homonuclear diatomic molecule is nonpolar and remains so when the bond length changes in the vibration. Therefore, N_2 and O_2 molecules do not absorb IR radiation (IR-inactive). A heteronuclear diatomic molecule has a dipole moment in its equilibrium state, and vibration changes its dipole. Therefore, vibration of HCl and CO can be observed (IR-active). In a polyatomic molecule, whether a specific mode can be observed in the IR spectroscopy or not depends on whether the mode changes the dipole. Even a nonpolar molecule has some modes that can be observed. For example, in a CO_2 molecule, antisymmetric stretch displaces the center of the two O atoms relative to the C atom. As a result, the molecule acquires a dipole, and this mode can be observed at around $2300 \, \text{cm}^{-1}$. In contrast, symmetric stretch does not change the dipole, and it remains zero. Therefore, the symmetric stretch mode is not observed. Bend motions can be observed, since they change the dipole. In a water molecule, all of the three modes change the dipole of this highly polar molecule, and therefore can be observed. IR-inactive modes may be observed in Raman scattering.

3.6.4 Electronic Energy Levels

So far we have not listed the energy of electrons in a molecule, $\varepsilon_{\text{elec}}$, as a part of the energy of the molecule. We can include $\varepsilon_{\text{elec}}$, but will find that it is not necessary. For example, in a hydrogen atom, the energy levels of the electron are specified by a positive integer n, called the principal quantum number. The energy of the nth level is

$$\varepsilon_{\text{elec}}(n) = -13.6\text{eV} \times \frac{1}{n^2} \tag{3.66}$$

As $n \to \infty$, $\varepsilon_{\text{elec}}$ increases to zero, the energy level of a free electron. The 13.6 eV ($1\,\text{eV} = 1.602 \times 10^{-19}$ J) is the difference of the ground-state energy level and the level for the unbound electron, and is equal to the ionization energy of the H atom.

The difference of $\varepsilon_{\text{elec}}$ between the ground state and the first excited state ($n = 2$) is $\varepsilon_{\text{elec}}(2) - \varepsilon_{\text{elec}}(1) = 10.2\,\text{eV} = 1.63 \times 10^{-18}$ J. This energy is about 400 times as large as the $k_B T$ at room temperature. Therefore, all of the H atoms are at the electronically ground state at room temperature. Equating $k_B T$ to 1.63×10^{-18} J gives the temperature for a large portion of H atoms to be in excited states. That temperature is $12\,000$ K, higher than the temperature of a white dwarf star.

3.6.5 Comparison of Energy Level Spacings

Table 3.9 summarizes the energies for the translation, rotation, and vibration of a diatomic molecule in quantum-mechanical expressions.

We compared the spacing between adjacent levels in each mode of motion with $k_B T$. Here we compare the three spacings. Figure 3.30 depicts the energy levels for $^{14}N_2$ in a box of $(10\,\text{cm})^3$ The vibrational levels are equally spaced by 4.69×10^{-20} J. The spacing for the rotational energy levels is a lot narrower. When zoomed, it looks like the one shown in the figure. The spacing between adjacent translation levels is even narrower.

The energy level of the molecule is specified by n_x, n_y, n_z, J, and n. For a given rotational energy level, there are many choices for n_x, n_y, and n_z. Each rotational level entails these narrowly spaced translational energy levels. For a given

Table 3.9 Energy levels in three modes of motion of a diatomic molecule. The quantum number and degeneracy are also listed.

Mode of motion	Energy levels	Quantum number	Degeneracy
Translation	$\dfrac{h^2}{8mL^2}\left(\dfrac{n_x^2}{L_x^2} + \dfrac{n_y^2}{L_y^2} + \dfrac{n_z^2}{L_z^2}\right)$	$n_x, n_y, n_z = 1, 2, 3, \ldots$	1
Rotation	$\dfrac{h^2}{8\pi^2 I}J(J + 1)$	$J = 0, 1, 2, \ldots$	$2J + 1$
Vibration	$h\nu\left(n + \dfrac{1}{2}\right)$	$n = 0, 1, 2, \ldots$	1

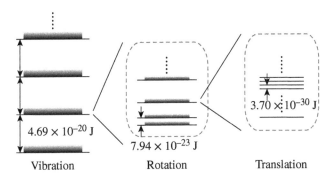

4.69 × 10⁻²⁰ J → 4.69×10^{-20} J

7.94 × 10⁻²³ J → 7.94×10^{-23} J

3.70 × 10⁻³⁰ J → 3.70×10^{-30} J

Vibration   Rotation   Translation

Figure 3.30 Hierarchy of energy levels. Spacings of the energy levels are compared for vibration, rotation, and translation of a $^{14}N_2$ molecule in a box of $(10\,\text{cm})^3$. Shaded bands represent the fine structures by the other modes of motion.

vibrational energy level, there are many possibilities for J plus n_x, n_y, and n_z. In Figure 3.30, a shaded band on each rotational level represents the fine structure by the translation. Likewise, a shaded band on each vibrational level represents the fine structure by the rotation and translation.

Problems

3.1 *Variance of bond stretch.* Suppose r in Eq. (3.14) is distributed with a normal distribution $f(r)$ given as

$$f(r) = \left(\frac{\kappa}{2\pi k_B T} \right)^{1/2} \exp\left(-\frac{\kappa}{2k_B T}(r - r_0)^2 \right)$$

(1) What is the mean of r?
(2) What is the mean of $(r - r_0)^2$?
(3) If $\kappa = 500\,\text{N m}^{-1}$, what is the root-mean-square (RMS) displacement $\langle (r - r_0)^2 \rangle^{1/2}$ at 300 K?

3.2 *Moment of inertia.* Apply the definition of the moment of inertia by Eq. (3.35) to the diatomic molecule and show that I is given by Eq. (3.23).

3.3 *Spring force constant of a bond.* Estimate the force constant of the spring for each of the molecules listed in Table 3.8.

3.4 *Isotopes.* The characteristic frequency of vibration in $^{35}Cl_2$ is 1.678×10^{13} Hz. What is the characteristic frequency of $^{35}Cl^{37}Cl$?

How about $^{37}\text{Cl}_2$? Assume that the force constant of the spring is shared by the three isotope forms of Cl_2.

3.5 *Translational energy.* Equation (3.57) applies to all gas molecules. In Section 3.6.1, we considered hydrogen, the lightest molecule. Does the narrow spacing ($\Delta\varepsilon_{trans}/\varepsilon_{trans}$ is small) apply with a greater margin for a heavy molecule such as O_2? Consider how $\Delta\varepsilon_{trans}/\varepsilon_{trans}$ depends on the mass of the molecule.

4

Statistical Mechanics

We learn several fundamental concepts of statistical mechanics – states, phase space, ensemble, and partition function – in this chapter. There are a few assumptions or hypotheses in statistical mechanics, and from them we can derive all the thermodynamics functions such as pressure, heat capacity, free energy, and chemical potential. The expression of the pressure gives the equation of state. When you finish this chapter, you will have learned all the necessary tools for applying statistical mechanics to different chemical systems.

4.1 Basic Assumptions, Microcanonical Ensembles, and Canonical Ensembles

4.1.1 Basic Assumptions

Statistical mechanics is founded on a few assumptions stated in simple words. Starting with these assumptions, various laws of thermodynamics that we learned in undergraduate physical chemistry courses are derived. So far, no phenomena that go against the laws have been discovered. That is why we accept the basic assumptions of statistical mechanics as the first principles, just as the Schrödinger equation is accepted in quantum mechanics. Recall that, in the physical chemistry courses, many of the laws were given, not derived.

In the first eight sections of this chapter, we consider a closed system that consists of N particles – a fixed number of particles. A closed system can exchange heat and work with its surroundings. The particles are molecules, atoms, electrons, etc. They can be in any phase – vapor, liquid, or solid. The system's physical dimension is of laboratory scale, and therefore N is huge, close to Avogadro's number.

In the nonstatistical version of thermodynamics, the system would be a box containing some continuous medium, and it does not involve the concept of

Statistical Thermodynamics: Basics and Applications to Chemical Systems, First Edition. Iwao Teraoka.
© 2019 John Wiley & Sons, Inc. Published 2019 by John Wiley & Sons, Inc.
Companion website: www.wiley.com/go/Teraoka_StatsThermodynamics

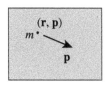

Figure 4.1 A system consisting of one monatomic molecule can be specified by its position **r** and linear momentum **p**.

molecules. In contrast, statistical mechanics pays attention to each of the many molecules and starts with a detailed description of the molecules.

Let us start with the simplest system, a system of one monatomic molecule ($N = 1$). The state of the system is specified by the position $\mathbf{r} = (x, y, z)$ and linear momentum $\mathbf{p} = (p_x, p_y, p_z) = m\mathbf{v}$ of the molecule, where m is the mass of the molecule, and $\mathbf{v}$ the velocity (Figure 4.1). With **r** and **p**, we specify a **microscopic state** (in short, a **microstate**, or just a **state**) of the system. In the one-particle system, its microstate is specified by six real numbers.

We can add more details to the specification. The latter can also include the electronic state – which atomic orbitals are occupied. If the molecule is diatomic, the description needs more information, namely, the state of rotation around its center of mass and the state of vibration, i.e. bond length change with time. The details beyond **r** and **p** are considered in Chapter 5.

When the system consists of N monatomic molecules, its microstate is specified by the positions and linear momenta of all the particles: $\mathbf{r}_1, \mathbf{r}_2,..., \mathbf{r}_N$ for the positions, and $\mathbf{p}_1, \mathbf{p}_2,..., \mathbf{p}_N$ for the linear momenta. Each of $\mathbf{r}_1, \mathbf{r}_2,..., \mathbf{r}_N$, $\mathbf{p}_1, \mathbf{p}_2,...,$ and $\mathbf{p}_N$ has x, y, z components, and therefore a total $6N$ variables are needed to specify the state.

We go back to a system with $N = 1$. First, we simplify the system further by allowing the particle to move only in the x direction. Now, the state is specified by x and p_x only. We introduce a function $f(x, p_x)$ such that $f(x, p_x)\mathrm{d}x\,\mathrm{d}p_x$ represents the probability for the particle to be between x and $x + \mathrm{d}x$, and have the linear momentum between p_x and $p_x + \mathrm{d}p_x$, where $\mathrm{d}x$ and $\mathrm{d}p_x$ are small. The $f(x, p_x)\mathrm{d}x\,\mathrm{d}p_x$ is a joint probability for the two random variables, x and p_x, and $f(x, p_x)$ is called the probability density. In Figure 4.2, $f(x, p_x)$ is represented by the shade in a gray scale.

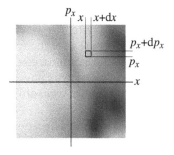

Figure 4.2 The probability to find the particle in a small rectangle between x and $x + \mathrm{d}x$ and between p_x and $p_x + \mathrm{d}p_x$ is indicated by the gray scale of the rectangle.

For a system with $N = 1$ in three dimensions, the probability density function is $f(\mathbf{r}, \mathbf{p}) = f(x, y, z; p_x, p_y, p_z)$. It is defined so that $f(\mathbf{r}, \mathbf{p})\mathrm{d}\mathbf{r}\, \mathrm{d}\mathbf{p} = f(x, y, z; p_x, p_y, p_z)\mathrm{d}x\, \mathrm{d}y\, \mathrm{d}z\, \mathrm{d}p_x\, \mathrm{d}p_y\, \mathrm{d}p_z$ represents the probability for the particle to be in a box defined by x and $x + \mathrm{d}x$, y and $y + \mathrm{d}y$, z and $z + \mathrm{d}z$, p_x and $p_x + \mathrm{d}p_x$, p_y and $p_y + \mathrm{d}p_y$, and p_z and $p_z + \mathrm{d}p_z$ in the x, y, z, p_x, p_y, p_z space, respectively.

In a system of $N = 2$, the function is $f(\mathbf{r}_1, \mathbf{r}_2; \mathbf{p}_1, \mathbf{p}_2)$. In a system of N particles, it is $f(\mathbf{r}_1, \mathbf{r}_2, ..., \mathbf{r}_N; \mathbf{p}_1, \mathbf{p}_2, ..., \mathbf{p}_N)$.

It is convenient to introduce a coordinate system called a **phase space**. Figure 4.3a illustrates a phase space for five particles in one dimension. Each point indicates where the particle is and with what linear momentum it is moving (arrow). For example, particle 1 is in $x > 0$, $p_x > 0$, and therefore moving in the $+x$ direction. Particle 2 is in $x > 0$, $p_x < 0$, and therefore moving in the $-x$ direction. The more distant the point is from the x axis in the phase space, the faster the speed. The p_x coordinate of the point tells where the particle will be at the next moment relative to the current position. We can put as many points as we wish in the phase space. Figure 4.3b is an example. It is a snapshot of many particles. When the snapshot is taken at another instance, the points will be at different positions in the phase space.

Now we move over to three dimensions. The phase space needs six coordinates $f(x, y, z; p_x, p_y, p_z)$ for one particle. The state of each particle in the system is represented by a point in the six-dimensional space. For a system of N particles in three dimensions, the six-dimensional phase space will have N points to specify the state.

Alternatively, we could use a $6N$-dimensional phase space for a system of N particles. A state of the system will be represented by one point in the $6N$-dimensional phase space. However, it is much more convenient to place N points in six-dimensional phase space than placing one point in the $6N$-dimensional phase space, and we will adopt the easier view.

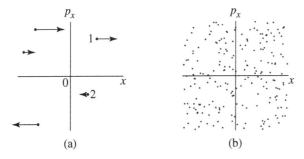

(a)   (b)

Figure 4.3 (a) Phase space in one dimension for five particles including 1 and 2. Each particle is moving with the linear momentum indicated by the arrow. (b) Phase space in one dimension for many particles. Each point represents a particle at a specific position moving with a specific linear momentum.

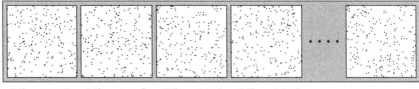

Microstate 1   Microstate 2   Microstate 3   Microstate 4

Figure 4.4 An ensemble is a collection of many microstates of a system. Each microstate is represented by a collection of points in the phase space.

Imagine many similar systems (replicas), each having the same N particles, and take many snapshots of the systems in the phase space. Each snapshot represents a microstate of the system. A collection of microstates is called an **ensemble** (Figure 4.4).

We can also follow the movement of the particles in one system, but over a long time. The record of their positions in the phase space at many instances will give us another collection, and the states in the collection will be similar to those in Figure 4.4. The equivalence of the latter collection to the ensemble is called **ergodicity**. In other words, a system will experience over a long time all possible microstates of the ensemble. The ergodicity is one of the assumptions of the statistical mechanics, and is usually stated as follows:

The ensemble average is equal to the long-time average   (S1)

So far, we have considered continuous states: x and p_x are continuous variables. Any small changes in x and p_x are possible, and the updated x and p_x make another point in the phase space. There are systems whose states are not continuous, but rather are discrete. In Section 3.6, we learned modes of motion whose states are discrete. The state of the vibration in a diatomic molecule is specified by a nonnegative integer, and the state of rotational motion is specified by another integer.

Let us consider a system of particles in which the energy of each particle is specified by a nonnegative integer J. You can imagine a hypothetical system of N diatomic molecules that can only rotate. If the system has only one particle, its state is specified by one integer J. In a system of two particles, the state is specified by two integers, J_1 and J_2. The state in a system of an N-particle system is specified by N integers, $J_1, J_2, \ldots, J_N$. The ensemble is a collection of these N integers. Snapshots of many similar systems will encompass all possibilities of the ensemble. Likewise, a given system will experience all the possibilities of these integers over a long time.

Once the state is specified, the system has a unique value for its energy E. We denote by E_k the energy of state k. In this book, we use index k, especially for a system of discrete states.

The hypothesis of statistical mechanics is stated as follows:

All microstates that have the same energy are equally probable   (S2)

This hypothesis can be rephrased as

All microstates of an isolated system occur with an equal probability

(S3)

Recall that the isolated system does not exchange heat or work with its surroundings, and therefore its energy does not change with time.

Three types of ensembles are often considered in statistical mechanics. In this section, we learn two of them, a microcanonical ensemble and a canonical ensemble.

4.1.2 Microcanonical Ensembles

The **microcanonical ensemble** is a collection of microstates with the same energy. The microcanonical ensemble has a unique value of E. According to (S3), all the states in a microcanonical ensemble have an equal probability. Here we look at a few examples.

A. Particle in a Box (1D)

Consider a particle of mass m that can move only in one direction in a box of length L (Figure 4.5a). As the particle hits the wall, it reflects the particle. The energy E of the particle is the kinetic energy, $E = p^2/(2m)$. In the phase space (Figure 4.5b), the trace of the particle consists of two lines at $p_x = \pm(2mE)^{1/2}$ parallel to the x axis. Since the particle moves at a constant speed in the box, the trace of the particle moves also at a constant speed in the phase space. Therefore, all the points on the two lines are equally probable. A particle with a large E has the two lines more distant from the x axis.

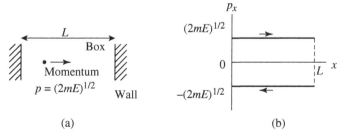

(a)   (b)

Figure 4.5 (a) Particle moving in the x direction in a box of length L. The linear momentum of the particle is $(2mE)^{1/2}$. (b) In the phase space, the particle moves along the two solid lines. The upper line is for moving in the $+x$ direction, the lower line for the $-x$ direction.

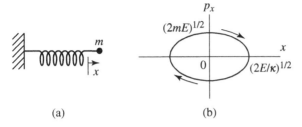

(a)   (b)

Figure 4.6 Classical harmonic oscillator in one dimension. (a) A particle of mass m connected to a spring of force constant κ. Displacement from the equilibrium position is x. (b) In the phase space, the harmonic oscillator moves along the ellipse.

B. Harmonic Oscillator (1D)

A particle of mass m is connected to a spring of force constant κ (Figure 4.6a). The energy E consists of the elastic energy and the kinetic energy:

$$\frac{1}{2}\kappa x^2 + \frac{p_x^2}{2m} = E \tag{4.1}$$

where x is the position of the particle relative to its equilibrium position. The elastic energy is given by a harmonic potential ($\propto x^2$), and therefore the particle is called a harmonic oscillator. Since this harmonic oscillator follows Newton's equation of motion, it is often called a classical harmonic oscillator, as opposed to the quantum-mechanical harmonic oscillator (or simply, harmonic oscillator) we learned about in Section 3.6. Equation (4.1) is rewritten to

$$\left[\frac{x}{(2E/\kappa)^{1/2}}\right]^2 + \left[\frac{p_x}{(2mE)^{1/2}}\right]^2 = 1 \tag{4.2}$$

which represents an ellipse that intersects the x axis at $x = \pm(2E/\kappa)^{1/2}$ and the p_x axis at $p_x = \pm(2mE)^{1/2}$ in the phase space (Figure 4.6b). The particle traces the ellipse clockwise. Problem 4.1 shows what rule the movement follows. A particle with a large E has a larger ellipse, but the ratio of the two axes remains the same.

C. Discrete Energy Levels

As an example of a microcanonical ensemble with discrete energy levels, we consider a system of three particles, each having the energy levels of $n\varepsilon$ ($n = 0, 1, 2,...$; $\varepsilon > 0$). The six microstates shown in Figure 4.7 have the same total energy, and that is 2ε. No other microstates have that total energy. Therefore, the six states in the figure represent a microcanonical ensemble. The hypothesis of the statistical mechanics says that the six states are equally probable.

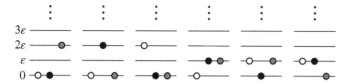

Figure 4.7 A system of three particles (O●◑). The energy level of each particle is $n\varepsilon$ ($n = 0$, 1, 2,...). This figure shows a microcanonical ensemble that has the energy of 2ε.

4.1.3 Canonical Ensembles

The microcanonical ensemble has a well-defined energy. We now consider closed systems that have a well-defined temperature. A collection of such systems is called a **canonical ensemble**, and most of applications of statistical mechanics consider a canonical ensemble. Most closed systems are equilibrated with the surroundings that have a well-defined temperature, rather than being isolated.

We consider a closed system at a constant temperature T, made possible by being in contact with a heat reservoir of constant temperature T (Figure 4.8). The heat reservoir supplies heat to the system or deprives the system of heat to keep the system's temperature at T. For the temperature to be held constant, the energy of the system may need to change with time. Alternatively, we can imagine many similar replica systems in contact with the heat reservoir. All such systems have the same temperature T, but the energy may be different from system to system.

In the next section, we prove that the probability of the system having a specific energy E decreases with an increasing E in the canonical ensemble. Here, we look at a canonical-ensemble version of the two examples with continuous energy that we looked at in Section 4.1.2.

A. Particles in a Box (1D)

If the particles are gas molecules, their values of p_x at a given temperature will be distributed around $p_x = 0$, since p_x and $-p_x$ are equally probable. As $|p_x|$ increases, the number of particles having that large $|p_x|$ decreases. The distribution of the particles in the phase space will look like the one indicated in Figure 4.9a; the darker the point, the higher the probability.

Figure 4.8 System in contact with a reservoir of heat at temperature T. The system exchanges heat with the reservoir to keep the temperature of the system always at T. As a result, the energy of the system E_k may change with time.

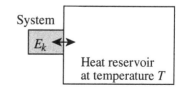

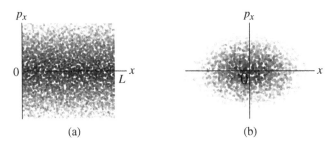

p_x p_x

0 $\dfrac{}{L}\, x$ x

(a) (b)

Figure 4.9 Snapshot of a phase space for a canonical ensemble of (a) particles moving in one dimension and (b) classical harmonic oscillators. The gray scale represents the number of particles.

B. Classical Harmonic Oscillators (1D)

A canonical ensemble of classical harmonic oscillators will also have the highest probability at around the origin, $x = 0$ and $p_x = 0$. The probability decreases with an increasing distance from the origin (see Figure 4.9b).

Now we raise a question: What is the probability f for the canonical ensemble to have the energy E? Unlike the microcanonical ensemble, E changes from system to system, and for a given system, from time to time. However, when we apply the hypothesis of statistical mechanics (S2), we find that the dependence of f on $\mathbf{r}_1,\ldots,\mathbf{p}_N$ is through E only. It means that, if two systems have the same energy, their probabilities are identical:

$$\text{If } E(\mathbf{r}_1,\ldots,\mathbf{p}_N) = E(\mathbf{r}_1',\ldots,\mathbf{p}_N'), \text{ then } f(\mathbf{r}_1,\ldots,\mathbf{p}_N) = f(\mathbf{r}_1',\ldots,\mathbf{p}_N').$$

There are many different values of $\mathbf{r}_1,\ldots,\mathbf{p}_N$ that have the same E. Therefore, we can rewrite f as

$$f(\mathbf{r}_1,\ldots,\mathbf{p}_N) = f(E(\mathbf{r}_1,\ldots,\mathbf{p}_N)) \tag{4.3}$$

to indicate that f depends on $\mathbf{r}_1,\ldots,\mathbf{p}_N$ through its E.

For a system with discrete states, the probability P_k of microstate k is determined by its energy E_k. Different microstates having the same energy (as we saw in Figure 4.7) have the same probability. In other words, the probability of a microstate is determined by the energy of the microstate. We denote by $P(E_k)$ the probability for state k that has energy E_k:

$$P_k = P(E_k) \tag{4.4}$$

Note that the probability P_k is for the microstate. It is not that P_k is the probability for the system to have the energy E. If there are six microstates with the same energy, then the probability for the system to have that energy is six times the probability for the system to be in one of the microstates.

Now that we have learned that the probability of a microstate is determined by its energy, we want to find how $f(E)$ depends on E, and how $P(E_k)$ depends on E_k.

As P_k is the probability, the sum of P_k with respect to k must be equal to 1:

$$1 = \sum_k P(E_k) \tag{4.5}$$

One simple possibility is easily rejected for a system of N monatomic molecules in a box of volume V. It $f(E)$ were a constant, it would lead to

$$\int_V d\mathbf{r}_1 \cdots \int_{-\infty}^{\infty} d\mathbf{p}_N f(E(\mathbf{r}_1, \ldots, \mathbf{p}_N)) = \text{constant} \times \int_V d\mathbf{r}_1 \cdots \int_{-\infty}^{\infty} d\mathbf{p}_N = +\infty \tag{4.6}$$

The integral by $\mathbf{r}_i$ is finite, but the integral by $\mathbf{p}_i$ is not. It means that $f(E) = \text{constant}$ cannot be normalized. For the integral to be finite, f must decrease to zero as either $|p_{1x}|$, $|p_{1y}|$, $|p_{1z}|$, $|p_{2x}|$, ..., or $|p_{Nz}|$ increases. Since $E = (2m)^{-1}\Sigma_i \mathbf{p}_i{}^2$, $f(E)$ must decrease to zero with an increasing E.

4.2 Probability Distribution in Canonical Ensembles and Partition Functions

4.2.1 Probability Distribution

We found that $P(E_k) = \text{constant}$ is not allowed in the canonical ensemble, and $P(E_k)$ must decrease quickly with an increasing E_k. In this section, we use a simple but effective method to find how $P(E_k)$ depends on E_k.

For that purpose, we consider two systems A and B in equilibrium with a common **heat reservoir** (Figure 4.10). We denote by $P_A(E_A)$ the probability of finding A in one of the (micro)states with energy E_A. The probability is for the state, not for the energy. Likewise, we denote by $P_B(E_B)$ the probability of finding B in one of the states with energy E_B.

Then, we combine the two systems to form a composite system A + B, while retaining a wall between A and B. Each of A and B is a closed system. We denote

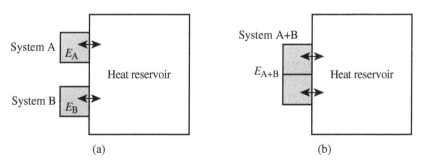

Figure 4.10 (a) System A and system B are independently in contact with a reservoir of heat at temperature T. (b) A composite system of A + B is in contact with the reservoir.

by $P_{A+B}(E_{A+B})$ the probability of finding A + B in one of the states with energy E_{A+B}.

Since A and B are macroscopic, the interaction between a particle in A and a particle in B is a negligible part of E_A or E_B; we proved it in Section 3.5. Thus,

$$E_{A+B} = E_A + E_B \qquad (4.7)$$

It follows that

$$P_{A+B}(E_{A+B}) = P_{A+B}(E_A + E_B) \qquad (4.8)$$

For the composite system to have the energy of E_{A+B}, the A part must have the energy of E_A, and the B part the energy of E_B (imagine that A is the rotational part and B the vibrational part of a diatomic molecule). This statement is written as

$$P_{A+B}(E_A + E_B) = P_A(E_A)p_B(E_B) \qquad (4.9)$$

Obviously, $P_A(E_A)P_B(E_B)$ is a joint probability. Furthermore, $P_A(E_A)$ and $P_B(E_B)$ are independent of each other. The independence can be easily understood, since bringing A and B into contact with each other does not change the state of A or the state of B; they had the same temperature before the contact. The added contact between A and B does not change the fact that A has the temperature of the heat reservoir and B has the same temperature.

We now differentiate Eq. (4.9) by E_A:

$$P'_{A+B}(E_A + E_B) = P'_A(E_A)p_B(E_B) \qquad (4.10)$$

where the prime denotes a derivative by the argument. The right-hand side is a straightforward derivative of a function of E_A by E_A, but the left-hand side used the chain rule:

$$\frac{\partial f(x+y)}{\partial x} = \frac{df(x+y)}{d(x+y)} \frac{\partial(x+y)}{\partial x} \qquad (4.11)$$

Likewise, we differentiate Eq. (4.9) by E_B to obtain

$$P'_{A+B}(E_A + E_B) = P_A(E_A)p'_B(E_B) \qquad (4.12)$$

Equations (4.10) and (4.12) share the left-hand side. Therefore,

$$\frac{P'_A(E_A)}{P_A(E_A)} = \frac{P'_B(E_B)}{P_B(E_B)} \qquad (4.13)$$

This equation is remarkable. The left-hand side is a function of function of E_A and the right-hand side is a function of E_B, which is possible only when Eq. (4.13) is equal to a constant (independent of E_A or E_B). Let the constant be $-\beta$, i.e.

$$\frac{P'_A(E_A)}{P_A(E_A)} = -\beta \qquad (4.14)$$

which is rewritten to

$$\frac{dP_A(E_A)}{dE_A} = -\beta P_A(E_A) \tag{4.15}$$

This differential equation is easily solved as

$$P_A(E_A) = C_A \exp(-\beta E_A) \tag{4.16}$$

where C_A is a constant of E_A, but is different for each system. Likewise,

$$P_B(E_B) = C_B \exp(-\beta E_B) \tag{4.17}$$

Note that β is common to A and B. In fact, it is a property of the heat reservoir, and therefore is a function of T. In Section 4.4, we find what exactly β is. At least, we know that β must be positive. Otherwise, the sum of $P_A(E_A)$ would not be finite when E_A approaches $+\infty$.

We now drop A in Eq. (4.16):

$$P(E) = Ce^{-\beta E} \tag{4.18}$$

$P(E)$ is the probability that a system, in equilibrium with a heat reservoir at temperature T, is at one of the microstates with the energy E.

The constant C depends on the details of the system. It is also a normalization constant.

4.2.2 Partition Function for a System with Discrete States

For discrete states, the normalization condition is

$$\sum_k P(E_k) = \sum_k C \exp(-\beta E_k) = C \sum_k \exp(-\beta E_k) \tag{4.19}$$

The sum must be equal to 1. Therefore,

$$C = \frac{1}{\sum \exp(-\beta E_k)} \tag{4.20}$$

We are reminded that the sum is taken with respect to k, not E.

Here we introduce an important quantity which is nearly always denoted by Z: $Z = 1/C$ or

$$Z \equiv \sum_k \exp(-\beta E_k) \tag{4.21}$$

Then,

$$P(E_k) = \frac{1}{Z} \exp(-\beta E_k) \tag{4.22}$$

The function Z is called the **partition function**. A German word, *Zustandssumme*, coined by Max Planck, is the origin of the symbol. *Zustand* means a state, and *summe* is a sum. The *s* between the two words is inserted to construct a new noun from the two nouns. Zustandssumme is pronounced as *tsu shtants zumme*. It means the sum of states. We need to keep this definition in mind. The English word, partition function, does not carry that meaning. The sum is taken with respect to the states, and all possible states must be included.

Let us calculate the partition function for a few simple systems with discrete energy levels. The energy level spacing is ε (>0).

A. Single Particle with Three Energy Levels at $-\varepsilon$, 0, and ε
    There are only three states. The sum of $e^{-\beta E}$ is

$$Z = e^{\beta \varepsilon} + 1 + e^{-\beta \varepsilon} = 1 + 2\cosh \beta \varepsilon \tag{4.23}$$

Here it is assumed that all of the three states are nondegenerate.

B. Single Particle with Energy Levels $E = 0, \varepsilon, 2\varepsilon, \ldots$
    There are an infinite number of states. The energy of the nth state ($n = 0, 1, 2, \ldots$) is $n\varepsilon$. The sum of $e^{-\beta E}$ is an infinite series:

$$Z = \sum_{n=0}^{\infty} e^{-\beta n \varepsilon} = \frac{1}{1 - e^{-\beta \varepsilon}} \tag{4.24}$$

See Eq. (A.22) in Appendix A for the calculation. Again, we assumed non-degeneracy. The probability of the nth state, $P_n = P(E_n)$ is

$$P_n = \frac{1}{Z} e^{-\beta n \varepsilon} = (1 - e^{-\beta \varepsilon}) e^{-\beta n \varepsilon} \tag{4.25}$$

Figure 4.11 shows a plot of P_n for $\beta \varepsilon = 0.2$. The ground state is the most probable, and P_n decreases with an increasing n.

C. Three Particles, Each with Energy Levels at 0 or ε
    The state of the three-particle system is specified by three integers, n_1, n_2, and n_3. Each of the three integers can be either 0 or 1, so there are $2^3 = 8$ states. The energy of the ith particle ($i = 1, 2, 3$) is $n_i \varepsilon$, and therefore the energy of the system is $E = (n_1 + n_2 + n_3)\varepsilon$. Table 4.1 is a complete list of the states, arranged in the order of an increasing E.

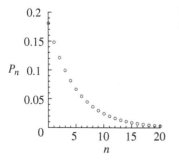

Figure 4.11 Probability P_n of the nth state, plotted as a function of n. $P_n = (1 - e^{-\beta \varepsilon}) e^{-\beta n \varepsilon}$ with $\beta \varepsilon = 0.2$.

Table 4.1 Complete list of states for a system of three particles, each with energy 0 or ε.

E_k/ε	n_1	n_2	n_3	$\exp(-\beta E_k)$
0	0	0	0	1
1	1	0	0	$e^{-\beta\varepsilon}$
1	0	1	0	$e^{-\beta\varepsilon}$
1	0	0	1	$e^{-\beta\varepsilon}$
2	1	1	0	$e^{-2\beta\varepsilon}$
2	1	0	1	$e^{-2\beta\varepsilon}$
2	1	1	0	$e^{-2\beta\varepsilon}$
3	1	1	1	$e^{-3\beta\varepsilon}$

The total energy E reduced by ε, the state of the three particles, and the exponential factor in the partition function are listed.

The table allows us to calculate the partition function:

$$Z = 1 + 3e^{-\beta\varepsilon} + 3e^{-2\beta\varepsilon} + e^{-3\beta\varepsilon} = (1 + e^{-\beta\varepsilon})^3 \tag{4.26}$$

The last equality indicates that the partition function of the three-particle system is a cube of the partition function of a one-particle system that has the energy levels of 0 and ε. Another way to get this partition function is shown here. The sum with respect to the state is now a threefold sum with respect to n_1, n_2, and n_3:

$$Z = \sum_{n_1=0}^{1}\sum_{n_2=0}^{1}\sum_{n_3=0}^{1} \exp(-\beta(n_1 + n_2 + n_3)\varepsilon)$$

$$= \sum_{n_1=0}^{1}\exp(-\beta n_1 \varepsilon) \sum_{n_2=0}^{1}\exp(-\beta n_2 \varepsilon) \sum_{n_3=0}^{1}\exp(-\beta n_3 \varepsilon) = \left(\sum_{n=0}^{1} e^{-\beta n\varepsilon}\right)^3$$

$$= (1 + e^{-\beta\varepsilon})^3 \tag{4.27}$$

4.2.3 Partition Function for a System with Continuous States

Next, we move to a system with continuous states. For the $f(\mathbf{r}_1,\ldots,\mathbf{p}_N) = f(E(\mathbf{r}_1,\ldots,\mathbf{p}_N))$, we can write the E dependence as

$$f(E(\mathbf{r}_1,\ldots,\mathbf{p}_N)) = C_{\text{cont}} \exp(-\beta E(\mathbf{r}_1,\ldots,\mathbf{p}_N)) \tag{4.28}$$

where a constant C_{cont} is determined from the normalization requirement:

$$\int_V d\mathbf{r}_1 \cdots \int_{-\infty}^{\infty} d\mathbf{p}_N f(E(\mathbf{r}_1,\ldots,\mathbf{p}_N)) = C_{\text{cont}} \int_V d\mathbf{r}_1 \cdots$$

$$\int_{-\infty}^{\infty} d\mathbf{p}_N \exp(-\beta E(\mathbf{r}_1,\ldots,\mathbf{p}_N)) = 1 \tag{4.29}$$

As we did for the discrete states, we introduce the partition function by

$$Z = \int_V d\mathbf{r}_1 \cdots \int_{-\infty}^{\infty} d\mathbf{p}_N \exp(-\beta E(\mathbf{r}_1, \ldots, \mathbf{p}_N)) \tag{4.30}$$

However, this definition of Z has a dimension, and that is $(\text{length})^{3N} \times (\text{momentum})^{3N}$, unlike the discrete-state version. Later we learn that dividing Eq. (4.30) by h^{3N}, where h is Planck's constant, matches the partition function calculated using the continuous states and the partition function for the discrete-state version. Therefore, we redefine Z as

$$Z = \frac{1}{h^{3N}} \int_V d\mathbf{r}_1 \cdots \int_{-\infty}^{\infty} d\mathbf{p}_N \exp(-\beta E(\mathbf{r}_1, \ldots, \mathbf{p}_N)) \tag{4.31}$$

for a system of continuous states.

The simplest example of a system with continuous states is a gas consisting of N noninteracting monatomic molecules. The energy of the system is equal to the sum of the energy of the ith molecule, ε_i:

$$E = \sum_{i=1}^{N} \varepsilon_i = \varepsilon_1 + \varepsilon_2 + \cdots + \varepsilon_N \tag{4.32}$$

Expressing E as a sum of the energy of individual particles is a hallmark of an **ideal** system. For this system,

$$\begin{aligned}
Z &= \frac{1}{h^{3N}} \int_V d\mathbf{r}_1 \cdots \int_{-\infty}^{\infty} d\mathbf{p}_N \exp(-\beta(\varepsilon_1 + \varepsilon_2 + \cdots + \varepsilon_N)) \\
&= \frac{1}{h^3} \int_V d\mathbf{r}_1 \int_{-\infty}^{\infty} d\mathbf{p}_1 \exp(-\beta\varepsilon_1) \times \frac{1}{h^3} \int_V d\mathbf{r}_2 \int_{-\infty}^{\infty} d\mathbf{p}_2 \exp(-\beta\varepsilon_2) \\
&\quad \times \cdots \times \frac{1}{h^3} \int_V d\mathbf{r}_N \int_{-\infty}^{\infty} d\mathbf{p}_N \exp(-\beta\varepsilon_N)
\end{aligned} \tag{4.33}$$

Let us introduce a **single-molecule partition function** Z_1:

$$Z_1 \equiv \frac{1}{h^3} \int_V d\mathbf{r} \int_{-\infty}^{\infty} e^{-\beta\varepsilon} d\mathbf{p} \tag{4.34}$$

where ε is one of ε_i's. Since all the molecules are identical,

$$Z = Z_1^N \tag{4.35}$$

If the molecules are monatomic with mass m,

$$\varepsilon = \frac{\mathbf{p}^2}{2m} \tag{4.36}$$

and

$$Z_1 = \frac{1}{h^3} \int_V d\mathbf{r} \int_{-\infty}^{\infty} d\mathbf{p} \left(-\beta \frac{\mathbf{p}^2}{2m} \right)$$

$$= \frac{V}{h^3} \int_{-\infty}^{\infty} \exp\left(-\beta \frac{p_x^2}{2m} \right) dp_x \int_{-\infty}^{\infty} \exp\left(-\beta \frac{p_y^2}{2m} \right) dp_y$$

$$\int_{-\infty}^{\infty} \exp\left(-\beta \frac{p_z^2}{2m} \right) dp_z \tag{4.37}$$

The integrals are calculated as

$$\int_{-\infty}^{\infty} \exp\left(-\beta \frac{p_x^2}{2m} \right) dp_x = \int_{-\infty}^{\infty} \exp\left(-\beta \frac{p_y^2}{2m} \right) dp_y$$

$$= \int_{-\infty}^{\infty} \exp\left(-\beta \frac{p_z^2}{2m} \right) dp_z = \left(\frac{2\pi m}{\beta} \right)^{1/2} \tag{4.38}$$

Therefore,

$$Z_1 = \frac{V}{h^3} \left(\frac{2\pi m}{\beta} \right)^{3/2} = V \left(\frac{2\pi m}{\beta h^2} \right)^{3/2} \tag{4.39}$$

Thus, the partition function for the gas system of N monatomic molecules is obtained as

$$Z = V^N \left(\frac{2\pi m}{\beta h^2} \right)^{3N/2} \tag{4.40}$$

The partition function is calculated for a given number of particles (recall the system is closed), say N, and a given β (or the temperature; we learn $\beta \sim T^{-1}$ in Section 4.4). The volume V of the system is also given, as we saw in this example. It means that Z is a function of V, β (or T), and N. In the system of noninteracting monatomic molecules, $Z = Z_1^N$ and $Z_1 \propto V\beta^{-3/2}$. The greater the V or T, the greater the partition function.

On the surface, Z is simply a normalization constant. However, Z is a powerful function: We will learn to use Z to calculate all sorts of thermodynamic functions such as pressure, internal energy, heat capacity, free energy, entropy, and chemical potential. All procedures start with obtaining an explicit expression of Z or a similar function in statistical mechanics.

4.2.4 Energy Levels and States

Equation (4.22) indicates that $P(E_k)$ peaks for the state of the lowest energy. It does not necessarily mean, however, that the most likely energy level of the system is the lowest. The $P(E_k)$ is the probability of the state, whereas the most likely energy level refers to the probability for the system to have that energy.

The difference is in the number of states that have the same energy, also known as **degeneracy**. The following rule explains the difference:

$$\left(\begin{array}{c}\text{probability that the}\\\text{system has energy } E\end{array}\right) = \left(\begin{array}{c}\text{number of states}\\\text{with energy } E\end{array}\right) \times P(E) \tag{4.41}$$

In Eq. (4.41), $P(E)$ is the probability for the system to be at one of the states that have the energy E. We call E the energy level to include all the states with the same energy. In other words, there may be more than one state at the energy level.

If we denote by $P_{\text{level}}(E)$ the probability that the system is at energy level E, then

$$P_{\text{level}}(E) = \sum_{E_k=E} P(E_k) \tag{4.42}$$

by definition. If $\Omega(E) = $ number of states with energy E, then the sum runs over $\Omega(E)$ states and

$$P_{\text{level}}(E) = \Omega(E)P(E) \tag{4.43}$$

Figure 4.12 illustrates the difference between $P(E)$ and $P_{\text{level}}(E)$. If $\Omega(E)$ increases exponentially with an increasing E, then the P_{level} may see a peak at some value of E.

With the introduction of the degeneracy, we have a second way to write Z for a discrete system. In the second method, the sum is with respect to the energy level. Let n be the index for the energy levels, and g_n the degeneracy for the level E_n:

$$g_n = \Omega(E_n) \tag{4.44}$$

Then, Z can be expressed in two ways:

$$Z = \sum_{k:\text{states}} \exp(-\beta E_k) = \sum_n g_n \exp(-\beta E_n) \tag{4.45}$$

In Section 4.2.2, we looked at three systems to calculate their partition functions. In examples A and B, all the energy levels had $g_n = 1$. In example C, the

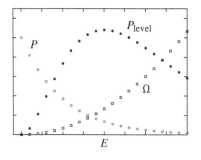

Figure 4.12 Probability of the state, P, number of states, Ω, and the probability of energy level, P_{level}, are plotted as a function of the energy E. The plots are an example.

system has four energy levels, 0, ε, 2ε, and 3ε. For the first and last energy levels, $g_n = 1$. For the second and third energy levels, $g_n = 3$.

If, in example A, the degeneracy is 1, 2, and 3 for the energy levels 0, ε, and 2ε, respectively, then

$$Z = \sum_{n=0}^{2}(n+1)\,e^{-\beta n\varepsilon} = 1 + 2e^{-\beta\varepsilon} + 3e^{-2\beta\varepsilon} \tag{4.46}$$

If a system consists of two or more particles and each of them has many energy levels, calculation of Z using Eq. (4.45) may appear difficult. Let us consider a system consisting of five noninteracting particles. The energy ε_i of the ith particle ($i = 1, 2, 3, 4, 5$) is given as

$$\varepsilon_i = n_i\varepsilon_0 \tag{4.47}$$

where $\varepsilon_0 > 0$, and $n_i = 0, 1, 2, 3,\ldots$ specifies the state of the ith particle. The total energy E of the system is

$$E = \varepsilon_1 + \varepsilon_2 + \varepsilon_3 + \varepsilon_4 + \varepsilon_5 = \varepsilon_0(n_1 + n_2 + n_3 + n_4 + n_5) \tag{4.48}$$

The energy measured in the unit of ε_0 is a sum of five nonnegative integers. The lowest energy level of the system is 0, when all of n_i are equal to 0. There is only one state for this energy level. The next energy level is ε_0, when one of n_i is equal to 1 and the rest is 0, for example, $n_1 = 1$, $n_2 = n_3 = n_4 = n_5 = 0$. Since the particle that has $n_i = 1$ can be any of the five particles, and the rest is automatic, there are $_5C_1$ states for this second lowest energy level. Table 4.2 lists typical states, the number of states, and $\Omega(E)$ for the five lowest energy levels. For the energy level of $2\varepsilon_0$, for example, two patterns of n_i are possible, and a typical state and the number of states are listed for each pattern. How the number of states is calculated is also indicated in the table. The degeneracy for $E = 2\varepsilon_0$ is $_5C_1 + _5C_2 = 15$. Some in Table 4.2 use a permutation $_nP_m$ which counts the number of arranging m particles out of n labeled particles, paying attention to the order.

From Table 4.2, Z is calculated as follows:

$$Z = 1\cdot\exp(-\beta\varepsilon_0 0) + 5\cdot\exp(-\beta\varepsilon_0 1) + 15\cdot\exp(-\beta\varepsilon_0 2)$$
$$+ 35\cdot\exp(-\beta\varepsilon_0 3) + 70\cdot\exp(-\beta\varepsilon_0 4) + \cdots \tag{4.49}$$

Adding the terms, however, becomes increasingly difficult with an increasing energy level.

There is a simpler way to calculate Z, that is to use the fact that each particle independently assumes $\varepsilon_i = n_i\varepsilon_0$. First, we note that the state of the system is specified by the five integers, so $k = (n_1, n_2, n_3, n_4, n_5)$:

$$Z = \sum_{n_1,n_2,\ldots,n_5}\exp(-\beta\varepsilon_0(n_1 + n_2 + n_3 + n_5)) \tag{4.50}$$

Table 4.2 Typical states, the number of states, and the degeneracy for each energy level (the lowest five levels) in a system of five noninteracting particles.

E/ε_0	Typical state n_1	n_2	n_3	n_4	n_5	Number of states	$\Omega(E)$
0	0	0	0	0	0	1	1
1	1	0	0	0	0	$_5C_1 = 5$	5
2	2	0	0	0	0	$_5C_1 = 5$	15
	1	1	0	0	0	$_5C_2 = 10$	
3	3	0	0	0	0	$_5C_1 = 5$	35
	2	1	0	0	0	$_5P_2 = 20$	
	1	1	1	0	0	$_5C_3 = 10$	
4	4	0	0	0	0	$_5C_1 = 5$	70
	3	1	0	0	0	$_5P_2 = 20$	
	2	2	0	0	0	$_5C_2 = 10$	
	2	1	1	0	0	$_5C_1 \cdot_4C_2 = 30$	
	1	1	1	1	0	$_5C_4 = 5$	

The exponential function can be decomposed into five factors:

$$\exp(-\beta\varepsilon_0(n_1 + n_2 + n_3 + n_4 + n_5))$$
$$= \exp(-\beta\varepsilon_0 n_1)\exp(-\beta\varepsilon_0 n_2)\exp(-\beta\varepsilon_0 n_3)\exp(-\beta\varepsilon_0 n_4)\exp(-\beta\varepsilon_0 n_5) \tag{4.51}$$

Therefore, Z is a product of five series:

$$Z = \sum_{n_1=0}^{\infty}\exp(-\beta\varepsilon_0 n_1)\sum_{n_2=0}^{\infty}\exp(-\beta\varepsilon_0 n_2)\sum_{n_3=0}^{\infty}\exp(-\beta\varepsilon_0 n_3)$$
$$\times \sum_{n_4=0}^{\infty}\exp(-\beta\varepsilon_0 n_4)\sum_{n_5=0}^{\infty}\exp(-\beta\varepsilon_0 n_5) \tag{4.52}$$

The five sums are identical. Each sum is a series of a common ratio:

$$\sum_{n_1=0}^{\infty}\exp(-\beta\varepsilon_0 n_1) = \frac{1}{1 - \exp(-\beta\varepsilon_0)} \tag{4.53}$$

Therefore,

$$Z = [1 - \exp(-\beta\varepsilon_0)]^{-5} \tag{4.54}$$

Figure 4.13 shows a plot of Z as a function of $\beta\varepsilon_0$.

The simple method we have just learned allows us to calculate Z, but does not give an explicit expression of $\Omega(E)$. Here we learn a method to

Figure 4.13 Partition function Z plotted as a function of $\beta\varepsilon_0$ for a system of five independent particles. The y axis is in a logarithmic scale.

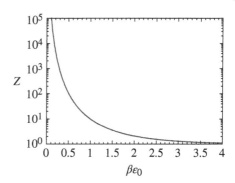

calculate $\Omega(E)$. Since our method gives a general formula for $\Omega(E)$ in a system consisting of N noninteracting particles, each having $\varepsilon_i = n_i\varepsilon_0$ with $n_i = 0, 1, 2,\ldots$, we move to the N-particle system. The energy of the system is $E = \varepsilon_1 + \varepsilon_2 + \varepsilon_3 + \cdots + \varepsilon_N = \varepsilon_0(n_1 + n_2 + n_3 + \cdots + n_N)$. The partition function of the N-particle system is

$$Z = [1 - \exp(-\beta\varepsilon_0)]^{-N} \tag{4.55}$$

We return to the five-particle system for now. To make $E/\varepsilon_0 = 3$, for example, the five nonnegative integers must add up to 3. A possible breakup of three into five nonnegative integers is equivalent to arranging three crosses and four bars, and assign the numbers of crosses in the five sections demarcated by the four bars to n_1, n_2, n_3, n_4, and n_5. We place four bars, since there are five particles $(4 = 5 - 1)$. Figure 4.14 shows two examples of breaking up of three into five parts.

Now we consider the system of N particles. To break up $E/\varepsilon_0 = n$ into N parts, we arrange $N - 1$ bars and n crosses. The number of the arrangements is

$$_{N-1+n}C_n = \frac{(N - 1 + n)!}{(N - 1)!n!} \tag{4.56}$$

In other words, we pick up $N - 1$ out of $N - 1 + n$ symbols, and assign bars to the $N - 1$ symbols and crosses to the rest. The following table is generated for the example of five particles (Table 4.3).

Problem 4.6 confirms that the partition function calculated using Eqs. (4.45) with (4.56) is identical to Eq. (4.55).

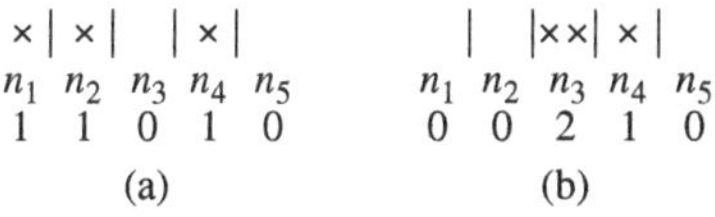

×	×		×	
n_1	n_2	n_3	n_4	n_5
1	1	0	1	0

(a)

		××	×	
n_1	n_2	n_3	n_4	n_5
0	0	2	1	0

(b)

Figure 4.14 An arrangement of three crosses and four bars can be mapped into breaking up three into five nonnegative integers. (a) $n_1 = 1, n_2 = 1, n_3 = 0, n_4 = 1$, and $n_5 = 0$. (b) $n_1 = 0, n_2 = 0, n_3 = 2, n_4 = 1$, and $n_5 = 0$.

Table 4.3 Number of states for each energy level (the lowest nine levels) in a system of five noninteracting particles.

$E/\varepsilon_0 = n$	$\Omega = {}_{N-1+n}C_n$
0	${}_4C_0 = 1$
1	${}_5C_1 = 5$
2	${}_6C_2 = 15$
3	${}_7C_3 = 35$
4	${}_8C_4 = 70$
5	${}_9C_5 = 126$
6	${}_{10}C_6 = 210$
7	${}_{11}C_7 = 330$
8	${}_{12}C_8 = 495$

4.3 Internal Energy

The first thermodynamic variable we learn to calculate from the partition function Z is the internal energy U. In classical thermodynamics, U is some extensive, macroscopic quantity that increases when the surroundings gives heat to the system or does work to the system. The law that describes this exchange is called the first law of thermodynamics. The general formula to calculate U from Z, derived subsequently, tells us what constitutes the internal energy microscopically.

Note that U is different from the system's energy E_k at a specific time. The energy is different from system to system among those replica systems in contact with a shared reservoir of constant temperature (Figure 4.15a). There is a

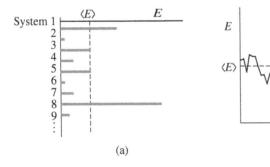

(a)   (b)

Figure 4.15 (a) Snapshot of energy E in replica systems in contact with a shared heat reservoir. The mean value $\langle E \rangle$ is indicated by a dashed line. (b) Time trace of the energy E of a given system in contact with a heat reservoir.

variability in E_k. The difference is caused by k; how E_k depends on k is shared by different replicas of the ensemble. Another view is that the state of a given system changes with time, and so does the energy (Figure 4.15b). In the ensemble average as well as in the long-time average, E_k fluctuates around its mean $\langle E \rangle$.

The fluctuations in k are not chaotic. They follow a rule we learned in the preceding section: The distribution of E_k follows the probability distribution $P(E_k)$ which depends on E_k as given by Eq. (4.22). Therefore, the ensemble average of the energy, $\langle E \rangle$, should be expressed as

$$\langle E \rangle = \sum_k E_k P(E_k) = \frac{1}{Z} \sum_k E_k \exp(-\beta E_k) \tag{4.57}$$

On the right-hand side, each term of the series contains E_k, which can be generated by differentiating $\exp(-\beta E_k)$ by β:

$$\frac{\partial}{\partial \beta} \sum_k \exp(-\beta E_k) = -\sum_k E_k \exp(-\beta E_k) \tag{4.58}$$

where the order of differentiation and summation has been interchanged. Thus, we find

$$\langle E \rangle = -\frac{1}{Z} \frac{\partial}{\partial \beta} \sum_k \exp(-\beta E_k) = -\frac{1}{Z} \frac{\partial Z}{\partial \beta} = -\frac{\partial \ln Z}{\partial \beta} \tag{4.59}$$

where the identity, $d \ln Z = Z^{-1} dZ$, was used. Since $\langle E \rangle$ is equivalent to the classical U, the general formula is written as

$$U = -\frac{1}{Z} \frac{\partial Z}{\partial \beta} = -\frac{\partial \ln Z}{\partial \beta} \tag{4.60}$$

As Z was calculated for a system of given T, V, and N, U is a function of T, V, and N.

4.4 Identification of β

So far, we have been treating β as something related to T, but we have not found how they are related to each other. In this section, we learn that

$$\beta = \frac{1}{k_B T} \tag{4.61}$$

We apply the general formula we obtained in the preceding section (Eq. (4.60)) to the partition function of N monatomic molecules. We obtained the partition function of the system in Section 4.2.3 (see Eq. (4.40)). The first step is to write the natural logarithm of Z:

$$\ln Z = N \ln V + \frac{3N}{2} \ln \frac{2\pi m}{\beta h^2} \tag{4.62}$$

Differentiating by β leads to

$$\frac{\partial}{\partial \beta} \ln Z = -\frac{3N}{2\beta} \tag{4.63}$$

Thus, we find

$$U = \frac{3N}{2\beta} \tag{4.64}$$

We know that, for n moles of monatomic gas ($n = N/N_A$, where N_A is the Avogadro's number; $Nk_B = nR$),

$$U = \frac{3}{2}nRT = \frac{3}{2}Nk_B T \tag{4.65}$$

Comparison of these two equations leads to Eq. (4.61). We used a system of monatomic molecules to derive this universal identity. Now we can use this identity in other systems including a perfect gas of diatomic molecules and quantum-mechanical systems.

By now, you may have noticed that the probability of a state given by Eq. (4.22) is identical to the **Boltzmann factor**. Usually, the latter is stated as follows: The ratio of the probability for the system to be in one of the states with energy E_1 to the probability to be in one of the states with energy E_2 is

$$\frac{P(E_1)}{P(E_2)} = \exp(-\beta(E_1 - E_2)) = \exp\left(-\frac{E_1 - E_2}{k_B T}\right) \tag{4.66}$$

Note again that the probability refers to the state, not to the energy level.

In Section 4.2.2, we looked at a plot of P_n, the probability for the nth state in a single-particle system with energy level $n\varepsilon$. Figure 4.11 was drawn for $\beta\varepsilon = 0.2$. Having learned that $\beta = (k_B T)^{-1}$, we examine here the states of the particle at low and high temperatures. Figure 4.16 shows a plot of P_n for $\beta\varepsilon = 0.1$ and 2. The plot for $\beta\varepsilon = 0.1$ decreases more slowly with an increasing n than does the P_n for $\beta\varepsilon = 2$. At sufficiently high temperatures ($\beta\varepsilon \to 0$), the plot will be nearly horizontal, indicating that all the states are equally probable, although each P_n is nearly zero. The plot for $\beta\varepsilon = 2$ is dominated by P_0. In the low-temperature

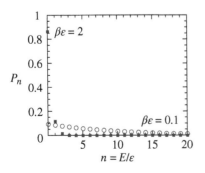

Figure 4.16 Distribution of normalized P_n at two temperatures, $\beta\varepsilon = 0.1$ (circle) and 2 (square) in a system of one particle with energy levels $n\varepsilon$ ($n = 0, 1, 2, \ldots$).

limit ($\beta\varepsilon \to \infty$), $P_n = 0$ except $P_0 = 1$. The particle is exclusively at the lowest energy level (ground state). A finite thermal energy excites the particle to higher energy levels.

4.5 Equipartition Law

We prove here the **equipartition law**: The internal energy is $\frac{1}{2}k_B T$ per variable in the expression of the energy of the classical system. It is required that the expression contains the variable in a quadratic form.

We first look at a classical expression of the energy. For an ideal gas of N monatomic molecules, E consists of $3N$ quadratic terms:

$$E = \sum_{i=1}^{N} \frac{\mathbf{p}_i^2}{2m} = \sum_{i=1}^{N} \left(\frac{p_{ix}^2}{2m} + \frac{p_{iy}^2}{2m} + \frac{p_{iz}^2}{2m} \right) \tag{4.67}$$

For a system of N noninteracting harmonic oscillators, E consists of $6N$ quadratic terms:

$$E = \sum_{i=1}^{N} \left(\frac{1}{2}\kappa\mathbf{r}_i^2 + \frac{\mathbf{p}_i^2}{2m} \right) = \sum_{i=1}^{N} \left(\frac{\kappa}{2}x_i^2 + \frac{\kappa}{2}y_i^2 + \frac{\kappa}{2}z_i^2 + \frac{p_{ix}^2}{2m} + \frac{p_{iy}^2}{2m} + \frac{p_{iz}^2}{2m} \right) \tag{4.68}$$

We pick up one component from E as cs^2 and write the rest as E':

$$E = cs^2 + E' \tag{4.69}$$

Here, the prime does not denote the derivative. For example, in the ideal gas system, if we choose $s = p_{jy}$, then $c = (2m)^{-1}$, and

$$E' = \sum_{i\neq j}^{N} \frac{\mathbf{p}_i^2}{2m} + \frac{p_{jx}^2}{2m} + \frac{p_{jz}^2}{2m} \tag{4.70}$$

In the system of harmonic oscillators, if we choose $s = z_j$, then $c = \kappa/2$, and

$$E' = \sum_{i\neq j}^{N} \left(\frac{1}{2}\kappa\mathbf{r}_i^2 + \frac{\mathbf{p}_i^2}{2m} \right) + \frac{\kappa}{2}x_j^2 + \frac{\kappa}{2}y_j^2 + \frac{p_j^2}{2m} \tag{4.71}$$

The partition function is an integral of $e^{-\beta E}$ by $\mathbf{r}_1,\ldots, \mathbf{p}_N$. With s singled out, the integral is expressed as

$$Z = \int_{-\infty}^{\infty} \exp(-\beta c s^2)ds \times Z' = \left(\frac{\pi}{\beta c} \right)^{1/2} Z' \tag{4.72}$$

where

$$Z' = \int \cdots \int e^{-\beta E'} \, d(\text{remaining variables}) \tag{4.73}$$

We proceed as usual. The logarithm of Z is

$$\ln Z = \frac{1}{2} \ln \frac{\pi}{\beta c} + \ln Z' \tag{4.74}$$

The internal energy is

$$U = -\frac{\partial}{\partial \beta} \ln Z = \frac{1}{2\beta} + U' = \frac{1}{2} k_B T + U' \tag{4.75}$$

where

$$U' = -\frac{\partial}{\partial \beta} \ln Z' \tag{4.76}$$

is the contribution to the internal energy from the remaining variables.

Now we have found that one variable contributes $\frac{1}{2} k_B T$ to the U of the system, proving the equipartition law. Table 4.4 summarizes U for the system of N noninteracting monatomic molecules and the system of N independent harmonic oscillators. The energy per particle is $\frac{3}{2} k_B T$ in the gas, but it is $3 k_B T$ in the harmonic oscillator. Each degree of freedom in the harmonic oscillator has two variables: $\mathbf{r}_i$ and $\mathbf{p}_i$. Per dimension, the contribution is $k_B T$.

The situation is different in a system of quantum-mechanical harmonic oscillators. We learned in Section 3.6 that the nth energy level ε_n is

$$\varepsilon_n = (n + \frac{1}{2}) h\nu \tag{4.77}$$

where $n = 0, 1, 2,\dots$ The single-particle partition function Z_1 is calculated as

$$Z_1 = \sum_{n=0}^{\infty} e^{-(n+1/2)\beta h\nu} = \frac{e^{-\beta h\nu/2}}{1 - e^{-\beta h\nu}} = \frac{1}{2 \sinh(\beta h\nu/2)} \tag{4.78}$$

Then, the partition function Z of the system of N harmonic oscillators is

$$Z = [2 \sinh(\beta h\nu/2)]^{-N} \tag{4.79}$$

The logarithm of Z is

$$\ln Z = -N \ln 2 - N \ln(\sinh(\beta h\nu/2)) \tag{4.80}$$

Table 4.4 Internal energy U and the energy U per particle in a gas of monatomic molecules and a system of harmonic oscillators.

System	Number of variables	U	U per particle
Ideal gas of N monatomic molecules	$3N$	$\frac{3}{2} N k_B T$	$\frac{3}{2} k_B T$
N noninteracting harmonic oscillators	$6N$	$3 N k_B T$	$3 k_B T$

Then, U is calculated as

$$U = -\frac{\partial}{\partial\beta}\ln Z = N\frac{\cosh(\beta h\nu/2)}{\sinh(\beta h\nu/2)}\frac{h\nu}{2} = \frac{Nh\nu}{2}\frac{1}{\tanh(\beta h\nu/2)} \tag{4.81}$$

Per oscillator, the internal energy is

$$\frac{U}{N} = \frac{h\nu}{2}\frac{1}{\tanh(\beta h\nu/2)} \tag{4.82}$$

At low temperatures, $\frac{1}{2}\beta h\nu \gg 1$ and

$$\frac{U}{N} \cong \frac{h\nu}{2} \tag{4.83}$$

In the low-temperature limit, the harmonic oscillator is at the ground state. At high temperatures, $\tanh(\beta h\nu/2) \cong \beta h\nu/2$, and

$$\frac{U}{N} \cong \frac{h\nu}{2}\frac{2}{\beta h\nu} = \frac{1}{\beta} = k_B T \tag{4.84}$$

Thus, we find that the classical harmonic oscillator is equivalent to the high-temperature asymptote of the quantum-mechanical harmonic oscillator.

4.6 Other Thermodynamic Functions

In Section 4.3, we learned how to calculate U from Z. In this section, we learn how to calculate the entropy, pressure, Helmholtz free energy, and chemical potential.

We start with the partition function $Z(\beta, V)$ for a closed system of N particles. The total derivative of $\ln Z$ is

$$d\ln Z = \frac{\partial\ln Z}{\partial\beta}d\beta + \frac{\partial\ln Z}{\partial V}dV \tag{4.85}$$

With Eq. (4.60), it is rewritten to

$$d\ln Z = -U\,d\beta + \frac{\partial\ln Z}{\partial V}dV \tag{4.86}$$

The total derivative of βU is

$$d(\beta U) = U\,d\beta + \beta\,dU \tag{4.87}$$

Adding these two equations side by side, we obtain

$$d(\ln Z + \beta U) = \beta\,dU + \frac{\partial\ln Z}{\partial V}dV \tag{4.88}$$

which is rearranged into

$$dU = k_B T d(\ln Z + \beta U) - k_B T\frac{\partial\ln Z}{\partial V}dV \tag{4.89}$$

Compare this equation with the law of thermodynamics (first and second laws combined for reversible changes):

$$dU = T\,dS - p\,dV \tag{4.90}$$

The latter equation means that the energy increase is equal to sum of the heat given to the system and the work done to the system. In a reversible process, the given heat is equal to $T\,dS$. Comparison of the last two equations leads to

$$dS = k_B d(\ln Z + \beta U) \tag{4.91}$$

$$p = k_B T \frac{\partial \ln Z}{\partial V} \tag{4.92}$$

The second equation is called the **equation of state**.

Equation (4.91) defines the entropy S in the canonical ensemble. Upon integration, the equation leads to

$$S = k_B(\ln Z + \beta U) + \text{constant} \tag{4.93}$$

where the constant depends on N and the external field (electric, magnetic, etc.) if it is present. As is the case with the free energy, what matters is a difference of S or a change of S, and therefore the constant does not matter. For now, we set constant $= 0$:

$$S = k_B(\ln Z + \beta U) \tag{4.94}$$

Then, the Helmholtz free energy F, defined as $F = U - TS$, is expressed using Z as

$$F = -k_B T \ln Z \tag{4.95}$$

Again, the constant is neglected here; but it does not matter, since what counts is a change in the free energy.

Recall $dF = -S\,dT - p\,dV$. It means that

$$S = -\left(\frac{\partial F}{\partial T}\right)_V \tag{4.96}$$

$$p = -\left(\frac{\partial F}{\partial V}\right)_T \tag{4.97}$$

We can confirm consistency by calculating S and P using the expression of F (Eq. (4.95)). The results are identical to those in Eqs. (4.94) and (4.92).

The chemical potential μ is obtained by differentiating F by N:

$$\mu = \left(\frac{\partial F}{\partial N}\right)_{T,V} = -k_B T \frac{\partial \ln Z}{\partial N} \tag{4.98}$$

Here is a summary of formulas for thermodynamic functions we can derive from Z:

$$U = -\frac{\partial \ln Z}{\partial \beta} \tag{4.60}$$

$$p = k_B T \frac{\partial \ln Z}{\partial V} \qquad (4.92)$$

$$F = -k_B T \ln Z \qquad (4.95)$$

$$S = k_B (\ln Z + \beta U) \qquad (4.94)$$

$$\mu = -k_B T \frac{\partial \ln Z}{\partial N} \qquad (4.98)$$

From Eq. (4.95), we obtain

$$Z = e^{-\beta F} \qquad (4.99)$$

Then, the probability of a state in the canonical ensemble is expressed as

$$P(E) = Z^{-1} e^{-\beta E} = e^{\beta(F-E)} \qquad (4.100)$$

Now we apply the formulas to a few simple systems.

A. Gas of N Noninteracting Monatomic Molecules
The system has N independent monatomic molecules of mass m in volume V. We obtained the partition function for the system and its natural logarithm (Eqs. (4.40) and (4.62)). Applying the formula for p gives

$$p = k_B T \frac{\partial}{\partial V} \left(N \ln V + \frac{3N}{2} \ln \frac{2\pi m}{\beta h^2} \right) = \frac{N k_B T}{V} \qquad (4.101)$$

This equation of state is identical to $pV = nRT$. The gas that follows this equation of state is called a perfect gas. We derived this equation of state assuming that the molecules do not interact with each other (ideal). Thus, we have proved that the ideal gas is perfect. We consider how the interactions change the equation of state in Chapter 7.

The Helmholtz free energy is calculated as

$$F = -N k_B T \left(\ln V + \frac{3}{2} \ln \frac{2\pi m}{\beta h^2} \right) \qquad (4.102)$$

With Eq. (4.65) that gives U, we obtain the entropy through $S = (U - F)/T$ as

$$S = N k_B \left(\ln V + \frac{3}{2} \ln \frac{2\pi m}{\beta h^2} + \frac{3}{2} \right) \qquad (4.103)$$

The chemical potential of each molecule is

$$\mu = -k_B T \left(\ln V + \frac{3}{2} \ln \frac{2\pi m}{\beta h^2} \right) \qquad (4.104)$$

Equation (4.104) is incorrect. The correct expression of μ must wait for Chapter 6 in which we learn about indistinguishability. The expressions of S and F will be also different.

B. Single Particle with Energy Levels $E = n\varepsilon$ ($n = 0, 1, 2,\ldots; \varepsilon > 0$)

Earlier, we calculated the partition function (Eq. (4.24)). Our next job is to write the natural logarithm of Z:

$$\ln Z = -\ln(1 - e^{-\beta\varepsilon}) \tag{4.105}$$

Then, the internal energy, free energy, and entropy are calculated as

$$U = \frac{\partial}{\partial\beta}\ln(1 - e^{-\beta\varepsilon}) = \frac{\varepsilon e^{-\beta\varepsilon}}{1 - e^{-\beta\varepsilon}} = \frac{\varepsilon}{e^{\beta\varepsilon} - 1} \tag{4.106}$$

$$F = k_B T \ln(1 - e^{-\beta\varepsilon}) \tag{4.107}$$

$$S = k_B \left[\frac{\beta\varepsilon}{e^{\beta\varepsilon} - 1} - \ln(1 - e^{-\beta\varepsilon}) \right] \tag{4.108}$$

From Eq. (4.106), we can calculate the heat capacity:

$$C_V = \frac{\partial U}{\partial T} = -k_B\beta^2 \frac{\partial}{\partial\beta} \frac{\varepsilon}{e^{\beta\varepsilon} - 1} = k_B\beta^2 \frac{\varepsilon^2 e^{\beta\varepsilon}}{(e^{\beta\varepsilon} - 1)^2} = \frac{k_B(\beta\varepsilon)^2}{2(\cosh\beta\varepsilon - 1)} \tag{4.109}$$

where the identity

$$\frac{\partial}{\partial T} = -k_B\beta^2 \frac{\partial}{\partial\beta} = -\frac{1}{k_B T^2} \frac{\partial}{\partial\beta} \tag{4.110}$$

was used.

It is interesting to see how U and C_V change with T at low and high temperatures. First, at low temperatures, $\beta\varepsilon \gg 1$. Then, $e^{\beta\varepsilon} \gg 1$ and

$$U \cong \varepsilon e^{-\beta\varepsilon} = \varepsilon \exp\left(-\frac{\varepsilon}{k_B T}\right) \tag{4.111}$$

$$C_V \cong k_B(\beta\varepsilon)^2 e^{-\beta\varepsilon} = k_B \left(\frac{\varepsilon}{k_B T}\right)^2 \exp\left(-\frac{\varepsilon}{k_B T}\right) \tag{4.112}$$

In the low-temperature limit, both quantities are zero. These two equations retain T dependence. Therefore, they are the low-temperature asymptotes. Second, at high temperatures, $\beta\varepsilon \ll 1$. Then, $e^{\beta\varepsilon} \cong 1 + \beta\varepsilon$ and $\cosh\beta\varepsilon = 1 + \frac{1}{2}(\beta\varepsilon)^2 + \frac{1}{24}(\beta\varepsilon)^4$, and we obtain

$$U \cong \frac{\varepsilon}{\beta\varepsilon} = k_B T \tag{4.113}$$

$$C_V \cong \frac{k_B(\beta\varepsilon)^2}{2\left[\frac{1}{2}(\beta\varepsilon)^2 + \frac{1}{24}(\beta\varepsilon)^4\right]} \cong k_B \left[1 - \frac{1}{12}\left(\frac{\varepsilon}{k_B T}\right)^2\right] \tag{4.114}$$

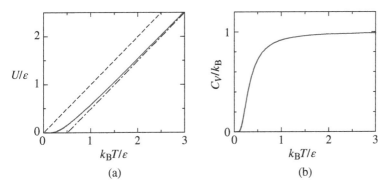

Figure 4.17 Plot of (a) reduced internal energy U/ε and (b) reduced heat capacity C_V/k_B, plotted as a function of reduced temperature $k_B T/\varepsilon$. In (a), the dashed line represents the leading term of the high-temperature asymptote. The dash-dotted line includes the next leading term.

The high-temperature limit of C_V is k_B, but U diverges to $+\infty$; there is no high-temperature limit. These two equations are the high-temperature asymptotes.

Figure 4.17a shows a plot of U/ε as a function of $k_B T/\varepsilon$. Dividing U by ε reduces it to a dimensionless quantity. Likewise, $k_B T/\varepsilon$ is a reduced temperature that is dimensionless. Figure 4.17b is a plot of reduced heat capacity C_V/k_B as a function of the reduced temperature. The curves agree with the low- and high-temperature asymptotes we obtained in the abovementioned discussion. One thing to note is that the high-temperature asymptote of U (Eq. (4.113)), indicated by the dashed line in Figure 4.17a, does not run close to the solid line. The reason is that the approximation in Eq. (4.113) is not sufficient. We need one more term, as shown here.

$$U \cong \frac{\varepsilon}{\beta\varepsilon + \frac{1}{2}(\beta\varepsilon)^2} = \frac{1}{\beta}\left(1 - \frac{1}{2}\beta\varepsilon\right) = k_B T - \frac{\varepsilon}{2} \tag{4.115}$$

This equation, indicated by a dash-dotted line, runs close to the curve.

4.7 Another View of Entropy

In general chemistry and physical chemistry classes, you would have learned that the entropy represents randomness of the system. The randomness was vaguely defined, at best, such as random orientation of water molecules, a greater freedom to move, etc. In this section, we take a second look at the entropy to interpret it in statistical mechanics.

Let us start by listing relevant equations we have learned:

$$S = k_B(\ln Z + \beta U) \tag{4.94}$$

$$1 = \sum_k P(E_k) \tag{4.5}$$

$$U = \langle E \rangle = \sum_k E_k P(E_k) \tag{4.57}$$

Combining these equations leads to

$$S = k_B \left(\ln Z \sum_k P(E_k) + \beta \sum_k E_k P(E_k) \right) = k_B \sum_k (\ln Z + \beta E_k) P(E_k)$$
$$\tag{4.116}$$

where

$$\ln Z + \beta E_k = \ln[Z \exp(\beta E_k)] = \ln \frac{1}{P(E_k)} = -\ln P(E_k) \tag{4.117}$$

Then,

$$S = -k_B \sum_k P(E_k) \ln P(E_k) \tag{4.118}$$

Recall the uncertainty we defined in Section 2.4:

$$H = -\sum_k P(E_k) \ln P(E_k) \tag{4.119}$$

Thus, we find that the entropy is just a constant times the uncertainty of microstates:

$$S = k_B H \tag{4.120}$$

and the constant is the Boltzmann constant.

Consider a microcanonical ensemble. All the microstates of the system have the same energy. If there are W microstates in the system, $P(E_k) = W^{-1}$, and we obtain

$$S = -k_B \sum_{k=1}^{W} \frac{1}{W} \ln \frac{1}{W} = k_B \ln W \tag{4.121}$$

In the white tombstone of Ludwig Boltzmann in the Central Cemetery (Zentralfriedhof) of Vienna, "$S = k \log W$" is inscribed in gold. A photograph on the cover of this book shows the tomb. Before he died, Boltzmann instructed this inscription. His grave is not far from the graves of great composers such as Ludwig van Beethoven, Franz Schubert, and Johannes Brahms.

In Section 4.4, we compared $P_n = P(n\varepsilon)$ for two values of $\beta\varepsilon$ in a system of a single particle with energy levels $n\varepsilon$. The higher the temperature, the broader the distribution of P_n. In the sum of $P_n \times (-\ln P_n)$ with respect to n, $-P_n \ln P_n$ remains large at large n, when the temperature is high, and therefore the greater entropy.

4.8 Fluctuations of Energy

In Section 4.3, we learned how to calculate the mean energy $\langle E \rangle$ from the partition function. The energy E fluctuates from system to system among replicas, and, for a given system, from time to time. In this section, we learn how small the fluctuations of the energy are. Our goal is to estimate the root-mean-square (RMS) deviation $\langle \Delta E^2 \rangle^{1/2}$, where $\Delta E = E - \langle E \rangle$.

Let us start with listing some relevant equations we already know.

$$\langle E \rangle = \sum_k E_k P(E_k) = \frac{1}{Z} \sum_k E_k \exp(-\beta E_k) \tag{4.57}$$

$$\langle E \rangle = U = -\frac{1}{Z} \frac{\partial Z}{\partial \beta} = -\frac{\partial \ln Z}{\partial \beta} \tag{4.60}$$

We calculate $\langle \Delta E^2 \rangle$ using the identity:

$$\langle \Delta E^2 \rangle = \langle E^2 \rangle - \langle E \rangle^2 \tag{4.122}$$

The variance is equal to the mean of the square (aka second moment) minus the square of the mean, as we learned in Section 2.4. In the same way as $\langle E \rangle$ was calculated, $\langle E^2 \rangle$ is calculated according to

$$\langle E^2 \rangle = \sum_k E_k^2 P(E_k) = \frac{1}{Z} \sum_k E_k^2 \exp(-\beta E_k)$$

$$= \frac{1}{Z} \frac{\partial^2}{\partial \beta^2} \sum_k \exp(-\beta E_k) = \frac{1}{Z} \frac{\partial^2 Z}{\partial \beta^2} \tag{4.123}$$

Then,

$$\langle \Delta E^2 \rangle = \frac{1}{Z} \frac{\partial^2 Z}{\partial \beta^2} - \left(\frac{1}{Z} \frac{\partial Z}{\partial \beta} \right)^2 \tag{4.124}$$

The following identity allows easy calculation of the right-hand side.

$$\frac{\partial^2 \ln Z}{\partial \beta^2} = \frac{1}{Z} \frac{\partial^2 Z}{\partial \beta^2} - \left(\frac{1}{Z} \frac{\partial Z}{\partial \beta} \right)^2 \tag{4.125}$$

This identity can be easily confirmed by differentiating Eq. (4.59) by β. Thus,

$$\langle \Delta E^2 \rangle = \frac{\partial^2 \ln Z}{\partial \beta^2} \tag{4.126}$$

This equation can be rewritten into an expression with thermodynamic variables. Differentiating Eq. (4.59) by β and we obtain

$$-\frac{\partial U}{\partial \beta} = \frac{\partial^2 \ln Z}{\partial \beta^2} \tag{4.127}$$

Therefore,

$$\langle \Delta E^2 \rangle = -\frac{\partial U}{\partial \beta} \tag{4.128}$$

Since $-\partial/\partial\beta = k_B T^2 \partial/\partial T$ (see Eq. (4.110)),

$$\langle \Delta E^2 \rangle = k_B T^2 \frac{\partial U}{\partial T} = k_B T^2 C_V \qquad (4.129)$$

This is the general formula we were looking for.

Now we compare $\langle \Delta E^2 \rangle^{1/2}$ and $\langle E \rangle$ for a perfect gas of monatomic molecules. Since $C_V = \frac{3}{2}Nk_B$,

$$\langle \Delta E^2 \rangle^{1/2} = (3N/2)^{1/2} k_B T \qquad (4.130)$$

The ratio of the standard deviation to the mean of the energy is

$$\langle \Delta E^2 \rangle^{1/2} / \langle E \rangle = (3N/2)^{-1/2} \qquad (4.131)$$

For $N = 10^{18}$, the coefficient of variation is $\sim 10^{-9}$. The fluctuation in the energy is essentially negligible. Although we allow for energy variations in the canonical ensemble, they are nearly zero. The situation will be different, if N is small.

4.9 Grand Canonical Ensembles

As the canonical ensemble was right for a closed system, there is a different type of ensemble that is right for an open system of volume V at temperature T. The open system allows an exchange of particles with the surroundings on top of an exchange of heat and work (Figure 4.18).

A room with an open door is an open system, if the walls are heat-conductive. Here, we look at another open system important in chemistry experiments. That is the volume in a solution used in a light scattering experiment (Figure 4.19). A collimated (or focusing) laser beam travels the solution in

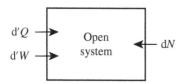

Figure 4.18 An open system exchanges heat, work, and particles with its surroundings to reach an equilibrium.

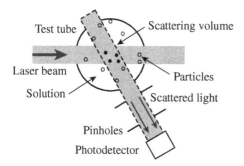

Figure 4.19 An example of an open system. Top view of a light scattering experiment. The intersection of a laser beam path and the volume subtended by two pinholes defines the scattering volume. Solute and solvent molecules can enter or leave the scattering volume.

a test tube. Illuminated molecules, including solvent molecules, can scatter light. A pair of pinholes is placed between the test tube and the photodetector to restrict the molecules that scatter photons to those within a cylindrical volume subtended by the pinholes, indicated by the dashed lines in the figure. The intersection of this cylinder and the laser beam path is called the scattering volume, since only those photons scattered by the molecules within that volume can reach the photodetector. Obviously, there are no walls for the volume: Solute and solvent molecules are free to enter the volume and leave the volume, as long as they satisfy incompressibility requirement (no volume change). As a result of this movement, the number of solute molecules in the scattering volume fluctuates, while the rest of the solution in the test tube serves as a reservoir for the volume.

Sampling an open system is also the case with UV absorption spectroscopy and fluorescence spectroscopy of a solution in a cuvette. Only those molecules within the volume illuminated by incident light contribute to the absorption and fluorescence, while the rest of the solution serves as a reservoir.

As the difference of the temperature between the system and the reservoir determines the flow of heat, the transfer of particles is dictated by a difference of chemical potential μ. If the particles in the system have a higher chemical potential compared with the particles in the reservoir, the mass transfer will occur from the system to the reservoir, assuming that the system is thermally equilibrated with the reservoir. The transfer will continue until the exodus lowers μ within the system to match the one in the reservoir.

In the open system, T and μ are specified (also V, if it is relevant). The reservoir is now a reservoir of heat and particles (see Figure 4.20).

When the system adds a particle, its energy changes, which may change the temperature. Then, the system reestablishes the equilibrium with the heat reservoir to hold its temperature at the specified value.

In Section 4.1, we introduced hypothesis (S2) for the microcanonical ensemble. In that statement, the condition that the number of particles be fixed was implicit. Here, we restate the hypothesis:

All microstates with the same energy and the same number of particles are equally probable   (S4)

Figure 4.20 An open system is in contact with a reservoir of heat and particles. The energy and the number of particles of the system fluctuate with time.

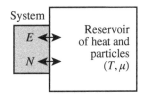

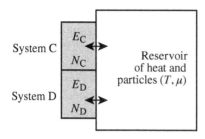

System C

E_C

N_C

System D

E_D

N_D

Reservoir
of heat and
particles (T, μ)

Figure 4.21 Systems C and D are in contact with a reservoir of heat and particles. The two systems are in contact with each other.

Now the probability P of the state depends on its energy E_k and the number of particles N, i.e. $P = P(E_k, N)$. We know how P depends on E_k. Here, we want to find how P depends on N.

We bring two systems C and D into contact with a reservoir of heat and particles. Each of the two systems is equilibrated with the reservoir independently. Then, the two systems will share T and μ. Therefore, bringing the two systems into contact with each other will not change the states of the two systems (Figure 4.21).

The total energy of the composite system, C + D, can be written as $E_C + E_D$, and the number of particles in the system is $N_C + N_D$. The probability for the combined system to have $E_C + E_D$ and $N_C + N_D$ is equal to the joint probability for system C to have E_C and N_C and for system D to have E_D and N_D. The two probabilities are independent, and therefore

$$P_{C+D}(E_C + E_D, N_C + N_D) = P_C(E_C, N_C)P_D(E_D, N_D) \tag{4.132}$$

As we did for the canonical ensemble, we differentiate this equation by E_C and E_D to obtain

$$\frac{\partial P_C}{\partial E_C}P_D = P_C\frac{\partial P_D}{\partial E_D} \tag{4.133}$$

Likewise, differentiating Eq. (4.132) by N_C and N_D leads to

$$\frac{\partial P_C}{\partial N_C}P_D = P_C\frac{\partial P_D}{\partial N_D} \tag{4.134}$$

As before, Eq. (4.133) is rewritten to

$$\frac{\partial \ln P_C}{\partial E_C} = \frac{\partial \ln P_D}{\partial E_D} = -\beta \tag{4.135}$$

since equating a function of E_C to a function of E_D requires that they be a constant. Equation (4.135) is solved as

$$\ln P_C = -\beta E_C + C_C(N_C) \tag{4.136}$$

where $C_C(N_C)$ is a constant of E_C, but can be a function of N_C.

Equation (4.134) is rewritten to

$$\frac{1}{P_C}\frac{\partial P_C}{\partial N_C} = \frac{1}{P_D}\frac{\partial P_D}{\partial N_D} \tag{4.137}$$

The left-hand side is a function of N_C, and the right-hand side is a function of N_D. Equating the two functions is possible only when they are a constant. Let the constant be $\beta\mu$:

$$\frac{\partial \ln P_C}{\partial N_C} = \frac{\partial \ln P_D}{\partial N_D} = \beta\mu \tag{4.138}$$

Thus introduced, μ is a constant. Later in this section, we prove that μ introduced in this way is identical to the chemical potential that we know. With Eq. (4.136),

$$\frac{dC_C}{dN_C} = \beta\mu \tag{4.139}$$

which is solved as

$$C_C = \beta\mu N_C + D_C \tag{4.140}$$

where D_C is a constant of N_C. Thus,

$$P_C(E_C, N_C) = \exp(D_C)\exp(\beta\mu N_C - \beta E_C) \tag{4.141}$$

Obviously, $\exp(D_C)$ is just a normalization constant. Eliminating subscripts, we have

$$P(E, N) = \frac{1}{\mathcal{Z}}e^{\beta\mu N - \beta E} \tag{4.142}$$

where $\mathcal{Z}$ is obtained from the normalization condition.

For a system with discrete energy levels,

$$\mathcal{Z} = \sum_{N=0}^{\infty}\sum_{k}\exp(\beta\mu N - \beta E_k) = \sum_{N=0}^{\infty}e^{\beta\mu N}\sum_{k}\exp(-\beta E_k) = \sum_{N=0}^{\infty}e^{\beta\mu N}Z_N \tag{4.143}$$

where Z_N is the partition function of the N-particle canonical ensemble.

For a system of continuous energy, we have

$$f(E(\mathbf{r}_1, \ldots, \mathbf{p}_N), N) = \frac{1}{\mathcal{Z}}\exp(\beta\mu N - \beta E(\mathbf{r}_1, \ldots, \mathbf{p}_N)) \tag{4.144}$$

where

$$\mathcal{Z} = \sum_{N=0}^{\infty}e^{\beta\mu N}\frac{1}{h^{3N}}\int_V d\mathbf{r}_1\cdots\int_{-\infty}^{\infty}d\mathbf{p}_N\exp(-\beta E(\mathbf{r}_1, \ldots, \mathbf{p}_N)) = \sum_{N=0}^{\infty}e^{\beta\mu N}Z_N \tag{4.145}$$

Equations (4.142) and (4.144) are called a grand canonical distribution. The function $\mathcal{Z}$ is called a **grand canonical partition function** or **grand partition function**, for short. The ensemble consisting of open systems similar to the one we are considering is called a **grand canonical ensemble**. The grand partition function is a function of T, V, and μ. We will use a different font of Z for the grand partition function.

Statistical averages under the grand canonical distribution are calculated in a way similar to those for the canonical ensemble. The general formula for the mean of a function $g(E, N)$ is

$$\langle g(E, N) \rangle = \frac{1}{\mathcal{Z}} \sum_{N=0}^{\infty} \sum_{k} g(E_k, N) \exp(\beta \mu N - \beta E_k) \tag{4.146}$$

for an ensemble with discrete energy levels. A similar expression can be written for an ensemble with continuous energy. The mean of E, E^2, N, and N^2 are calculated as

$$\langle E \rangle = \frac{1}{\mathcal{Z}} \sum_{N=0}^{\infty} \sum_{k} E_k \exp(\beta \mu N - \beta E_k) \tag{4.147}$$

$$\langle E^2 \rangle = \frac{1}{\mathcal{Z}} \sum_{N=0}^{\infty} \sum_{k} E_k^2 \exp(\beta \mu N - \beta E_k) \tag{4.148}$$

$$\langle N \rangle = \frac{1}{\mathcal{Z}} \sum_{N=0}^{\infty} \sum_{k} N \exp(\beta \mu N - \beta E_k) \tag{4.149}$$

$$\langle N^2 \rangle = \frac{1}{\mathcal{Z}} \sum_{N=0}^{\infty} \sum_{k} N^2 \exp(\beta \mu N - \beta E_k) \tag{4.150}$$

Statistical mechanics offers a toolbox for calculating thermodynamic functions. It often happens that calculating them from $\mathcal{Z}$ is simpler compared with calculating them from Z. Table 4.5 compares the two ensembles.

Now we find formulas to calculate thermodynamic functions from $\mathcal{Z}$. The procedure is lengthier compared with the one for the canonical ensemble. You can skip the details to jump to the list of formulas in a box, if you are comfortable with trusting them without going through the derivation yourself.

Table 4.5 Comparison of canonical and grand canonical ensembles.

Ensemble	System	Independent variables	Partition function
Canonical	Closed	V, T, N	$Z = \sum_{k} \exp(-\beta E_k)$
Grand canonical	Open	V, T, μ	$\mathcal{Z} = \sum_{N=0}^{\infty} e^{\beta \mu N} \sum_{k} \exp(-\beta E_k)$

Consider the total derivative of $\ln \mathcal{Z}$. Since $\mathcal{Z}$ is a function of β, V, and μ,

$$\mathrm{d}\ln\mathcal{Z} = \frac{\partial\ln\mathcal{Z}}{\partial\beta}\mathrm{d}\beta + \frac{\partial\ln\mathcal{Z}}{\partial V}\mathrm{d}V + \frac{\partial\ln\mathcal{Z}}{\partial\mu}\mathrm{d}\mu \tag{4.151}$$

where

$$\frac{\partial\ln\mathcal{Z}}{\partial\mu} = \frac{1}{\mathcal{Z}}\frac{\partial\mathcal{Z}}{\partial\mu} = \frac{1}{\mathcal{Z}}\sum_{N=0}^{\infty}\sum_{k}\beta N\exp(\beta\mu N - \beta E_k) = \beta\langle N\rangle \tag{4.152}$$

and

$$\frac{\partial\ln\mathcal{Z}}{\partial\beta} = \frac{1}{\mathcal{Z}}\frac{\partial\mathcal{Z}}{\partial\beta} = \frac{1}{\mathcal{Z}}\sum_{N=0}^{\infty}\sum_{k}(\mu N - E_k)\exp(\beta\mu N - \beta E_k) = \mu\langle N\rangle - \langle E\rangle \tag{4.153}$$

With these two equations, Eq. (4.151) is rewritten to

$$\mathrm{d}\ln\mathcal{Z} = (\mu\langle N\rangle - \langle E\rangle)\mathrm{d}\beta + \frac{\partial\ln\mathcal{Z}}{\partial V}\mathrm{d}V + \beta\langle N\rangle\mathrm{d}\mu \tag{4.154}$$

Separately, we take the total derivative of $(\mu\langle N\rangle - \langle E\rangle)\beta$:

$$\mathrm{d}[(\mu\langle N\rangle - \langle E\rangle)\beta] = (\mu\langle N\rangle - \langle E\rangle)\mathrm{d}\beta + \beta\mathrm{d}(\mu\langle N\rangle - \langle E\rangle) \tag{4.155}$$

where

$$\mathrm{d}(\mu\langle N\rangle - \langle E\rangle) = \mu\mathrm{d}\langle N\rangle + \langle N\rangle\mathrm{d}\mu - \mathrm{d}\langle E\rangle \tag{4.156}$$

From these two equations, we get

$$\mathrm{d}[(\mu\langle N\rangle - \langle E\rangle)\beta] = (\mu\langle N\rangle - \langle E\rangle)\mathrm{d}\beta + \beta\langle N\rangle\mathrm{d}\mu - \beta\mathrm{d}\langle E\rangle + \beta\mu\,\mathrm{d}\langle N\rangle \tag{4.157}$$

The first two terms on the right-hand side appear also in Eq. (4.154). Therefore,

$$\mathrm{d}\ln\mathcal{Z} = \mathrm{d}[(\mu\langle N\rangle - \langle E\rangle)\beta] + \frac{\partial\ln\mathcal{Z}}{\partial V}\mathrm{d}V + \beta\mathrm{d}\langle E\rangle - \beta\mu\,\mathrm{d}\langle N\rangle \tag{4.158}$$

which is rewritten to

$$\mathrm{d}[\ln\mathcal{Z} + (\langle E\rangle - \mu\langle N\rangle)\beta] = \frac{\partial\ln\mathcal{Z}}{\partial V}\mathrm{d}V + \beta\mathrm{d}\langle E\rangle - \beta\mu\mathrm{d}\langle N\rangle \tag{4.159}$$

Dividing Eq. (4.159) by β and rearranging terms, we obtain

$$\mathrm{d}\langle E\rangle = k_{\mathrm{B}}T\mathrm{d}[\ln\mathcal{Z} + (\langle E\rangle - \mu\langle N\rangle)\beta] - k_{\mathrm{B}}T\frac{\partial\ln\mathcal{Z}}{\partial V}\mathrm{d}V + \mu\mathrm{d}\langle N\rangle \tag{4.160}$$

The right-hand side lines up the three components that may increase the internal energy: the heat given to the system, the work done on the system, and the chemical energy by mass transfer (see Figure 4.22).

The energy balance for an open system is given as

$$\mathrm{d}U = T\,\mathrm{d}S - p\,\mathrm{d}V + \mu\,\mathrm{d}N \tag{4.161}$$

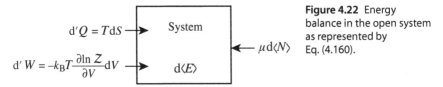

$$\mathrm{d}'Q = T\mathrm{d}S$$

$$\mathrm{d}'W = -k_\mathrm{B}T\frac{\partial \ln \mathcal{Z}}{\partial V}\mathrm{d}V$$

System

$\mu\,\mathrm{d}\langle N\rangle$

$\mathrm{d}\langle E\rangle$

Figure 4.22 Energy balance in the open system as represented by Eq. (4.160).

in thermodynamics. Compare this equation with Eq. (4.160), and we obtain the following relationships:

$$S = k_\mathrm{B}[\ln \mathcal{Z} + (\langle E\rangle - \mu\langle N\rangle)\beta] = k_\mathrm{B}\ln \mathcal{Z} + (\langle E\rangle - \mu\langle N\rangle)/T \qquad (4.162)$$

$$p = k_\mathrm{B}T\frac{\partial \ln \mathcal{Z}}{\partial V} \qquad (4.163)$$

Other thermodynamic functions can also be calculated. For instance,

$$F = U - TS = \langle E\rangle - k_\mathrm{B}T[\ln \mathcal{Z} + (\langle E\rangle - \mu\langle N\rangle)\beta] = \mu\langle N\rangle - k_\mathrm{B}T\ln \mathcal{Z} \qquad (4.164)$$

The following box recaps the formulas we have obtained for the grand canonical ensemble.

$$\langle N\rangle = \frac{1}{\beta}\frac{\partial \ln \mathcal{Z}}{\partial \mu} \qquad (4.165)$$

$$\langle E\rangle = \mu\langle N\rangle - \frac{\partial \ln \mathcal{Z}}{\partial \beta} \qquad (4.166)$$

$$S = k_\mathrm{B}[\ln \mathcal{Z} + (\langle E\rangle - \mu\langle N\rangle)\beta] = k_\mathrm{B}\ln \mathcal{Z} + (\langle E\rangle - \mu\langle N\rangle)/T \qquad (4.162)$$

$$p = k_\mathrm{B}T\frac{\partial \ln \mathcal{Z}}{\partial V} \qquad (4.163)$$

Now we calculate the fluctuations in the number of molecules in the grand canonical ensemble. Equation (4.150) indicates that $\langle N^2\rangle$ is calculated using

$$\langle N^2\rangle = \frac{1}{\beta^2}\frac{1}{\mathcal{Z}}\frac{\partial^2 \mathcal{Z}}{\partial \mu^2} \qquad (4.167)$$

Since $\beta^{-1}\partial \mathcal{Z}/\partial \mu = \mathcal{Z}\langle N\rangle$,

$$\langle N^2\rangle = \frac{1}{\beta}\frac{1}{\mathcal{Z}}\frac{\partial}{\partial \mu}\mathcal{Z}\langle N\rangle = \langle N\rangle\frac{1}{\beta}\frac{1}{\mathcal{Z}}\frac{\partial \mathcal{Z}}{\partial \mu} + \frac{1}{\beta}\frac{\partial \langle N\rangle}{\partial \mu} = \langle N\rangle^2 + \frac{1}{\beta}\frac{\partial \langle N\rangle}{\partial \mu} \qquad (4.168)$$

Thus, we find that the variance $\langle \Delta N^2\rangle$ is calculated as

$$\langle \Delta N^2\rangle = \langle N^2\rangle - \langle N\rangle^2 = \frac{1}{\beta}\frac{\partial \langle N\rangle}{\partial \mu} = \frac{1}{\beta^2}\frac{\partial^2 \ln \mathcal{Z}}{\partial \mu^2} \qquad (4.169)$$

Later in Section 6.6, we find that

$$\langle \Delta N^2 \rangle^{1/2} / \langle N \rangle \cong \langle N \rangle^{-1/2} \tag{4.170}$$

for a system of gas consisting of N molecules. Although particles are free to enter the system or leave, the number of particles in the system does not change much. This situation is similar to the small coefficient of variation for the energy in the canonical ensemble.

We do not look at examples of applying the tools of the grand canonical ensemble here. The application requires that we understand another important concept – indistinguishability of particles. We learn the latter in Chapter 6. We also learn how to apply the grand canonical ensemble in that chapter.

4.10 Cumulants of Energy

We return to the canonical ensemble. If you are familiar with probability theory and statistics, you may have speculated that $-\partial \ln Z / \partial \beta$ might be the first cumulant of the energy and $\partial^2 \ln Z / \partial \beta^2$ the second cumulant. Your guess is right. In this section, we prove that they represent the first two cumulants. The higher-order cumulants are obtained in the same way, namely, by calculating the higher-order derivatives $\partial^n \ln Z / \partial \beta^n$ ($n = 3, 4, \ldots$).

The first cumulant is the mean of the random variable, and the second cumulant is the variance. A statistical distribution is characterized by the cumulants. For example, a normal distribution has only the first two cumulants; the remaining cumulants are zero. The question we ask is: Does $P(E)$ approximately follow a normal distribution? If that is the case, it will facilitate calculation of various thermodynamic quantities.

Consider a function of t defined as

$$f(t) \equiv \sum_k \exp((t - \beta)E_k) \tag{4.171}$$

where t is a parameter, not time. Taylor expansion of $f(t)$ is explicitly written as

$$f(t) = \sum_k \exp(-\beta E_k) + t \sum_k E_k \exp(-\beta E_k)$$
$$+ \frac{t^2}{2} \sum_k E_k^2 \exp(-\beta E_k) + \frac{t^3}{6} \sum_k E_k^3 \exp(-\beta E_k) + \cdots \tag{4.172}$$

Dividing both sides by $f(0) = \sum \exp(-\beta E_k)$ leads to

$$\frac{f(t)}{f(0)} = 1 + t\langle E \rangle + \frac{t^2}{2}\langle E^2 \rangle + \frac{t^3}{6}\langle E^3 \rangle + \frac{t^4}{24}\langle E^4 \rangle + \cdots \tag{4.173}$$

We thus find that $f(t)/f(0)$ is a moment generating function. The logarithm of the moment generating function is called a cumulant generating function:

$$\ln \frac{f(t)}{f(0)} = 1 + t\kappa_1 + \frac{t^2}{2}\kappa_2 + \frac{t^3}{6}\kappa_3 + \frac{t^4}{24}\kappa_4 + \cdots \tag{4.174}$$

It can be proved (see, for example, Wikipedia) that the coefficients κ_1, κ_2, κ_3, and κ_4 are related to the distribution of E as listed here.

$$\kappa_1 = \langle E \rangle \tag{4.175}$$

$$\kappa_2 = \langle (E - \langle E \rangle)^2 \rangle = \langle \Delta E^2 \rangle \tag{4.176}$$

$$\kappa_3 = \langle \Delta E^3 \rangle \tag{4.177}$$

$$\kappa_4 = \langle \Delta E^4 \rangle - 3\langle \Delta E^2 \rangle^2 \tag{4.178}$$

Equation (4.174) indicates that κ_n's are the coefficients in the Taylor expansion of $\ln f(t)$. In Eq. (4.171), differentiating $\ln f(t)$ by t, followed by setting t to zero, is equivalent to differentiating $\ln f(0) = \ln Z$ by $-\beta$. Therefore,

$$\kappa_n = \left. \frac{\partial^n \ln f(t)}{\partial t^n} \right|_{t=0} = \frac{\partial^n \ln Z}{\partial(-\beta)^n} = (-1)^n \frac{\partial^n \ln Z}{\partial \beta^n} \tag{4.179}$$

Table 4.6 lists the cumulant coefficients κ_n for $n = 1, 2, 3,$ and 4.

We are familiar with the first two equations in the table. The third cumulant has a name, and the reduced cumulant $\langle \Delta E^3 \rangle / \langle \Delta E^2 \rangle^{3/2}$ is called a skewness. It can be positive or negative. The shape of the distribution of E is shown for positive and negative skewness together with a curve for the normal distribution ($\langle \Delta E^3 \rangle = 0$) in Figure 4.23. A positive skewness raises the right tail and curtails the left tail.

Likewise, the fourth cumulant has its name. A reduced quantity $\langle \Delta E^4 \rangle / \langle \Delta E^2 \rangle^2 - 3$ is called a kurtosis, and it can be positive or negative. The shape of the distribution of E is shown for positive and negative kurtosis ($\kappa_3 = 0$; $\kappa_4 \neq 0$) together with a curve for the normal distribution ($\kappa_3 = \kappa_4 = 0$) in Figure 4.24. A positive kurtosis enhances the tails at both ends of the

Table 4.6 First four cumulants of ln Z.

n	$\kappa_n = (-1)^n \partial^n \ln Z / \partial \beta^n$
1	$\langle E \rangle = -\dfrac{\partial \ln Z}{\partial \beta}$
2	$\langle \Delta E^2 \rangle = \dfrac{\partial^2 \ln Z}{\partial \beta^2}$
3	$\langle \Delta E^3 \rangle = -\dfrac{\partial^3 \ln Z}{\partial \beta^3}$
4	$\langle \Delta E^4 \rangle - 3\langle \Delta E^2 \rangle^2 = \dfrac{\partial^4 \ln Z}{\partial \beta^4}$

Figure 4.23 Examples of energy distribution with a positive skewness (dash-dotted line) and a negative skewness (dashed line). The solid line is a normal distribution with zero skewness.

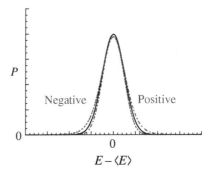

Figure 4.24 Examples of energy distribution with a positive kurtosis (dash-dotted line) and a negative kurtosis (dashed line). The solid line is a normal distribution with zero kurtosis.

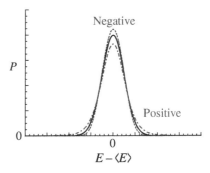

distribution compared with the normal distribution. A negative κ_4 makes the peak sharper.

In Section 4.8, we calculated $\langle \Delta E^2 \rangle$ for a gas of N noninteracting, monatomic molecules. We rewrite Eq. (4.129) into a function of β:

$$\frac{\partial^2 \ln Z}{\partial \beta^2} = \frac{C_V}{k_B \beta^2} \tag{4.180}$$

Further differentiation of this equation by β leads to the cumulants of E shown in Table 4.7. The skewness and kurtosis are also listed. We thus find that the skewness decreases with an increasing N as $N^{-1/2}$, and kurtosis decreases as

Table 4.7 Second, third, and fourth cumulants of E in a gas of monatomic molecules.

n	κ_n	$\kappa_n / \kappa_2^{n/2}$
2	$\langle \Delta E^2 \rangle = \dfrac{C_V}{k_B \beta^2}$	
3	$\langle \Delta E^3 \rangle = \dfrac{2 C_V}{k_B \beta^3}$	$\langle \Delta E^3 \rangle / \langle \Delta E^2 \rangle^{3/2} = 2\left(\dfrac{k_B}{C_V}\right)^{1/2} = 2\left(\dfrac{2}{3N}\right)^{1/2}$
4	$\langle \Delta E^4 \rangle - 3\langle \Delta E^2 \rangle^2 = \dfrac{6 C_V}{k_B \beta^4}$	$\langle \Delta E^4 \rangle / \langle \Delta E^2 \rangle^2 - 3 = 6\dfrac{k_B}{C_V} = \dfrac{4}{N}$

N^{-1}. Both are essentially zero when N is the number of molecules in a system of a laboratory bench top scale. Therefore, $P(E)$ for a system of noninteracting monatomic molecules with a sufficiently large N can be approximated by a normal distribution, and its distribution is extremely narrow. This is a part of the central limit theorem.

Problems

4.1 *Phase space of a harmonic oscillator.* Inspection of Eq. (4.2) indicates that x and p_x are represented by a shared parameter θ as $x = (2E/\kappa)^{1/2}\cos\theta$ and $p_x = (2mE)^{1/2}\sin\theta$. Show that θ changes with time at a constant rate.

4.2 *Particle with three states.* Consider a system of single particles that can be at energy levels 0, ε, and 2ε. The degeneracy is one for all the levels.
(1) At what value of $\beta\varepsilon$ does the probability of finding the particle at level 0 become equal to 0.5?
(2) At the temperature found in (1), what is the probability of finding the particle at energy level 2ε?

4.3 *Canonical ensemble, $N = 1$ and 2, two states.* We consider a particle that has only two states, represented by ↑ and ↓. Their energy levels are $-\varepsilon$ and ε, respectively, where $\varepsilon > 0$. When the system of one particle is in thermal equilibrium with a reservoir, the probability that the particle is in the two states is found to be

$$P(\uparrow) = 0.8, P(\downarrow) = 0.2$$

(1) Determine the constants C and β in $P = Ce^{-\beta E}$.
(2) Consider a system made up of two of these (independent) particles. The system is coupled with the same reservoir. Complete the following table for the remaining states.

State	Energy E	Probability
↑ ↑	-2ε	0.64

(3) What is the probability that the composite system has zero energy?
(4) Show that the probabilities of the states in the table you prepared in (2) satisfy

$$P = De^{-\beta' E}$$

(5) What is the relationship between (D, β') and (C, β) that was defined in (1)?

(6) Which of these constants depend on the system?
(7) Determine the value of U for the composite system.

4.4 *System specified by two integers.* What is the partition function for a system of a single particle whose energy ε is specified by two nonnegative integers n_1 and n_2 as $\varepsilon = n_1\varepsilon_1 + n_2\varepsilon_2$? The degeneracy is one for all states. $\varepsilon_1 > 0, \varepsilon_2 > 0$.

4.5 *System of degeneracy $n + 1$.* What is the partition function for a system of a single particle that has energy $n\varepsilon$ and degeneracy $n + 1$, where $n = 0, 1, 2, \ldots$?

4.6 *Calculation of Z through g_n.* Show that the partition function calculated using Eq. (4.45) with $g_n = {}_{N-1+n}C_n$ is identical to Eq. (4.55).

4.7 *Three energy levels, mean energy.* Consider a system of a single particle that can be at energy levels of ε, 0, and $-\varepsilon$ ($\varepsilon > 0$) with degeneracies 1, 2, and 1, respectively.
(1) What is the partition function?
(2) The partition function is identical to the one for a system consisting of two distinguishable, but otherwise identical particles. What are the energy levels and their degeneracies for each particle?
(3) Calculate the heat capacity.
(4) Draw a sketch for the plot of $\langle E \rangle / \varepsilon$ as a function of $\beta\varepsilon$.
(5) What is the high-temperature asymptote of $\langle E \rangle$?
(6) What is the state in the low-temperature limit?
(7) Draw a sketch for the plot of C_V/k_B as a function of $\beta\varepsilon$.

4.8 *Constant in energy level.* Calculate the internal energy U, Helmholtz free energy F, and entropy S for a system of N independent particles, each with two energy levels ε_0 and $\varepsilon_0 + \Delta\varepsilon$ ($\Delta\varepsilon > 0$) to show that a nonzero ε_0 simply adds a constant to each of U and F, and there is no effect on S.

4.9 *Entropy calculation as an uncertainty.* For a system of one particle with energy levels $n\varepsilon$ ($n = 0, 1, 2, \ldots$; $\varepsilon > 0$), the probability P_n for the nth state is

$$P_n = (1 - e^{-\beta\varepsilon})e^{-n\beta\varepsilon}$$

(1) Use the formula $S = -k_B \sum P_n \ln P_n$ to calculate the entropy S.
(2) What is the low-temperature asymptote of S? How about the high-temperature asymptote?
(3) Draw a sketch for the plot of S/k_B as a function of $k_B T/\varepsilon$.

4.10 *Entropy vs temperature.* There are two systems of a single particle, C and D. In system C, the energy levels are $E = n\varepsilon$, where $n = 0, 1, 2,...$ In system D, the energy levels are $E = 2n\varepsilon$ where $n = 0, 1, 2,...$ The two systems share $\varepsilon > 0$. Compare the entropies of the two systems.

4.11 *Grand canonical ensemble of monatomic gas.* Find the grand partition function of ideal gas of N monatomic molecules of mass m in volume V with chemical potential μ. Then, calculate the pressure and express it as a function of $\langle N \rangle$. [The grand partition function you will obtain here is wrong, since it neglects the indistinguishability of the molecules. See Problem 6.7. Nevertheless, solve this problem.]

4.12 *Complexation.* Consider a solution of molecule C of chemical potential μ. When a small amount of molecule D is added to the solution, some of the C molecules bind to D molecules. A molecule D has two sites, and each of them can bind a C molecule. The energy of the D molecule is ε_1 when a C molecule is at one of the two sides and ε_2 when two C molecules are at the two sides, where the zero level of the energy is for the D molecule without a C molecule.
(1) What is the grand partition function of a molecule D?
(2) Let n be the number of C molecules bound to a D molecule. What is $\langle n \rangle$?

4.13 *Cumulants of E.* In a system consisting of N noninteracting particles, each with energy $n\varepsilon$ ($n = 0, 1, 2,...$; $\varepsilon > 0$), the partition function is given as (Eq. (4.55))

$$Z = (1 - e^{-\beta\varepsilon})^{-N}$$

(1) Calculate $\langle E \rangle$, $\langle \Delta E^2 \rangle$, $\langle \Delta E^3 \rangle$, and $\langle \Delta E^4 \rangle - 3\langle \Delta E^2 \rangle^2$.
(2) What are the low-temperature asymptotes of $\langle E/\varepsilon \rangle$, $\langle \Delta E^2/\varepsilon^2 \rangle$, $\langle \Delta E^3/\varepsilon^3 \rangle$, and $\langle \Delta E^4/\varepsilon^4 \rangle - 3\langle \Delta E^2/\varepsilon^2 \rangle^2$?
(3) From the results in (2), find what distribution E/ε follows.
(4) Explain why E/ε follows that distribution at low temperatures.

5

Canonical Ensemble of Gas Molecules

Having learned the partition function Z of a canonical ensemble and how to obtain thermodynamic functions from Z, we apply these tools to an ideal gas in this chapter. In Section 5.1, we apply the canonical distribution function to find a distribution of the velocity of molecules. In Section 5.2, we consider the thermodynamics of an ideal gas using classical representation of the energy. In Section 5.3, we consider its quantum-mechanical version.

The thermodynamic functions we consider in this chapter are the internal energy and heat capacity only. We need to learn the indistinguishability in Chapter 7 before being able to correctly express the chemical potential.

In Section 5.4, we consider the most probable state for the rotation of a diatomic molecule. Section 5.5 considers conformations of a molecule.

5.1 Velocity of Gas Molecules

We apply the canonical distribution to find the velocity distribution of molecules in vapor phase. Molecules in a gas collide with each other to change their velocities. At equilibrium, a steady distribution of the velocity is established. It was James Clark Maxwell who first derived the correct distribution function in 1857, long before statistical mechanics was introduced or the concept of Boltzmann distribution was established.

Considering monatomic gas makes the derivation easy, since the molecules do not have rotational or vibrational motion. Since only the center-of-mass movement is involved in the velocity distribution, the same formulation applies to diatomic and other molecules, and the result of the velocity distribution remains the same. We adopt classical-mechanical representation of the translational energy here.

In Section 4.2.3, we derived the single-molecule partition function for a monatomic molecule of mass m in volume V:

$$Z_1 = \frac{1}{h^3} \int_V d\mathbf{r} \int_{-\infty}^{\infty} d\mathbf{p} \exp\left(-\beta \frac{\mathbf{p}^2}{2m}\right) = V\left(\frac{2\pi m}{\beta h^2}\right)^{3/2} \tag{4.39}$$

Statistical Thermodynamics: Basics and Applications to Chemical Systems, First Edition. Iwao Teraoka.
© 2019 John Wiley & Sons, Inc. Published 2019 by John Wiley & Sons, Inc.
Companion website: www.wiley.com/go/Teraoka_StatsThermodynamics

The canonical distribution $f(\mathbf{r}, \mathbf{p})$ is the probability density for the molecule to be at $\mathbf{r}$ and have a momentum $\mathbf{p}$:

$$f(\mathbf{r}, \mathbf{p}) = \frac{1}{Z_1 h^3} \exp\left(-\frac{\beta}{2m}\mathbf{p}^2\right) \tag{5.1}$$

Integrating $f(\mathbf{r}, \mathbf{p})$ by $\mathbf{r}$ leads to the probability density $f(\mathbf{p})$ for $\mathbf{p}$:

$$f(\mathbf{p}) = \int f(\mathbf{r}, \mathbf{p})d\mathbf{r} = \frac{V}{Z_1} \frac{1}{h^3} \exp\left(-\frac{\beta}{2m}\mathbf{p}^2\right)$$

$$= \left(\frac{\beta}{2\pi m}\right)^{3/2} \exp\left(-\frac{\beta}{2m}\mathbf{p}^2\right) \tag{5.2}$$

Now we convert $f(\mathbf{p})$ to $f(\mathbf{v})$, the distribution of the velocity $\mathbf{v} = \mathbf{p}/m$. Since $d\mathbf{p} = m^3 d\mathbf{v}$, equating $f(\mathbf{p})d\mathbf{p}$ to $f(\mathbf{v})d\mathbf{v}$ yields

$$f(\mathbf{v}) = \left(\frac{\beta m}{2\pi}\right)^{3/2} \exp\left(-\frac{\beta}{2}m v^2\right) \tag{5.3}$$

This distribution is called a **Maxwell distribution**. It is easy to confirm that $f(\mathbf{v})$ is normalized.

We can decompose $f(\mathbf{v})d\mathbf{v}$ into three components:

$$f(\mathbf{v})d\mathbf{v} = f_x(v_x)dv_x f_y(v_y)dv_y f_z(v_z)dv_z \tag{5.4}$$

where $f_x(v_x)$ represents the distribution of the x component v_x of the velocity:

$$f_x(v_x) = \left(\frac{\beta m}{2\pi}\right)^{1/2} \exp\left(-\frac{\beta}{2}m v_x^2\right) \tag{5.5}$$

for instance. We find that $f_x(v_x)$ is a normal distribution with zero mean and variance $(\beta m)^{-1}$ (see Figure 5.1), i.e.

$$\langle v_x \rangle = 0 \tag{5.6}$$

$$\langle v_x^2 \rangle = \frac{1}{\beta m} = \frac{k_B T}{m} \tag{5.7}$$

Equation (5.3) indicates that $f(\mathbf{v})$ depends on $\mathbf{v}$ through its magnitude $v = |\mathbf{v}|$ only. In other words, the distribution is isotropic. Then, we can use the following identity,

$$f(\mathbf{v})4\pi v^2\, dv = f_v(v)dv \tag{5.8}$$

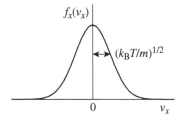

Figure 5.1 Probability distribution of the x component of the velocity of a gas molecule. The distribution is a normal distribution with zero mean and variance $k_B T/m$.

Figure 5.2 Distribution of speed v of a gas molecule. The speed is reduced by $(\beta m)^{-1/2}$. The most probable speed v_m, mean speed $\langle v \rangle$, and the root-mean-square velocity $\langle \mathbf{v}^2 \rangle^{1/2}$ are indicated by dashed vertical lines.

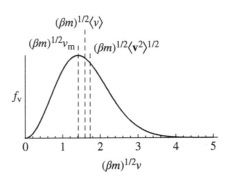

to convert $f(\mathbf{v})$ into $f_v(v)$, the distribution of the speed v. The result is

$$f_v(v) = 4\pi v^2 \left(\frac{\beta m}{2\pi} \right)^{3/2} \exp\left(-\frac{\beta}{2}mv^2 \right) \tag{5.9}$$

The identity (Eq. (5.8)) is derived from the difference between the volume of a sphere of radius $v + dv$ and that of a sphere of radius v, which is equal to $4\pi v^2 dv$. Figure 5.2 displays a plot of $f_v(v)$ as a function of $v(\beta m)^{1/2}$. The latter is the speed reduced by $(\beta m)^{-1/2}$. The function consists essentially of two factors – v^2 that increases with an increasing v and $\exp(-\beta mv^2/2)$ that decreases with v. The combined effect is a peaking at some value of v. The value of v that peaks at $f_v(v)$ is called the most probable speed and is denoted as v_m. From $df_v(v)/dv = 0$, v_m is obtained as

$$v_m = \left(\frac{2}{\beta m} \right)^{1/2} = \left(\frac{2k_B T}{m} \right)^{1/2} \tag{5.10}$$

We can also calculate the mean of v (mean speed), $\langle v \rangle$:

$$\langle v \rangle = \int_0^\infty v f_v(v) dv = 4\pi \left(\frac{\beta m}{2\pi} \right)^{3/2} \int_0^\infty v^3 \exp\left(-\frac{\beta}{2}mv^2 \right) dv$$

$$= \left(\frac{8}{\pi \beta m} \right)^{1/2} = \left(\frac{8k_B T}{\pi m} \right)^{1/2} \tag{5.11}$$

Figure 5.2 also shows the positions of v_m and $\langle v \rangle$. Since $\langle v \rangle$ is weighed by v, it is greater than the peak position v_m.

The mean square of the velocity is calculated as

$$\langle \mathbf{v}^2 \rangle = 3\langle v_x^2 \rangle = 3 \int_{-\infty}^\infty v_x^2 f_x(v_x) dv_x = \frac{3}{\beta m} \tag{5.12}$$

Its square root is another measure of the average speed:

$$\langle \mathbf{v}^2 \rangle^{1/2} = \left(\frac{3}{\beta m} \right)^{1/2} = \left(\frac{3k_B T}{m} \right)^{1/2} \tag{5.13}$$

We can confirm that $\langle v^2 \rangle = \langle \mathbf{v}^2 \rangle$. The greater weight of v in the integral for calculating $\langle v^2 \rangle^{1/2}$ compared with $\langle v \rangle$ makes $\langle \mathbf{v}^2 \rangle^{1/2}$ larger than $\langle v \rangle$. The results of v_m, $\langle v \rangle$, and $\langle \mathbf{v}^2 \rangle^{1/2}$ are the same for molecules that are not monatomic. The ratio of the three measures of the average speed is

$$v_m : \langle v \rangle : \langle \mathbf{v}^2 \rangle^{1/2} = \sqrt{2} : \sqrt{\frac{8}{\pi}} : \sqrt{3} \tag{5.14}$$

The average velocities are expressed as a function of temperature and mass of the molecule. We can estimate the mean velocities for any molecule in the vapor phase. For example, the mass of an O_2 molecule is $m = 32.00\,\text{g mol}^{-1}/(6.022 \times 10^{-23}\,\text{mol}^{-1}) = 5.316 \times 10^{-26}\,\text{kg}$. At $300\,\text{K}$,

$$\left(\frac{k_B T}{m}\right)^{1/2} = \left(\frac{1.381 \times 10^{-23}\,\text{J K}^{-1} \times 300\,\text{K}}{5.316 \times 10^{-26}\,\text{kg}}\right)^{1/2} = 279\,\text{m s}^{-1} \tag{5.15}$$

The three averages are $v_m = 396\,\text{m s}^{-1}$, $\langle v \rangle = 445\,\text{m s}^{-1}$, and $\langle v^2 \rangle^{1/2} = 484\,\text{m s}^{-1}$. These velocities are supersonic. Our skin is bombarded with these molecules that collide at these speeds. However, we do not feel any pain. That is because the mass of the molecule is so small that the momentum transfer (or the change of linear momentum upon collision to the skin) is negligible. The skin does not feel a lot of pressure.

Another question we may ask is: Why do we not hear the boom from a shock wave that is typical of an object flying at a supersonic speed? If the air we breathe were filled with such booms, our ears would not survive. The answer is as follows.

The propagation of sound as a wave assumes that the medium it travels is continuous and featureless. There is no place for molecules. In contrast, when we consider the velocity distribution of molecules in vapor, we place them in vacuum that does not transmit a density wave. Therefore, the sound theory does not apply. The shock wave is generated when a macroscopic object flies at a supersonic speed in air or another vapor.

5.2 Heat Capacity of a Classical Gas

In this section, we use the platform of canonical ensemble to calculate the heat capacity of an ideal gas. We consider a system consisting of N noninteracting molecules. Therefore, the partition function Z of the system is given as $Z = Z_1{}^N$, where Z_1 is the single-molecule partition function. We consider three models in classical mechanics to describe monatomic and diatomic molecules.

For each model, we will calculate U and then the heat capacity at constant volume (isocholic), C_V. The isobaric heat capacity C_p is related to C_V by

$$C_p = C_V + N k_B \tag{5.16}$$

as heating a gas at a constant pressure results in thermal expansion. The work the system does to the surroundings is $p\,dV = Nk_B\,dT$ for a temperature increase dT. The heat given to the system is partly consumed to do the work.

5.2.1 Point Mass

A monatomic molecule is modeled as a point mass. In Section 4.4, we derived expressions of U for a system of monatomic molecules of mass m in volume V:

$$U = \frac{3}{2}Nk_B T \tag{4.65}$$

Then, the heat capacities are calculated as

$$C_V = \frac{3}{2}Nk_B, \quad C_p = \frac{5}{2}Nk_B \tag{5.17}$$

A monatomic molecule has the center-of-mass translation as the only component of kinetic energy. Equation (5.13) gives

$$N\frac{m}{2}\langle \mathbf{v}^2 \rangle = \frac{3}{2}Nk_B T \tag{5.18}$$

that is equal to U.

The point-mass model is also appropriate for diatomic and polyatomic molecules at extremely low temperatures. The translational motion is the only allowed mode of motion when the temperature is too low for the molecule to rotate.

5.2.2 Rigid Dumbbell

A diatomic molecule with a fixed bond length can move its center of mass and rotate. Since the mass is at both ends of the bond, the model is called a rigid dumbbell model (see Figure 5.3). When we consider a quantum-mechanical version of the gas system in the next section, we find that a diatomic molecule at room temperature is approximately described by this model, unless the molecule is as heavy as I_2.

In Section 3.2.1, we learned that the kinetic energy of rotation is given as $\frac{1}{2}I(\omega_1^2 + \omega_2^2)$, where I is the moment of inertia and ω_1 and ω_2 are angular velocities in two orthogonal directions of rotation (see Eq. (3.21)). As we did for translation, we rewrite the energy using angular momenta L_1 and L_2. They are related to ω_1 and ω_2 by

$$L_i = I\omega_i \tag{5.19}$$

Figure 5.3 Rigid dumbbell model for a diatomic molecule. The dumbbell can move its center of mass and rotate, but the distance between the two masses does not change.

in a way similar to the one **p** is related to **v**. Classically, the kinetic energy of rotation is given as $(L_1^2 + L_2^2)/(2I)$. The kinetic energy ε of a rigid dumbbell consists of center-of-mass translation and rotation:

$$\varepsilon = \frac{\mathbf{p}^2}{2M} + \frac{\mathbf{L}^2}{2I} \tag{5.20}$$

where M is the mass of the dumbbell, and $\mathbf{L} = [L_1, L_2]^{\mathrm{T}}$. Note that **p** has three components, whereas **L** has two.

We can construct a phase space for rotation. A point in the space is specified by the orientation Ω ($=\theta, \phi$ in spherical polar coordinates) and the angular momenta. The single-molecule partition function is given as

$$Z_1 = V \left(\frac{2\pi m}{\beta h^2} \right)^{3/2} \frac{1}{h^2} \int d\Omega \int d\mathbf{L} \exp\left(-\beta \frac{\mathbf{L}^2}{2I} \right) \tag{5.21}$$

where $d\Omega = \sin\theta \, d\theta \, d\phi$. Division by h^2 makes Z_1 calculated here,

$$Z_1 = V \left(\frac{2\pi m}{\beta h^2} \right)^{3/2} \frac{8\pi^2 I}{\beta h^2} \tag{5.22}$$

agree with the high-temperature asymptote of quantum-mechanical calculation; see the next section. The division is again due to the uncertainty. With Eq. (5.22), $\ln Z$ of the system consisting of N rigid dumbbells is given as

$$\ln Z = N \ln Z_1 = N \ln V + \frac{3N}{2} \ln \frac{2\pi m}{\beta h^2} + N \ln \frac{8\pi^2 I}{\beta h^2} \tag{5.23}$$

Use of Eq. (4.60) gives

$$U = \frac{5N}{2\beta} = \frac{5}{2} N k_{\mathrm{B}} T \tag{5.24}$$

The heat capacities are then calculated as

$$C_V = \frac{5}{2} N k_{\mathrm{B}}, \quad C_p = \frac{7}{2} N k_{\mathrm{B}} \tag{5.25}$$

5.2.3 Elastic Dumbbell

Now we allow the bond length of the dumbbell to change. As we learn in the next section, this model is equivalent to the high temperature asymptote in quantum-mechanical treatment of a diatomic molecule. The elastic dumbbell shown in Figure 5.4 can move its center of mass, rotate, and vibrate.

Figure 5.4 Elastic dumbbell model for a diatomic molecule. The dumbbell can move its center of mass, rotate, and change the bond length.

In the elastic dumbbell model, both rotation and vibration are fully excited. The bond is a spring of force constant κ, instantaneous length r_{ab}, and equilibrium length r_0. We derived the expression for the energy in Section 3.2.1. The kinetic energy consists of three parts:

$$\varepsilon = \frac{\mathbf{p}^2}{2m} + \frac{\mathbf{L}^2}{2I} + \left(\frac{p_{ab}^2}{2\mu} + \frac{\kappa}{2}(r_{ab} - r_0)^2 \right) \tag{5.26}$$

where $p_{ab} = \mu \, dr_{ab}/dt$ is the linear momentum conjugate to r_{ab}. The phase space for the single elastic dumbbell has 12 dimensions – 6 from $\mathbf{r}$ and $\mathbf{p}$, 4 from Ω and $\mathbf{L}$, and 2 from r_{ab} and p_{ab}. The single-molecule partition function is expressed as

$$Z_1 = V \left(\frac{2\pi m}{\beta h^2} \right)^{3/2} \frac{8\pi^2 I}{\beta h^2} \frac{1}{h} \int_{-\infty}^{\infty} dp_{ab} \exp\left(-\beta \frac{p_{ab}^2}{2\mu} \right) \int_{-\infty}^{\infty} dr_{ab}$$

$$\exp\left(-\beta \frac{\kappa}{2}(r_{ab} - r_0)^2 \right) \tag{5.27}$$

which is calculated as

$$Z_1 = V \left(\frac{2\pi m}{\beta h^2} \right)^{3/2} \frac{8\pi^2 I}{\beta h^2} \frac{1}{h} \left(\frac{2\pi\mu}{\beta} \right)^{1/2} \left(\frac{2\pi}{\beta\kappa} \right)^{1/2} \tag{5.28}$$

Note that division by h is per degree of freedom or a pair of variable and its conjugate. Here, the pair is r_{ab} and p_{ab}. With $2\pi\nu = (\kappa/\mu)^{1/2}$ (Eq. (3.63)), Eq. (5.28) is simplified to

$$Z_1 = V \left(\frac{2\pi m}{\beta h^2} \right)^{3/2} \frac{8\pi^2 I}{\beta h^2} \frac{1}{\beta h\nu} \tag{5.29}$$

Then, for a system of N elastic dumbbells,

$$\ln Z = N \ln Z_1 = N \ln V + \frac{3N}{2} \ln \frac{2\pi m}{\beta h^2} + N \ln \frac{8\pi^2 I}{\beta h^2} + N \ln \frac{1}{\beta h\nu} \tag{5.30}$$

and

$$U = \frac{7N}{2\beta} = \frac{7}{2} N k_B T \tag{5.31}$$

The heat capacities are

$$C_V = \frac{7}{2} N k_B, \quad C_p = \frac{9}{2} N k_B \tag{5.32}$$

Table 5.1 summarizes U, C_V, and C_p for the three models. With rotation added to the center-of-mass motion, both C_V and C_p jump by $N k_B$, and with vibration added on top of that, C_V and C_p are further stepped up by $N k_B$. The heat capacity ratio

$$\gamma \equiv \frac{C_p}{C_V} \tag{5.33}$$

is also listed.

Table 5.1 Heat capacity C_V at a constant volume, heat capacity C_p at a constant pressure, and their ratio $\gamma = C_p/C_V$ for three models of gas molecules.

Model	C_V	C_p	γ
Point mass	$\frac{3}{2}Nk_B$	$\frac{5}{2}Nk_B$	$\frac{5}{3}$
Rigid dumbbell	$\frac{5}{2}Nk_B$	$\frac{7}{2}Nk_B$	$\frac{7}{5}$
Elastic dumbbell	$\frac{7}{2}Nk_B$	$\frac{9}{2}Nk_B$	$\frac{9}{7}$

Table 5.2 Heat capacity ratio of some diatomic and polyatomic gases.

Gas	γ	T(K)	Gas	γ	T (K)
He	1.666	300	Br_2	1.292	400
Ne	1.666	300	I_2	1.285	500
Ar	1.666	300	NO	1.386	300
H_2	1.405	300	CO	1.399	300
N_2	1.400	300	HCl	1.399	300
O_2	1.394	300	CO_2	1.288	300
Cl_2	1.324	300	NH_3	1.304	300

Temperature is also listed.
Source: Data from NIST Chemistry WebBook.

It is interesting to compare actual values of γ with the values of these models. Table 5.2 lists γ for some molecules together with the temperature (300 K for most of the molecules). For monatomic molecules, $\gamma = 1.666$ agrees with the model for the point mass. For light diatomic molecules (H_2, N_2, O_2, CO, NO, HCl), γ is close to 1.4, the value of the rigid dumbbell model. For heavy diatomic molecules (Cl_2, Br_2, I_2) and polyatomic molecules (CO_2, NH_3), γ is less than 1.4, and is between the value for the rigid dumbbell model and the one for the elastic dumbbell model.

5.3 Heat Capacity of a Quantum-Mechanical Gas

5.3.1 General Formulas

In the preceding section, we derived expressions of the heat capacity for three models of a molecule in classical mechanics. In principle, the energy of a molecule consists of three parts:

$$\varepsilon = \varepsilon_{\text{trans}} + \varepsilon_{\text{rot}} + \varepsilon_{\text{vib}} \tag{5.34}$$

Depending on the temperature, one or two of the three components of the energy may not be excited. The classical model assumes that each mode of motion within the model is fully excited, and that is why we needed three distinct models.

In contrast, the quantum-mechanical model we learn in this section applies to all temperatures, and we adopt a single model for the energy as given by Eq. (5.34). The partition function is a continuous function of temperature, and so is the heat capacity. In some ranges of the temperature, these expressions are approximated by the classical versions.

The state of a molecule that can translate, rotate, and vibrate is specified by indices for the translational state, rotational state, and vibrational state. We symbolically denote the indices as trans, rot, and vib. The single-molecule partition function Z_1 is expressed as

$$Z_1 = \sum_{\text{trans}} \sum_{\text{rot}} \sum_{\text{vib}} \exp(-\beta(\varepsilon_{\text{trans}} + \varepsilon_{\text{rot}} + \varepsilon_{\text{vib}})) \tag{5.35}$$

where the threefold summation is taken over all possible states of translation, rotation, and vibration. Equation (5.35) can be rewritten to

$$Z_1 = Z_{1,\text{trans}} Z_{1,\text{rot}} Z_{1,\text{vib}} \tag{5.36}$$

where

$$Z_{1,\text{trans}} = \sum_{\text{trans}} \exp(-\beta\varepsilon_{\text{trans}}) \tag{5.37}$$

$$Z_{1,\text{rot}} = \sum_{\text{rot}} \exp(-\beta\varepsilon_{\text{rot}}) \tag{5.38}$$

$$Z_{1,\text{vib}} = \sum_{\text{vib}} \exp(-\beta\varepsilon_{\text{vib}}) \tag{5.39}$$

Once Z_1 is expressed as a product of the three factors, the partition function of the N-molecule system is written as

$$Z = Z_1{}^N = Z_{1,\text{trans}}{}^N Z_{1,\text{rot}}{}^N Z_{1,\text{vib}}{}^N \tag{5.40}$$

and therefore

$$\ln Z = N \ln Z_{1,\text{trans}} + N \ln Z_{1,\text{rot}} + N \ln Z_{1,\text{vib}} \tag{5.41}$$

Then, we can write the internal energy U as the sum of three components:

$$U = U_{\text{trans}} + U_{\text{rot}} + U_{\text{vib}} \tag{5.42}$$

where

$$U_{\text{trans}} = -N \frac{\partial}{\partial \beta} \ln Z_{1,\text{trans}} \tag{5.43}$$

$$U_{\text{rot}} = -N \frac{\partial}{\partial \beta} \ln Z_{1,\text{rot}} \tag{5.44}$$

$$U_{vib} = -N \frac{\partial}{\partial \beta} \ln Z_{1,vib} \tag{5.45}$$

Likewise, the heat capacity C_V consists of three components:

$$C_V = C_{V,trans} + C_{V,rot} + C_{V,vib} \tag{5.46}$$

where

$$C_{V,trans} = \frac{\partial U_{trans}}{\partial T} \tag{5.47}$$

$$C_{V,rot} = \frac{\partial U_{rot}}{\partial T} \tag{5.48}$$

$$C_{V,vib} = \frac{\partial U_{vib}}{\partial T} \tag{5.49}$$

Here, we calculate $Z_{1,trans}$, $Z_{1,rot}$, and $Z_{1,vib}$ using quantum-mechanical models and then calculate $C_{V,trans}$, $C_{V,rot}$, and $C_{V,vib}$. We will limit ourselves to diatomic molecules.

5.3.2 Translation

In Section 3.6.1, we learned that ε_{trans} is specified by three positive integers, n_x, n_y, and n_z:

$$\varepsilon_{trans} = \frac{h^2}{8mL^2}(n_x^2 + n_y^2 + n_z^2) \tag{5.50}$$

where the gas molecules are contained in a box of $L \times L \times L$. Then,

$$Z_{1,trans} = \sum_{n_x=1}^{\infty} \sum_{n_y=1}^{\infty} \sum_{n_z=1}^{\infty} \exp\left(-\frac{\beta h^2}{8mL^2}(n_x^2 + n_y^2 + n_z^2)\right) \tag{5.51}$$

which is rewritten to

$$Z_{1,trans} = \sum_{n_x=1}^{\infty} \exp\left(-\frac{\beta h^2}{8mL^2}n_x^2\right) \sum_{n_y=1}^{\infty} \exp\left(-\frac{\beta h^2}{8mL^2}n_y^2\right)$$
$$\times \sum_{n_z=1}^{\infty} \exp\left(-\frac{\beta h^2}{8mL^2}n_z^2\right) \tag{5.52}$$

The difference among the three series is just the index. The three sums are identical, and therefore,

$$Z_{1,trans} = \left[\sum_{n=1}^{\infty} \exp\left(-\frac{\beta h^2}{8mL^2}n^2\right)\right]^3 \tag{5.53}$$

We cannot express the sum using a simple analytical expression, and therefore we need to adopt an approximation. Let us consider a series, $g(x)$ defined as

$$g(x) \equiv \sum_{n=1}^{\infty} \exp(-xn^2) \tag{5.54}$$

This function cannot be expressed with known functions of x. However, when x is small and therefore $\exp(-xn^2)$ remains nonnegligible until n reaches a large value, we can approximate the sum with an integral by n. Regarding n as a continuous variable, we obtain

$$\sum_{n=1}^{\infty} \exp(-xn^2) \cong \int_0^{\infty} \exp(-xn^2)dn = \frac{1}{2}\left(\frac{\pi}{x}\right)^{1/2} \tag{5.55}$$

In Figure 5.5, the area of gray vertical strips represents the sum of the series, and the area under the curve is equal to the integral. The two areas are close to each other, and a major difference is that the integral has an excess that is the integral of $\exp(-xn^2)$ from $n = 0$ to $n = \frac{1}{2}$. There are also minor differences due to a mismatch between the curve and the horizontal line at the top of each strip.

Since $x = \beta h^2/(8mL^2)$ in our series,

$$\sum_{n=1}^{\infty} \exp\left(-\frac{\beta h^2}{8mL^2}n^2\right) \cong \left(\frac{2\pi m}{\beta h^2}\right)^{1/2} L \tag{5.56}$$

With Eq. (5.53), we have

$$Z_{1,\text{trans}} \cong \left(\frac{2\pi m}{\beta h^2}\right)^{3/2} L^3 = V\left(\frac{2\pi m}{\beta h^2}\right)^{3/2} \tag{5.57}$$

which is identical to the Z_1 obtained classically (see Eq. (4.39)). The agreement justifies dividing the integral by h^3 in Eq. (4.34) that shows how to calculate Z_1 for the continuous energy.

The approximation of the sum by the integral will be good, if the excess of the half-width strip between 0 and $\frac{1}{2}$ is negligible compared with the whole

Figure 5.5 The sum of $\exp(-xn^2)$ is represented by the area of vertical strips of a unit width. The sum is approximated by the area under the curve.

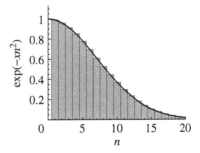

integral. The latter is the case if the distribution of $\exp(-xn^2)$ is broad, i.e.

$$x = \frac{\beta h^2}{8mL^2} \ll 1 \tag{5.58}$$

Here, we introduce a **thermal wavelength** λ_{th}:

$$\lambda_{th} \equiv \frac{h}{(2\pi m k_B T)^{1/2}} \tag{5.59}$$

Then, the requirement is rewritten to

$$\lambda_{th} \ll L \tag{5.60}$$

At 300 K, λ_{th} of a nitrogen molecule is 19 pm, a lot smaller compared with the linear dimension of the molecule. Since the volume of the container of the gas is sufficiently greater than $(19\,\mathrm{pm})^3$, the requirement is satisfied, and the integral gives a good approximation to the series. In other words, the classical representation of the center-of-mass movement is sufficient.

The internal energy and the heat capacity contributed by the translation motion are as follows.

$$U_{\mathrm{trans}} = -N\frac{\partial}{\partial \beta}\left(\ln V + \frac{3}{2}\ln\frac{2\pi m}{\beta h^2}\right) = \frac{3N}{2\beta} = \frac{3}{2}Nk_B T \tag{5.61}$$

$$C_{V,\mathrm{trans}} = \frac{\partial}{\partial T}\frac{3}{2}Nk_B T = \frac{3}{2}Nk_B \tag{5.62}$$

5.3.3 Rotation

In Section 3.6.2, we learned that $\varepsilon_{\mathrm{rot}}$ is specified by a nonnegative integer J:

$$\varepsilon_{\mathrm{rot}} = \frac{h^2}{8\pi^2 I}J(J+1) \tag{3.58}$$

and the Jth level has a degeneracy of $2J + 1$. Therefore,

$$Z_{1,\mathrm{rot}} = \sum_{J=0}^{\infty}(2J+1)\exp\left(-\frac{\beta h^2}{8\pi^2 I}J(J+1)\right) \tag{5.63}$$

For convenience, we introduce a characteristic temperature for rotation, Θ_{rot}, defined as

$$\Theta_{\mathrm{rot}} \equiv \frac{h^2}{8\pi^2 I k_B} \tag{5.64}$$

Then,

$$Z_{1,\mathrm{rot}} = \sum_{J=0}^{\infty}(2J+1)\exp\left(-\frac{\Theta_{\mathrm{rot}}}{T}J(J+1)\right) \tag{5.65}$$

Figure 5.6 The sum of $(2J+1)\exp(-J(J+1)$ $\Theta_{rot}/T)$ is represented by the area of vertical strips of a unit width. The sum is approximated by the area under the curve.

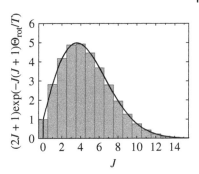

As we did for the translation, we approximate the sum of the series to an integral by J:

$$Z_{1,rot} \cong \int_0^\infty (2J+1)\exp\left(-\frac{\Theta_{rot}}{T}J(J+1)\right)dJ \tag{5.66}$$

Figure 5.6 compares the integral with the sum of the series. The integral is short of a half-width strip (-0.5 to 0), and there is a mismatch between the curve and the top of the strip.

With $2J+1$ in the integrand, we can calculate the integral as

$$Z_{1,rot} \cong \left[-\frac{T}{\Theta_{rot}}\exp\left(-\frac{\Theta_{rot}}{T}J(J+1)\right)\right]_{J=0}^{J=\infty} = \frac{T}{\Theta_{rot}} = \frac{8\pi^2 I}{\beta h^2} \tag{5.67}$$

The approximation is good, if the integrand in Eq. (5.66) does not decrease until J becomes large, i.e. $T \gg \Theta_{rot}$, that is a high temperature. At $T \gg \Theta_{rot}$, the internal energy and the heat capacity contributed by the rotation are as follows:

$$U_{rot} = -N\frac{\partial}{\partial\beta}\ln\frac{8\pi^2 I}{\beta h^2} = \frac{N}{\beta} = Nk_B T \tag{5.68}$$

$$C_{V,rot} = \frac{\partial}{\partial T}Nk_B T = Nk_B \tag{5.69}$$

The contribution is identical to the difference between the rigid dumbbell model and the point-of-mass model.

Table 5.3 lists Θ_{rot} for some diatomic molecules. Except the lightest molecule H_2, the room temperature is sufficiently higher than Θ_{rot}. Therefore, the high-temperature approximations in Eqs. (5.66)–(5.69) apply.

We can improve the approximation at high temperatures using the Euler–Maclaurin formula. See Problem 5.1 for details [7]. The result is

$$Z_{1,rot} = \frac{T}{\Theta_{rot}}\left[1 + \frac{1}{3}\frac{\Theta_{rot}}{T} + \frac{1}{15}\left(\frac{\Theta_{rot}}{T}\right)^2 + \cdots\right] \tag{5.70}$$

Table 5.3 Characteristic temperature of rotation, Θ_{rot}, and the characteristic temperature of vibration, Θ_{vib}, for some diatomic molecules.

Molecule	Θ_{rot} (K)	Θ_{vib} (K)
H_2	87.6	6331
N_2	2.88	3393
O_2	2.08	2239
F_2	1.27	1283
Cl_2	0.351	805.3
HCl	15.2	4303
CO	2.78	3122
NO	2.45	2740

Then,

$$\ln Z_{1,rot} = \ln \frac{T}{\Theta_{rot}} + \frac{1}{3}\frac{\Theta_{rot}}{T} + \frac{1}{90}\left(\frac{\Theta_{rot}}{T}\right)^2 + \cdots \tag{5.71}$$

which gives

$$U_{rot} = Nk_B T^2 \frac{\partial}{\partial T}\ln Z_{1,rot} = Nk_B T - \frac{1}{3}Nk_B\Theta_{rot} - \frac{1}{45}Nk_B\frac{\Theta_{rot}^2}{T} + \cdots \tag{5.72}$$

$$C_{V,rot} = Nk_B + \frac{1}{45}Nk_B\frac{\Theta_{rot}^2}{T^2} + \cdots \tag{5.73}$$

At low temperatures, Eq. (5.63) is dominated by the first few terms in the series:

$$Z_{1,rot} = 1 + 3\exp\left(-2\frac{\Theta_{rot}}{T}\right) + 5\exp\left(-6\frac{\Theta_{rot}}{T}\right) + \cdots \tag{5.74}$$

Its logarithm is

$$\ln Z_{1,rot} = 3\exp\left(-2\frac{\Theta_{rot}}{T}\right) - \frac{9}{2}\exp\left(-4\frac{\Theta_{rot}}{T}\right) + \cdots \tag{5.75}$$

Then,

$$U_{1,rot} = k_B T^2 \frac{\partial}{\partial T}\ln Z_{1,rot}$$

$$= k_B\Theta_{rot}\left[6\exp\left(-2\frac{\Theta_{rot}}{T}\right) - 18\exp\left(-4\frac{\Theta_{rot}}{T}\right) + \cdots\right] \tag{5.76}$$

Therefore,

$$\frac{C_{V,rot}}{Nk_B} = 12\left(\frac{\Theta_{rot}}{T}\right)^2\exp\left(-2\frac{\Theta_{rot}}{T}\right)\left[1 - 6\exp\left(-2\frac{\Theta_{rot}}{T}\right) + \cdots\right] \tag{5.77}$$

The low-temperature limit of $C_{V,rot}/(Nk_B)$ is 0.

Figure 5.7 Heat capacity due to rotation, $C_{V,\text{rot}}$, reduced by Nk_B, is plotted as a function of temperature T reduced by Θ_{rot}. The dashed line represents the high-temperature limit.

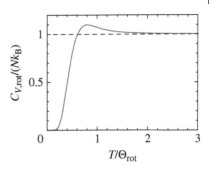

Figure 5.7 shows a plot of $C_{V,\text{rot}}/(Nk_B)$ as a function of T/Θ_{rot}. The transition from the low-temperature limit ($=0$) to the high-temperature limit ($=1$) occurs at around $T/\Theta_{\text{rot}} = 0.4$. Note an overshoot: $C_{V,\text{rot}}$ approaches the high-temperature limit from above as T increases. The main component of the overshoot is $(1/45)(\Theta_{\text{rot}}/T)^2$, and the overshoot is close to 0.1 at its peak.

5.3.4 Vibration

As opposed to contorted calculations of $Z_{1,\text{trans}}$ and $Z_{1,\text{rot}}$, calculation of $Z_{1,\text{vib}}$ is straightforward. We have already done it; see Eq. (4.78), which is reproduced here.

$$Z_{1,\text{vib}} = \sum_{n=0}^{\infty} \exp\left(-\beta h\nu\left(n+\frac{1}{2}\right)\right) = \frac{1}{2\sinh(\beta h\nu/2)} \tag{4.78}$$

This expression does not involve any approximation, unlike $Z_{1,\text{trans}}$ and $Z_{1,\text{rot}}$. We introduce the characteristic temperature of vibration, Θ_{vib}, defined as

$$\Theta_{\text{vib}} \equiv \frac{h\nu}{k_B} \tag{5.78}$$

Table 5.3 lists Θ_{vib} for some diatomic molecules. We find that $\Theta_{\text{vib}} \gg RT$ for all the molecules listed. At room temperature, $\exp(-\beta h\nu) \ll 1$, and therefore nearly all the diatomic molecules are at the vibrationally ground state.

With Θ_{vib}, $Z_{1,\text{vib}}$ is expressed as

$$Z_{1,\text{vib}} = \frac{1}{2\sinh(\Theta_{\text{vib}}/2T)} \tag{5.79}$$

The internal energy and the heat capacity contributed by the vibration are as follows:

$$U_{\text{vib}} = N\frac{\partial}{\partial\beta}\ln(2\sinh(\beta h\nu/2)) = \frac{Nh\nu}{2}\frac{1}{\tanh(\beta h\nu/2)} = \frac{Nh\nu}{2}\frac{1}{\tanh(\Theta_{\text{vib}}/2T)} \tag{5.80}$$

$$C_{V,\text{vib}} = \frac{\partial}{\partial T}\frac{Nh\nu}{2}\frac{1}{\tanh(\Theta_{\text{vib}}/2T)} = Nk_B\left[\frac{\Theta_{\text{vib}}/2T}{\sinh(\Theta_{\text{vib}}/2T)}\right]^2 \tag{5.81}$$

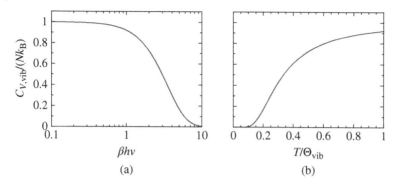

Figure 5.8 Reduced heat capacity by vibration, $C_{V,\mathrm{vib}}/(Nk_B)$, plotted as a function of (a) $\beta h\nu = \Theta_{\mathrm{vib}}/T$ and (b) T/Θ_{vib}.

Figure 5.8a shows a plot of $C_{V,\mathrm{vib}}/(Nk_B)$ as a function of $\beta h\nu = \Theta_{\mathrm{vib}}/T$, and Figure 5.8b shows a plot of the same quantity as a function of T/Θ_{vib}. The transition from the low-temperature limit ($=0$) to the high-temperature limit ($=1$) occurs at around $T/\Theta_{\mathrm{vib}} = 0.3$, and the curve of $C_{V,\mathrm{vib}}$ is sigmoidal. Unlike the plot of $C_{V,\mathrm{rot}}$, there is no overshoot.

5.3.5 Comparison with Classical Models

Now we can combine the three components of C_V in Sections 5.3.2–5.3.4. Figure 5.9 shows the result for an HCl molecule ($\Theta_{\mathrm{rot}} = 15.2\ \mathrm{K}$, $\Theta_{\mathrm{vib}} = 4303\ \mathrm{K}$). The temperature is in a logarithmic scale. The heat capacity increases in two steps.

At $T \ll \Theta_{\mathrm{rot}}$, $C_{V,\mathrm{vib}}/(Nk_B) = 3/2$, as only the translational motion is activated. The classical model of point mass is appropriate for this temperature range. At $\Theta_{\mathrm{rot}} \ll T \ll \Theta_{\mathrm{vib}}$, $C_{V,\mathrm{vib}}/(Nk_B) = 5/2$, as rotation is also fully excited. Room temperature belongs to this range of temperature. The classical model of the rigid dumbbell is right for this range. At $T \gg \Theta_{\mathrm{vib}}$, $C_{V,\mathrm{vib}}/(Nk_B) = 7/2$, as all of

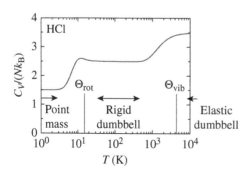

Figure 5.9 Heat capacity of HCl, reduced by Nk_B, plotted as a function of temperature T. The characteristic temperatures of rotation and vibration, Θ_{rot} and Θ_{vib}, are indicated by vertical lines. The arrows indicate the ranges of temperature in which the classical models of point mass, rigid dumbbell, and elastic dumbbell are effective.

the three modes of motion are fully activated. The classical model of the elastic dumbbell describes the diatomic molecule at these high temperatures.

Note that the normal boiling point of HCl is 188.1 K, and therefore we can observe the transition around Θ_{rot} only at very low pressures. Another limitation comes from thermal decomposition. At $T > 2000$ K or so, HCl is not stable. Therefore, we may not be able to see the transition due to the activation of vibrationally excited states.

Notice a diffusiveness of the transition in C_V when the temperature exceeds Θ_{vib}. Even in the logarithmic scale, the transition by vibration is more diffuse than is the transition by rotation.

In Table 5.2, the values of γ for Cl_2, Br_2, and I_2 are less than the value for the rigid dumbbell. The heavier the molecule, the greater the deviation. The reason for the deviation is the low Θ_{vib}, as the large reduced mass brings Θ_{vib} closer to the temperature of measurement.

Table 5.2 also lists γ for two polyatomic molecules, CO_2 and NH_3. CO_2 is a linear molecule, and the high-temperature limit of $C_{V,rot}$ is Nk_B, identical to the one for the diatomic molecule. If RT were sufficiently lower than Θ_{vib}, γ of CO_2 would be 1.4, but the actual value is a lot less. The reason is that Θ_{vib} is not sufficiently high. The bending modes (degeneracy = 2) of the molecule has $\bar{\nu} = 757$ cm^{-1} or $\nu = 1.578 \times 10^{13}$ Hz. For this mode, $\Theta_{vib} = 757$ K; RT is not sufficiently low to neglect the contribution of the mode to U and C_V.

Ammonia is a nonlinear molecule, and its rotation has three degrees of freedom. The high-temperature limit of $C_{V,rot}$ is $\frac{3}{2} Nk_B$. If RT is sufficiently lower than Θ_{vib}, γ of NH_3 would be 1.333, and the actual value is a slightly less. The deviation is a lot less compared with CO_2. The lowest-frequency vibration mode in NH_3 is N–H wagging (degeneracy = 1) at $\bar{\nu} = 1138$ cm^{-1} or $\nu = 3.147 \times 10^{13}$ Hz. For this mode, $\Theta_{vib} = 1639$ K, and therefore $C_{V,vib}$ is small at RT.

5.4 Distribution of Rotational Energy Levels

We learned that RT $\ll \Theta_{vib}$, and therefore nearly all diatomic molecules are at the vibrationally ground state at RT. The situation is different with the rotation. At room temperature, $\Theta_{rot} \ll T$, and therefore many rotational states are accessible by thermal excitation. Still, the energy level increases in proportion to $J(J + 1)$, and therefore states with a higher J are less populated. In fact, there is a most probable value of J. Here, we find which J is the most probable.

The probability for a diatomic molecule to be at energy level J is proportional to

$$g(J) \equiv (2J + 1) \exp\left(-J(J + 1)\frac{\Theta_{rot}}{T}\right) \tag{5.82}$$

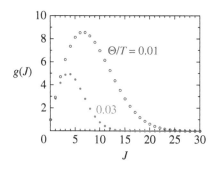

Figure 5.10 Plot of $g(J) = (2J + 1)$ $\exp(-J(J + 1)\Theta_{rot}/T)$ at $\Theta_{rot}/T = 0.01$ (open circles) and 0.03 (closed circles).

A plot of this function has a peak, since $2J + 1$ increases and $\exp(-J(J+1)$ $\Theta_{rot}/T)$ decreases with an increasing J. Figure 5.10 shows a plot of $g(J)$ for two values of Θ_{rot}/T. At each temperature, $g(J)$ peaks at a different value of J. The higher the temperature, the greater the J at the peak.

To find which J maximizes $g(J)$, we regard J as a continuous variable for now and calculate the derivative:

$$g'(J) = \left[2 - (2J + 1)^2 \frac{\Theta_{rot}}{T}\right] \exp\left(-J(J + 1)\frac{\Theta_{rot}}{T}\right) \tag{5.83}$$

We find that $g'(J) = 0$ at $J = J_m$ given by

$$2J_m + 1 = \sqrt{\frac{2T}{\Theta_{rot}}} \tag{5.84}$$

When $J < J_m$, $g'(J) > 0$, and therefore $g(J)$ maximizes at J_m, if J were continuous. However, J is discrete. We need to compare $g(J)$ at the two values of J closest to J_m. The two candidates are $[J_m]$ and $[J_m] + 1$, where $[x]$ is the greatest integer that does not exceed x. If $g([J_m]) > g([J_m] + 1)$, $[J_m]$ maximizes $g(J)$; if $g([J_m]) < g([J_m] + 1)$, $[J_m] + 1$ maximizes $g(J)$. For example, an HCl molecule has $\Theta_{rot} = 15.2$ K. At 300 K, $J_m = 6.28$. We then evaluate $g(J)$ for $J = 6$ and 7, and find that $g(6) > g(7)$. Thus, we find that $J = 6$ is the most probable state of rotation for HCl at 300 K.

5.5 Conformations of a Molecule

Many molecules exist in two or more conformations. For example, it is well known that cyclohexane has a boat conformation and a chair conformation. Figure 5.11 depicts the energy diagram for the two conformations of cyclohexane.

We have not considered the electronic energy levels of a molecule. Certainly, excitation of one of the electrons in the highest occupied molecular orbital to the lowest unoccupied molecular orbital requires a high energy. However, different conformations of a molecule, due to the electronic energy difference,

Figure 5.11 Conformational energy of cyclohexane. The lowest energy conformer is a chair. The boat conformation has a higher energy level.

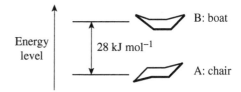

may have their energy level difference comparable to the vibrational energy difference. In this section, we consider the population of different conformations.

We consider a molecule in two conformations A and B. Let us denote by ε_A and ε_B the energy levels of the two conformations, and by g_A and g_B their degeneracies. The conformational part of the single-molecule partition function is written as

$$Z_{1,\text{conf}} = g_A \exp(-\beta\varepsilon_A) + g_B \exp(-\beta\varepsilon_B) \tag{5.85}$$

The overall partition function of a single molecule is

$$Z_1 = Z_{1,\text{trans}} Z_{1,\text{rot}} Z_{1,\text{vib}} Z_{1,\text{conf}} \tag{5.86}$$

Here, it is implicitly assumed that $Z_{1,\text{trans}}$, $Z_{1,\text{rot}}$, and $Z_{1,\text{vib}}$ are shared by the two conformations, and that is why we can factorize Z_1 into the four components. We can easily understand that $Z_{1,\text{trans}}$ is shared. We consider that the moment of inertia and normal modes are not much different for the two conformations.

The two conformations are interconvertible. For Eq. (5.86) to be effective, the interconversion must be a lot slower compared with the rotation. Otherwise, the rotation would occur for a conformation-averaged molecule.

The probability P_B for the molecule to be in conformation B is

$$P_B = \frac{g_B \exp(-\beta\varepsilon_B)}{Z_{1,\text{conf}}} = \frac{1}{1 + (g_A/g_B)\exp(\beta(\varepsilon_B - \varepsilon_A))} \tag{5.87}$$

For the two conformations of cyclohexane, $g_A = g_B$. From the energy level difference in Figure 5.11, we find that the probability for the molecule to be in the boat conformation at 300 K is $P_B = 1.34 \times 10^{-5}$. At 400 K, $P_B = 2.21 \times 10^{-5}$.

Figure 5.12 Energy diagram for different conformations of *n*-butane. The anti conformer has the lowest energy, and the two gauche conformers are at local minima of the energy.

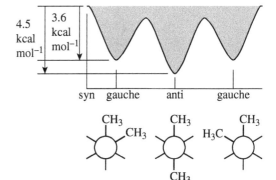

Now, we look at another example, n-butane. We regard the molecule as dimethyl-substituted ethane. Rotation around the C_2–C_3 bond in the molecule changes the energy, as illustrated in Figure 5.12. The plot of the energy has local minima for three angles of rotation.

We consider that the molecule is either in anti or one of the two gauche conformations (discrete energy levels). Let A = anti and B = gauche. Since $g_A/g_B = \frac{1}{2}$, we find that $P_B = 0.31$ at 300 K.

Problems

5.1 *High-temperature approximation of $C_{V,rot}$.* The high-temperature asymptote of $C_{V,rot}$ can be obtained using the Euler–Maclaurin formula:

$$\sum_{J=m}^{n} f(J) = \int_{m}^{n} f(J)dJ + \frac{f(m) + f(n)}{2} + \frac{1}{6}\frac{f'(n) - f'(m)}{2!}$$
$$- \frac{1}{30}\frac{f'''(n) - f'''(m)}{4!} + \cdots$$

For our series,

$$f(J) = (2J + 1)e^{-xJ(J+1)}$$

where $x = \Theta_{rot}/T$.

(1) Show that the formula reduces to

$$\sum_{J=0}^{\infty} (2J + 1)\exp(-xJ(J + 1)) = \frac{1}{x} + \frac{1}{3} + \frac{1}{15}x + \cdots$$

(2) Calculate $C_{V,rot}$ at high temperatures.

5.2 *Low-temperature approximation of $C_{V,rot}$.* This problem considers the low-temperature asymptote of U_{rot} and $C_{V,rot}$ for a system of N noninteracting diatomic molecules.

(1) Start with

$$Z_{1,rot} = \sum_{J=0}^{\infty} (2J + 1)\exp\left(-\frac{\Theta_{rot}}{T}J(J + 1)\right)$$
$$= \sum_{J=0}^{\infty} (2J + 1)\exp(-k_B\Theta_{rot}\beta J(J + 1))$$

to derive expressions for U_{rot} and $C_{V,rot}$. These expressions should be effective at all temperatures.

(2) Find how U_{rot} and $C_{V,rot}$ change with temperature T when T is sufficiently low.

5.3 *Rigid rotor.* A molecule that allows only to rotate is called a rigid rotor. We use the high-temperature asymptotes for the partition function:

$$Z = \frac{T}{\Theta_{\text{rot}}} \left[1 + \frac{1}{3} \frac{\Theta_{\text{rot}}}{T} + \frac{1}{15} \left(\frac{\Theta_{\text{rot}}}{T} \right)^2 + \cdots \right]$$

For a rigid rotor with Θ_{rot}, calculate $\langle E \rangle$ and $\langle \Delta E^2 \rangle$, where $E = k_B \Theta_{\text{rot}} J(J + 1)$

5.4 *Harmonic oscillator.* For a harmonic oscillator with characteristic frequency v,
(1) Calculate $\langle n \rangle$ and $\langle \Delta n^2 \rangle$.
(2) What are the high-temperature asymptotes of $\langle n \rangle$ and $\langle \Delta n^2 \rangle$?
(3) Draw a sketch for the plots of $\langle n \rangle$ and $\langle \Delta n^2 \rangle^{1/2}$ as a function of $k_B T/(hv)$ in a single chart.

5.5 *Most probable J.* Section 5.4 explains how to find the most probable J. Find the most probable J for $^{14}N_2$ and $^1H^{35}Cl$ at $T = 300$ K.

5.6 *Cylohexane.* Consider a system of N molecules of cyclohexane. Each molecule is either in a chair conformation or a boat conformation. Their energy levels are $\varepsilon_{\text{chair}}$ and $\varepsilon_{\text{boat}}$, respectively, and their degeneracies are one. For simplicity, we assume that the molecules are allowed only for the conformational changes, and do not move, rotate, or vibrate.
(1) The state of the system is specified by n, the number of the molecules in the chair conformation, where $n = 0, 1, 2, \ldots, N$. What is the partition function of the system?
(2) Look at the expression you obtained in (1) to derive a formula for $\langle n \rangle$.
(3) Apply the formula to calculate $\langle n \rangle$.

5.7 *n-Butane.* This problem continues on the discussion of n-butane in Section 5.5. Consider a system consisting of N molecules of butane in vapor phase.
(1) Calculate the part of the heat capacity contributed by the conformation, $C_{V,\text{conf}}$.
(2) Draw a sketch of $C_{V,\text{conf}}/(Nk_B)$ as a function of temperature for T between 0 and 100 °C. Pay attention to the values at 0 and 100 °C.
(3) Assume that n-butane remains as a stable gas over an extended range of temperature. Find the low- and high-temperature asymptotes of $C_{V,\text{conf}}/(Nk_B)$. Then, draw a sketch of $C_{V,\text{conf}}/(Nk_B)$ as a function of temperature for T over a broad range of temperature.

6

Indistinguishable Particles

This chapter introduces the concept of indistinguishability of particles. Although we have neglected it in the preceding chapters, the partition function must be calculated, taking into account the indistinguishability, if the particles of the system are indistinguishable.

First, we learn the concept of indistinguishability in Section 6.1. The next section describes how to express the partition function of a system consisting of N indistinguishable particles using a single-particle partition function. The expression is effective under a certain condition, which we examine in Section 6.3. Sections 6.4 and 6.5 look at ramifications of the indistinguishability. In the last section, we consider an open system of gas molecules using the concept of indistinguishability.

6.1 Distinguishable Particles and Indistinguishable Particles

So far, we have assumed, without explicitly stating, that all the particles are distinguishable and therefore can be numbered or labeled. It is as if each particle has its own face. Obviously, however, oxygen molecules are not distinguishable. Unlike humans (or animals), the molecules are not distinguishable unless their structures are different.

The particle may be an atom, a molecule, an electron, a wave, or a phonon. Table 6.1 lists examples of **distinguishable** particles and **indistinguishable** particles. For distinguishable particles, how to distinguish them is also described. Figure 6.1a shows molecules in a solid state and Figure 6.1b the molecules in vapor phase as examples of distinguishable particles and indistinguishable particles, respectively.

An N_2 molecule has features exactly identical to those owned by another N_2 molecule in the system. If the molecules move to change their center-of-mass positions, we cannot tell one particle from another. In a gas and a liquid, the

Statistical Thermodynamics: Basics and Applications to Chemical Systems, First Edition. Iwao Teraoka.
© 2019 John Wiley & Sons, Inc. Published 2019 by John Wiley & Sons, Inc.
Companion website: www.wiley.com/go/Teraoka_StatsThermodynamics

Table 6.1 Examples of distinguishable particles and indistinguishable particles.

Distinguishable particles		Indistinguishable particles
Particles	How to distinguish them	
Atoms in a crystal	Position within the crystal	N_2 molecules in a box
Repeat unit in a polymer chain	Position from the chain end	Water molecules in a beaker
Molecules immobilized on a surface	Position on the surface	Molecules adsorbed on a surface; movable after adsorption

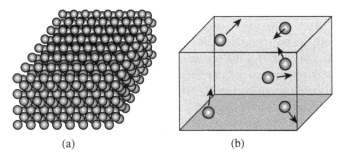

(a)   (b)

Figure 6.1 (a) Molecules in a solid state and (b) molecules in vapor phase. They are examples of distinguishable particles and indistinguishable particles, respectively.

particles can move and reach any corner of the volume. In a solid state of Ar molecules, in contrast, they do not change their positions within the crystal, allowing us to label each molecule.

The difference will vividly show up in systems of molecules adsorbed to a surface. If the adsorption results in fixed positions of the adsorbate, then we can distinguish the particles by the positions. Chemisorption belongs to such modes of adsorption. If the particles are mobile on the surface, they are indistinguishable. Likewise, if the particles can desorb and adsorb (in a dynamic equilibrium with the vapor adjacent to the surface), they are indistinguishable.

The indistinguishability makes the statistics different. We want to start with writing the partition function, but the rule we learned in Chapter 4 needs to be modified. Now we go back to the original definition of the partition function Z:

$$Z = \sum_{ms} \exp(-\beta E_{ms}) \tag{4.21}$$

where the series is taken over all possible microstates ms, and E_{ms} is the energy of the microstate. When calculating Z, each microstate must be distinctly different, regardless of whether the system consists of distinguishable particles or indistinguishable particles.

6.2 Partition Function of Indistinguishable Particles

We consider a system of N distinguishable particles and a system of N indistinguishable particles separately. We know how to express the partition function for the system of N distinguishable particles. Here we learn how to express the partition function for N indistinguishable particles using a single-particle partition function.

6.2.1 System of Distinguishable Particles

In the absence of interactions, we can write the energy E of the system as

$$E = \sum_{i=1}^{N} \varepsilon_i \qquad (6.1)$$

where the ith particle has energy ε_i ($i = 1, 2,\ldots, N$). Note that we can number all the particles in the system. The state "ms" of the N-particle system is specified by the states of the N particles (state s_i for the ith particle). For example, if the particle is an electron, and we are concerned about its spin, $s_i = \frac{1}{2}, -\frac{1}{2}$ (or up, down). The microstate of the N-particle system is specified by $\{s_1, s_2,\ldots, s_N\}$. Then, the sum by "ms" is represented by the N-fold sum with respect to $s_1, s_2,\ldots$, and s_N, and the partition function of the N-particle system is

$$Z = \sum_{s_1} \cdots \sum_{s_N} \exp\left(-\beta \sum_{i=1}^{N} \varepsilon_i \right) = \sum_{s_1} \exp(-\beta \varepsilon_1) \times \cdots \times \sum_{s_N} \exp(-\beta \varepsilon_N)$$

$$(6.2)$$

We have used only this formula so far. If, in addition, all the particles have the same energy levels and degeneracies,

$$Z = \left[\sum_{s_1} \exp(-\beta \varepsilon_1) \right]^N \qquad (6.3)$$

6.2.2 System of Indistinguishable Particles

Let us start by looking at a simple example. Consider a system of two indistinguishable particles, each capable of adopting one of three energy levels, 0, ε, and 2ε, where $\varepsilon > 0$. Both particles have the same structure of energy levels. Otherwise, the particles would be distinguishable. Figure 6.2a shows all possibilities

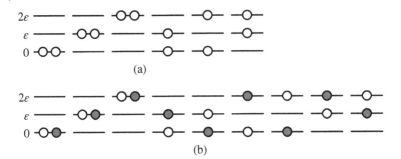

Figure 6.2 A system of two particles, each on one of three energy levels, 0, ε, and 2ε. (a) Indistinguishable particles, (b) distinguishable particles.

of distributing the two particles to the three energy levels. There are six in total. For reference, all possibilities of the arrangement, if the particles were distinguishable, are shown in Figure 6.2b. When the particles are distinguishable, the total count is greater. The difference grows rapidly with an increasing number of particles.

Each state in Figure 6.2a may be specified by n_1, n_2, and n_3, where n_s is the number of particles at the sth energy level. Note that s is not the index for a particle as in the ith particle for a system of distinguishable particles; rather, s is a label for the energy level of a particle. We call n_s "occupation number." Table 6.2 lists n_s of the three levels ($\varepsilon_1 = 0$, $\varepsilon_2 = \varepsilon$, and $\varepsilon_3 = 2\varepsilon$) for the six states in Figure 6.2a.

Now we move to a system of N indistinguishable particles. Let ε_s be the energy level for any one of identical particles, where $s = 1, 2, \ldots$, and occupation number n_s be the number of particles at energy level ε_s. Obviously, $n_1, n_2, \ldots$ satisfy the following equalities. The total number of particles in the system, N, is

$$N = \sum_s n_s \tag{6.4}$$

Table 6.2 Occupation numbers n_1, n_2, and n_3 for the six states in Figure 6.2a.

n_1	n_2	n_3
2	0	0
0	2	0
0	0	2
1	1	0
1	0	1
0	1	1

The energy E of the system is expressed as

$$E = \sum_s n_s \varepsilon_s \tag{6.5}$$

Specifying N and E does not necessarily specify the state of the system. Within the condition that N and E be specified, there may be different distributions of n_1, n_2,... In Table 6.2, the second and fifth rows have the same E.

Before considering the partition function for a system of N indistinguishable particles, we go back to the simple example in Table 6.2 to find its partition function:

$$Z = e^0 + e^{-2\beta\varepsilon} + e^{-4\beta\varepsilon} + e^{-\beta\varepsilon} + e^{-2\beta\varepsilon} + e^{-3\beta\varepsilon} = (1 + e^{-\beta\varepsilon} + e^{-2\beta\varepsilon})(1 + e^{-2\beta\varepsilon}) \tag{6.6}$$

This partition function is different from the one calculated for the distinguishable-particle version illustrated in Figure 6.2b:

$$Z_{\text{dist}} = (1 + e^{-\beta\varepsilon} + e^{-2\beta\varepsilon})^2 \tag{6.7}$$

The partition function for a system of indistinguishable particles cannot be expressed by the Nth power of a single-particle partition function. This restriction is inconvenient. However, when $N \gg 1$ and a certain condition is met, we can have a simple expression for the partition function of indistinguishable particles.

That condition is called nondegeneracy. This one is different from the one used in quantum mechanics. In the latter, nondegeneracy of a given energy level means that the degeneracy is one – only one state for that energy level. The nondegeneracy for the system of indistinguishable particles is defined as

$$[\text{number of states with } \varepsilon_s < k_B T] \gg N \tag{6.8}$$

It means that a lot more seats (states) are available for the particles to occupy compared with the number of particles. Imagine a ballpark that is very poorly attended or a nearly empty airplane. It is then not likely that two or more particles are in the same state for any of the available states. As illustrated in Figure 6.3, n_s is either 0 or 1 for nearly all s.

Now, we find how the partition function for a nondegenerate system consisting of N indistinguishable particles is related to the partition function of a single particle. We start with $N = 1$.

A. **$N = 1$.** For this system, it is guaranteed that n_s is either 0 or 1. A list of all possible states is

$$\{n_1, n_2, n_3, \ldots\} = \{1, 0, 0, \ldots\}, \{0, 1, 0, \ldots\}, \{0, 0, 1, \ldots\}, \tag{6.9}$$

Their energies are $\varepsilon(1)$, $\varepsilon(2)$, $\varepsilon(3)$, and so on. For clarity, we use $\varepsilon(s)$ for ε_s here. The partition function $Z(N = 1)$ is simply a single-particle partition

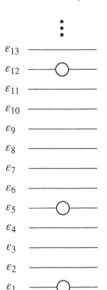

ε_{13}

ε_{12}

ε_{11}

ε_{10}

ε_9

ε_8

ε_7

ε_6

ε_5

ε_4

ε_3

ε_2

ε_1

Figure 6.3 Illustration of a nondegenerate system. Nearly all the energy levels are either empty or occupied by one particle. Doubly occupied levels are extremely rare.

function Z_1:

$$Z(N=1) = Z_1 = e^{-\beta\varepsilon(1)} + e^{-\beta\varepsilon(2)} + e^{-\beta\varepsilon(3)} + \cdots = \sum_{s=1}^{\infty} e^{-\beta\varepsilon(s)} \quad (6.10)$$

B. $N = 2$. The system has two types of microstates. One has each of two energy levels singly occupied, and the other type has one energy level doubly occupied:

$$\{n_1, n_2, n_3, n_4, \cdots\} = \{1, 1, 0, 0, \cdots\}, \{1, 0, 1, 0, \cdots\}, \{1, 0, 0, 1, \cdots\}, \cdots,$$
$$\{0, 1, 1, 0, \cdots\}, \{0, 1, 0, 1, \cdots\}, \cdots,$$
$$\{0, 0, 1, 1, \cdots\}, \cdots$$
$$\cdots,$$
$$\{2, 0, 0, 0, \cdots\}, \{0, 2, 0, 0, \cdots\}, \{0, 0, 2, 0, \cdots\}, \{0, 0, 0, 2, \cdots\}, \cdots \quad (6.11)$$

The partition function $Z(N=2)$ is

$$Z(N=2) = e^{-\beta[\varepsilon(1)+\varepsilon(2)]} + e^{-\beta[\varepsilon(1)+\varepsilon(3)]} + e^{-\beta[\varepsilon(1)+\varepsilon(4)]} + \cdots$$
$$+ e^{-\beta[\varepsilon(2)+\varepsilon(3)]} + e^{-\beta[\varepsilon(2)+\varepsilon(4)]} + \cdots$$
$$+ e^{-\beta[\varepsilon(3)+\varepsilon(4)]} + \cdots$$
$$+ \cdots$$
$$+ e^{-\beta 2\varepsilon(1)} + e^{-\beta 2\varepsilon(2)} + e^{-\beta 2\varepsilon(3)} + e^{-\beta 2\varepsilon(4)} + \cdots \quad (6.12)$$

Note that terms such as $e^{-\beta[\varepsilon(2)+\varepsilon(1)]}$ are not included, because $e^{-\beta[\varepsilon(2)+\varepsilon(1)]}$ is indistinguishable from $e^{-\beta[\varepsilon(1)+\varepsilon(2)]}$.

This partition function $Z(N = 2)$ looks similar to the square of Z_1:

$$
\begin{aligned}
Z_1^2 &= [e^{-\beta\varepsilon(1)} + e^{-\beta\varepsilon(2)} + e^{-\beta\varepsilon(3)} + \cdots]^2 \\
&= 2e^{-\beta[\varepsilon(1)+\varepsilon(2)]} + 2e^{-\beta[\varepsilon(1)+\varepsilon(3)]} + 2e^{-\beta[\varepsilon(1)+\varepsilon(4)]} + \cdots \\
&\quad + 2e^{-\beta[\varepsilon(2)+\varepsilon(3)]} + 2e^{-\beta[\varepsilon(2)+\varepsilon(4)]} + \cdots \\
&\quad\quad + 2e^{-\beta[\varepsilon(3)+\varepsilon(4)]} + \cdots \\
&\quad\quad\quad + \cdots \\
&\quad + e^{-\beta 2\varepsilon(1)} + e^{-\beta 2\varepsilon(2)} + e^{-\beta 2\varepsilon(3)} + e^{-\beta 2\varepsilon(4)} + \cdots
\end{aligned}
\tag{6.13}
$$

The difference between $Z(N = 2)$ and Z_1^2 is that states with two singly occupied levels are counted twice in Z_1^2. Counting of the states with a doubly occupied level is the same in $Z(N = 2)$ and Z_1^2. We can show that $Z(N = 2) \approx \frac{1}{2}Z_1^2$. Let M be the number of energy levels. Then, the total number of states for the first type (two singly occupied energy levels) is $\frac{1}{2}M(M-1)$, whereas the number of states for the second type (one doubly occupied energy level) is M. Since $M \gg 1$ from the assumption of nondegeneracy, the count of the first type far exceeds the count of the second type. Therefore,

$$
Z(N = 2) \cong \frac{1}{2}Z_1^2
\tag{6.14}
$$

C. $N = 3$. The system has three types of microstates. The first type has each of three energy levels singly occupied; the second type has a singly occupied energy level and a doubly occupied energy level; and the third type has one triply occupied energy level. The first type dominates in the partition function $Z(N = 3)$. As was the case with $Z(N = 2)$, $Z(N = 3)$ is similar to Z_1^3, but there is a difference. In the latter, terms consisting of permutations of $\varepsilon(1)$, $\varepsilon(2)$, and $\varepsilon(3)$ appear six times. The number is equal to 3!. Therefore,

$$
Z(N = 3) \cong \frac{1}{3!}Z_1^3
\tag{6.15}
$$

Equation (6.14) can be also expressed as $Z(N = 2) \approx Z_1^2/2!$. In general, we can approximate the N-particle partition function as

$$
Z(N) \cong \frac{1}{N!}Z_1^N
\tag{6.16}
$$

Recall that the partition function of N particles that are distinguishable, but otherwise identical, is Z_1^N. Thus, we have a simple rule for writing the partition function for the system of N particles that are indistinguishable and identical: Divide the distinguishable version of the partition function by $N!$, the number of permutations. Note that this protocol works only when $N \gg 1$ and the nondegeneracy condition is satisfied. Otherwise, we need to count all possible states as we did for $Z(N = 2)$.

Once we have obtained Z, we proceed to calculating its natural logarithm:

$$
\ln Z = -\ln N! + N \ln Z_1
\tag{6.17}
$$

Here we use Stirling's formula to approximate $\ln N!$. In most situations, a crude approximation, $\ln N! \approx N(\ln N - 1)$ is sufficient. See Appendix A.4. With the Stirling's formula,

$$\ln Z \cong N(\ln Z_1 - \ln N + 1) \tag{6.18}$$

6.3 Condition of Nondegeneracy

In this section, we consider what is required for the nondegeneracy to hold. We will find the requirement for a perfect gas in a box of cube of L^3. As the indistinguishability is associated with the translational movement of molecules, we pay attention to the translational states of molecules in the system and their energy levels. As we learned in Section 3.6, each translational state is specified by three positive integers n_x, n_y, and n_z. With these quantum numbers, the energy level of that state, ε_{trans}, is written as

$$\varepsilon_{trans} = \frac{h^2}{8mL^2}(n_x^2 + n_y^2 + n_z^2) \tag{5.50}$$

where $n_x = 1, 2, 3, \ldots$; $n_y = 1, 2, 3, \ldots$; $n_z = 1, 2, 3, \ldots$. It is convenient to introduce a vector $\mathbf{n}$ whose x, y, z components are n_x, n_y, and n_z, respectively:

$$\mathbf{n} = \begin{bmatrix} n_x \\ n_y \\ n_z \end{bmatrix} \tag{6.19}$$

Each set of n_x, n_y, and n_z represents a cubic lattice point in the Cartesian coordinates. Figure 6.4 shows such points with n_x, n_y, $n_z \leq 4$.

With $\mathbf{n}$, we can write ε_{trans} as

$$\varepsilon_{trans} = \frac{h^2}{8mL^2}\mathbf{n}^2 \tag{6.20}$$

Figure 6.4 Each of the spheres at lattice points has specific values of n_x, n_y, and n_z (>0) and represents a specific state of translational movement of a molecule.

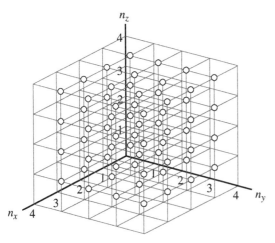

Figure 6.5 All the lattice points within one-eighth of a sphere of radius n_{max} represent states with energy less than $k_B T$.

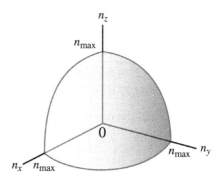

We want to find the number of states with $\varepsilon_{trans} < k_B T$ when the number is sufficiently large. Since ε_{trans} grows as $\mathbf{n}^2$ increases, we set the maximum of $\mathbf{n}$ at the threshold by equating Eq. (6.20) to $k_B T$. That is,

$$n_{max}{}^2 = \frac{8mL^2 k_B T}{h^2} \tag{6.21}$$

Different sets of n_x, n_y, and n_z with $|\mathbf{n}| < |\mathbf{n}_{max}| = n_{max}$ satisfy the requirement of energy level $< k_B T$. In Figure 6.5, the lattice points within one eighth of a sphere of radius n_{max} are those points.

The number of lattice points within the one-eighth sphere is

$$\frac{1}{8} \times \frac{4\pi}{3} n_{max}{}^3 = \frac{1}{8} \times \frac{4\pi}{3} \left(\frac{8mL^2 k_B T}{h^2} \right)^{3/2} = \frac{4\pi}{3} \left(\frac{2mk_B T}{h^2} \right)^{3/2} V \tag{6.22}$$

where $V = L^3$ is the volume of the box. The nondegeneracy condition is that this number be much greater than N, the number of molecules in the box:

$$\frac{4\pi}{3} \left(\frac{2mk_B T}{h^2} \right)^{3/2} V \gg N \tag{6.23}$$

Recall the definition of the thermal wavelength, $\lambda_T = h/(2\pi m k_B T)^{1/2}$, Eq. (5.59). With λ_T, the nondegeneracy condition is expressed as

$$\frac{4}{3\pi^{1/2}} \frac{V}{N} \gg \lambda_T^3 \tag{6.24}$$

where $4/(3\pi^{1/2})$ is close to 1. The above inequality is rewritten to

$$\lambda_T \ll (V/N)^{1/3} \tag{6.25}$$

where V/N is the volume of the box per molecule. That is, the nondegeneracy is that the mean distance between the nearest pair of molecules is a lot greater than the thermal wavelength.

To illustrate whether this condition can be easily met or not, let us consider a gas of helium at 300 K. Its thermal wavelength is

$$\lambda_T = \frac{6.626 \times 10^{-34}\text{J} \cdot \text{s}}{(2\pi \times 6.647 \times 10^{-27}\text{kg} \times 1.381 \times 10^{-23}\text{J K}^{-1} \times 300 \text{ K})^{1/2}}$$
$$= 5.037 \times 10^{-11}\text{m} \tag{6.26}$$

At $p = 10^5$ Pa, the mean distance between molecules is

$$\left(\frac{V}{N}\right)^{1/3} = \left(\frac{k_B T}{p}\right)^{1/3} = \left(\frac{1.381 \times 10^{-23}\text{J K}^{-1} \times 300 \text{ K}}{10^5 \text{ Pa}}\right)^{1/3}$$
$$= 3.460 \times 10^{-9}\text{m} \tag{6.27}$$

The inequality, Eq. (6.25), holds easily. The nondegeneracy condition will be satisfied, unless the pressure is excessively high or the temperature is low. For $(V/N)^{1/3}$ to be comparable to λ_T of helium at room temperature, the pressure must be more than 10^5 times as high. For heavier molecules including oxygen and nitrogen, the nondegeneracy requirement is satisfied with a greater margin.

6.4 Significance of Division by *N*!

6.4.1 Gas in a Two-Part Box

In Section 6.2, we learned that the partition function for a system of N indistinguishable particles can be obtained by dividing the distinguishable version of the partition function by $N!$ when $N \gg 1$. To better understand its significance, we consider a system of N indistinguishable monatomic molecules in a box consisting of parts A and B (see Figure 6.6). Their volumes are V_A and V_B. The molecules are distributed to the two parts separated by porous wall. The molecules can move from A to B, and from B to A.

Here, we obtain the partition function of the two-part system in two ways. Both employ classical, not quantum-mechanical, expressions of a single-molecule partition function.

A. Using the single-molecule partition function across the two parts
The single-molecule partition function Z_1 across the two-part box is expressed as

$$Z_1 = \int_{V_A + V_B} d\mathbf{r} \int d\mathbf{p} \exp\left(-\frac{\mathbf{p}^2}{2m}\beta\right) \tag{6.28}$$

Figure 6.6 Vapor-phase system consisting of parts A and B. Their volumes are V_A and V_B. The two parts are separated by porous wall that passes the molecules.

which is calculated as

$$Z_1 = (V_A + V_B) \left(\frac{2\pi m}{\beta h^2} \right)^{3/2} \tag{6.29}$$

With the nondegeneracy, the partition function of the system is written as

$$Z = \frac{Z_1^N}{N!} = \frac{1}{N!}(V_A + V_B)^N \left(\frac{2\pi m}{\beta h^2} \right)^{3N/2} \tag{6.30}$$

B. As a sum of products of the part-A partition function and the part-B partition function

We assign n indistinguishable molecules to part A and $N - n$ molecules to part B. The partition functions of the two parts, Z_A and Z_B, are

$$Z_A = \frac{1}{n!} \left[V_A \left(\frac{2\pi m}{\beta h^2} \right)^{3/2} \right]^n \tag{6.31}$$

$$Z_B = \frac{1}{(N-n)!} \left[V_B \left(\frac{2\pi m}{\beta h^2} \right)^{3/2} \right]^{N-n} \tag{6.32}$$

We can obtain the partition function of the two-part system by taking the sum of $Z_A Z_B$ over all possible values of n:

$$Z = \sum_{n=0}^{N} \frac{1}{n!(N-n)!} \left[V_A \left(\frac{2\pi m}{\beta h^2} \right)^{3/2} \right]^n \left[V_B \left(\frac{2\pi m}{\beta h^2} \right)^{3/2} \right]^{N-n} \tag{6.33}$$

We rewrite Eq. (6.33) into

$$Z = \frac{1}{N!} \left(\frac{2\pi m}{\beta h^2} \right)^{3N/2} \sum_{n=0}^{N} \frac{N!}{n!(N-n)!} V_A^n V_B^{N-n} \tag{6.34}$$

which is identical to Eq. (6.30) by the binomial theorem. We thus find that the division by the factorial of the number of molecules in each of the two parts allows a consistent expression of the partition function for the combined system. Note that the sum of the series on the right-hand side of Eq. (6.34) represents the partition function of the combined system filled with N distinguishable molecules, as $N!/[n!(N-n)!]$ is the number of ways to choose n molecules out of a total N distinguishable molecules.

6.4.2 Chemical Potential

The significance of the division by $N!$ is also evident in the chemical potential. Let us first calculate the chemical potential μ_{dist} in a system of N distinguishable monatomic molecules in volume V. Since the partition function Z_{dist} is given as

$$Z_{\text{dist}} = V^N \left(\frac{2\pi m}{\beta h^2} \right)^{3N/2} \tag{4.40}$$

the Helmholtz free energy F_{dist} is

$$F_{\text{dist}} = -k_B T \ln Z_{\text{dist}} = -N k_B T \left(\ln V + \frac{3}{2} \ln \frac{2\pi m}{\beta h^2} \right) \tag{4.102}$$

Then, μ_{dist} is calculated as

$$\mu_{\text{dist}} = \left(\frac{\partial F_{\text{dist}}}{\partial N} \right)_{V,T} = -k_B T \left(\ln V + \frac{3}{2} \ln \frac{2\pi m}{\beta h^2} \right) \tag{4.104}$$

We know that this expression of μ_{dist} is incorrect, as it is a function of an extensive variable V. A correct expression of the chemical potential should be a linear function of the logarithm of the density, N/V. Here we find that we can obtain a correct expression using the indistinguishable version:

$$Z_{\text{indist}} = \frac{V^N}{N!} \left(\frac{2\pi m}{\beta h^2} \right)^{3N/2} \tag{6.35}$$

$$F_{\text{indist}} = k_B T \ln N! - N k_B T \left(\ln V + \frac{3}{2} \ln \frac{2\pi m}{\beta h^2} \right)$$

$$\cong N k_B T \left(\ln N - 1 - \ln V - \frac{3}{2} \ln \frac{2\pi m}{\beta h^2} \right) \tag{6.36}$$

$$\mu_{\text{indist}} = k_B T \left(\ln \frac{N}{V} - \frac{3}{2} \ln \frac{2\pi m}{\beta h^2} \right) \tag{6.37}$$

As expected, μ_{indist} is a linear function of $\ln(N/V)$.

The pressure p, calculated from Eq. (6.35), is the same as Eq. (4.101):

$$p = k_B T \frac{\partial}{\partial V} \ln Z_{\text{indist}} = k_B T \frac{N}{V} \tag{6.38}$$

Then, the Gibbs free energy G_{indist} is

$$G_{\text{indist}} = F_{\text{indist}} + pV = N k_B T \left(\ln N - \ln V - \frac{3}{2} \ln \frac{2\pi m}{\beta h^2} \right) \tag{6.39}$$

which is identical to $N\mu_{\text{indist}}$, as required. Equation (6.37) is expressed also as

$$\frac{\mu_{\text{indist}}}{k_B T} = \ln p - \frac{5}{2} \ln(k_B T) - \frac{3}{2} \ln \frac{2\pi m}{h^2} \tag{6.40}$$

6.4.3 Mixture of Two Gases

How about a vapor-phase system consisting of molecules A and molecules B such as Ne and Ar? An A molecule and a B molecule can be distinguished, but molecules of A are indistinguishable among themselves. We can write the partition function for a one-part box of volume V having N_A monatomic molecules of A (here, N_A is not the Avogadro's number) and N_B monatomic molecules of B as

$$Z = \frac{V^{N_A}}{N_A!} \left(\frac{2\pi m_A}{\beta h^2} \right)^{3N_A/2} \frac{V^{N_B}}{N_B!} \left(\frac{2\pi m_B}{\beta h^2} \right)^{3N_B/2} \tag{6.41}$$

The division by the factorial is now by the product of two factorials. This rule can be extended to a mixture of three or more gases.

6.5 Indistinguishability and Center-of-Mass Movement

The indistinguishability is a result of particles' capability to change their center-of-mass positions. The movement makes a pair of particles indistinguishable from each other, since they can swap positions. If they cannot move and their positions are locked in a structure like a crystal, the particles can be distinguished by their positions within the structure. In contrast, a fluid consists of indistinguishable particles. A system of gas consists of indistinguishable molecules; so does a system of liquid.

It is conceptually wrong to consider a system of indistinguishable diatomic molecules that are allowed to rotate or vibrate, or do both, but not to move. The absence of movement automatically makes the molecules distinguishable. If you see problems asking for considering such a system, it is for practice purposes only. For the rest of this book, we do not explicitly say that molecules in vapor and liquid phases are indistinguishable, but it is being assumed.

6.6 Open System of Gas

Molecules in a system of gas are indistinguishable. In the preceding chapters, we neglected the indistinguishability to learn how the free energy, pressure, and entropy are expressed as a function of independent variables such as N, V, T and μ, V, T.

A canonical ensemble of N noninteracting molecules has now the partition function Z_N given as

$$Z_N = \frac{Z_1^N}{N!} \tag{6.42}$$

where Z_1 is the single-molecule partition function. Division by $N!$ changes the expression of F, S, and G, as we have seen in Section 6.4.2.

The changes due to $N!$ show up more acutely in the grand canonical ensemble and the thermodynamic functions derived from the grand partition function. In this section, we learn these changes.

The grand partition function $\mathcal{Z}$ is given as

$$\mathcal{Z} = \sum_{N=0}^{\infty} e^{\beta \mu N} Z_N \tag{6.43}$$

With Eq. (6.42), this equation is rewritten to

$$\mathcal{Z} = \sum_{N=0}^{\infty} e^{\beta \mu N} \frac{Z_1^N}{N!} = \exp(e^{\beta \mu} Z_1) \tag{6.44}$$

Therefore,

$$\ln \mathcal{Z} = e^{\beta \mu} Z_1 \tag{6.45}$$

Now, we consider a grand canonical ensemble of monatomic molecules with mass m in volume V. Since Z_1 is given by Eq. (4.39),

$$\ln \mathcal{Z} = V \left(\frac{2\pi m}{\beta h^2} \right)^{3/2} e^{\beta \mu} \tag{6.46}$$

We can use the formula, Eq. (4.152), to obtain the mean number of molecules, $\langle N \rangle$:

$$\langle N \rangle = \frac{1}{\beta} \frac{\partial \ln \mathcal{Z}}{\partial \mu} = \frac{1}{\beta} V \left(\frac{2\pi m}{\beta h^2} \right)^{3/2} \beta e^{\beta \mu} = \ln \mathcal{Z} \tag{6.47}$$

Note that $\langle N \rangle$ is equal to $\ln \mathcal{Z}$. Therefore, with Eq. (4.169),

$$\langle \Delta N^2 \rangle = \frac{1}{\beta} \frac{\partial \langle N \rangle}{\partial \mu} = \frac{1}{\beta} \frac{\partial \ln \mathcal{Z}}{\partial \mu} = \langle N \rangle \tag{6.48}$$

As μ increases in the reservoir the system is in contact with, $\langle N \rangle$ increases, but the increase is exponential. Recall that μ is a "logarithmic" concept; μ is a linear function of the logarithm of pressure.

The pressure p is calculated as

$$p = \frac{1}{\beta} \frac{\partial \ln \mathcal{Z}}{\partial V} = \frac{1}{\beta} \left(\frac{2\pi m}{\beta h^2} \right)^{3/2} e^{\beta \mu} = \frac{\ln \mathcal{Z}}{\beta V} \tag{6.49}$$

With Eq. (6.47), the equation of state is obtained as

$$pV = \frac{\langle N \rangle}{\beta} \tag{6.50}$$

The Nth term of the series in Eq. (6.44) is proportional to the probability for the open system to have N molecules. Therefore, the probability $P(N)$ is given as

$$P(N) = \frac{1}{\mathcal{Z}} e^{\beta \mu N} \frac{Z_1^N}{N!} = \frac{1}{\mathcal{Z}} \frac{1}{N!} (e^{\beta \mu} Z_1)^N \tag{6.51}$$

As we learned in Section 2.4, this $P(N)$ represents a Poisson distribution. Its mean and variance are equal to each other:

$$\langle N \rangle = \langle \Delta N^2 \rangle = e^{\beta \mu} Z_1 \tag{6.52}$$

With an increasing $\langle N \rangle$, the coefficient of variation decreases:

$$\frac{\langle \Delta N^2 \rangle^{1/2}}{\langle N \rangle} = \langle N \rangle^{-1/2} = (e^{\beta \mu} Z_1)^{-1/2} \tag{6.53}$$

The width of the distribution relative to its mean decreases with an increasing $\langle N \rangle$. If $\langle N \rangle = 10^{22}$, approximately equal to the number of molecules in a half-liter box at 1 bar and 300 K, $\langle \Delta N^2 \rangle^{1/2}/\langle N \rangle$ is 10^{-11}. Although the grand canonical ensemble allows N to vary, its distribution is extremely narrow.

As the box gets smaller, $\langle N \rangle$ decreases. For example, consider a box of $(10\,\text{nm})^3$. At 1 bar and 300 K, the box has $\langle N \rangle \cong 24$ and $\langle \Delta N^2 \rangle^{1/2}/\langle N \rangle \cong 0.2$.

Problems

6.1 *N particles, two energy levels.* Consider a system consisting of N indistinguishable particles, each capable of being at energy level $-\varepsilon$ or ε, where $\varepsilon > 0$. Do not assume $N \gg 1$.
   (1) Calculate the partition function and find $\langle E \rangle$.
   (2) What is the low-temperature limit of $\langle E \rangle$? What is the state of the system in the limit?
   (3) What is the high-temperature asymptote of $\langle E \rangle$? What is the state of the system in the limit?
   (4) Draw a sketch for the plot of $\langle E \rangle$ as a function of T.

6.2 *Five particles, two energy levels.* A system consists of five indistinguishable particles, each capable of adopting the energy of 0 and ε ($\varepsilon > 0$).
   (1) What is the partition function?
   (2) Find the heat capacity C_V.
   (3) What is the low-temperature limit of C_V?
   (4) What is the high-temperature asymptote of C_V?
   (5) Draw a sketch of C_V/k_B as a function of $k_B T/\varepsilon$.
   (6) What is the state of the system in the low-temperature limit?
   (7) What is the state of the system at high temperatures?
   (8) The way the equation in (2) was derived allows us to write an expression of C_V for N indistinguishable particles in place of five. Write the expression.
   (9) If $N \gg 1$ in (8), the expression can be simplified. Write that expression.

6.3 *Two particles, infinite number of energy levels.* Consider a system of two indistinguishable particles, each capable of adopting one of the energy levels $(0, \varepsilon, 2\varepsilon, \ldots,$ with $\varepsilon > 0)$.
   (1) What is the exact partition function Z?
   (2) Confirm that $Z_1^2/2$ is a good approximation to Z when the nondegeneracy condition is met.

6.4 *N particles, q energy levels.* Consider a system of N indistinguishable particles that can be at any of q equally spaced energy levels $(0, \varepsilon, 2\varepsilon, 3\varepsilon, \ldots,$ $(q-1)\varepsilon; \varepsilon > 0)$, on the condition that none of the levels accommodate two or more particles. Assume that $N < q$.
   (1) What is the state of the system in the limit of $T = 0$? What is the energy of the system, $\langle E \rangle$?
   (2) What is the state of the system in the limit of high temperature? What is $\langle E \rangle$?

6.5 *Gas in a two-part box.* Section 6.4 considered a two-part box that holds N indistinguishable monatomic molecules. The partition function was obtained as

$$Z = \sum_{n=0}^{N} \frac{1}{n!(N-n)!} \left[V_A \left(\frac{2\pi m}{\beta h^2} \right)^{3/2} \right]^n \left[V_B \left(\frac{2\pi m}{\beta h^2} \right)^{3/2} \right]^{N-n}$$

where part A has n molecules and part B has $N - n$. Answer the following questions:
   (1) Find a formula to calculate $\langle n \rangle$ and apply it to this system.
   (2) Find a formula to calculate $\langle \Delta n^2 \rangle$ and apply it to this system.
   (3) Suppose we change V_A. When does $\langle \Delta n^2 \rangle$ maximize? What is $\langle \Delta n^2 \rangle^{1/2} / \langle n \rangle$ at the maximum?

6.6 *Mixing of two gases.* A container has gas A in the left chamber of volume V_A and gas B in the right chamber of volume V_B. The number of molecules in the two chambers are N_A and N_B, respectively (N_A, $N_B \gg 1$; N_A is not the Avogadro's number here). Both chambers are in contact with a heat reservoir at temperature T. The nonconfigurational part of the single-molecule partition function of gas A is q_A $(= (2\pi m_A/\beta h^2)^{3/2})$; the one for gas B is q_B. The molecules do not interact with each other. We remove the partition that separates the two chambers to mix the two gases.

(1) Assume that gas A and gas B consist of the same molecules ($q_A = q_B = q$), and the molecules are distinguishable. How much does the free energy of the combined system change by removing the partition? Find the change ΔF. What is the entropy change ΔS?

(2) Assume that gas A and gas B consist of the same molecules, and the molecules are indistinguishable. What is ΔF? In what condition is $\Delta F = 0$? What is the entropy change ΔS?

(3) Which result of ΔS, (1) or (2), is reasonable? Justify your conclusion.

(4) Assume that gas A and gas B are different ($q_A \neq q_B$) and the molecules are indistinguishable within A or B. What is ΔF? What is ΔS?

6.7 *Open system of distinguishable particles.* If molecules were distinguishable in an open system of volume V in equilibrium with a reservoir of noninteracting, monatomic molecules of mass m and chemical potential μ at temperature T, what would be $\langle N \rangle$ and $\langle \Delta N^2 \rangle$? How does the coefficient of variation change as $\langle N \rangle$ increases?

6.8 *Molecules in a small box.* Consider a system of molecules in a system of a small box of volume V in contact with a reservoir of molecules with a chemical potential μ. The temperature is uniform throughout the system and the reservoir of the molecules. In the box, the energy of the molecule is lowered by ε, and therefore the single-molecule partition function is $e^{\beta \varepsilon} Vq$, where q is the nonconfigurational part of the single-molecule partition function. Calculate the grand partition function of the system and find $\langle N \rangle$ and $\langle \Delta N^2 \rangle$.

6.9 *Division by N!.* We learned in Section 6.4.2 that regarding molecules as distinguishable leads to an unreasonable expression of the chemical potential. Here, we will derive the division by $N!$ from thermodynamic requirement.

(1) To correct this problem, let us start with writing the partition function as

$$Z = \frac{V^N}{e^{f(N)}} \left(\frac{2\pi m}{\beta h^2} \right)^{3N/2}$$

where $f(N)$ is a function of N, not of V or T. Express μ using $f(N)$.

(2) What condition results from the requirement that the expression of μ be "reasonable"? Use that condition to find $f(N)$.

6.10 *Van der Waals gas.* Van der Waals equation of state,

$$\left(p + a \frac{N^2}{V^2} \right) (V - Nb) = \frac{N}{\beta}$$

where a and b are constants of N, V, and β, is known to give a good approximation to the state of a nonideal gas. The equation expresses p as a function of N, V, and β. Assume monatomic gas for simplicity. Interactions between molecules causes a and b to be nonzero.

(1) Use $p = \beta^{-1}\partial\ln Z/\partial V$ to find the expression for $\ln Z$. The constant of the integration should be determined by comparing the expression you have obtained (in the vanishing limit of a and b) with the $\ln Z$ of the ideal monatomic gas.

(2) What is U of the van der Waals gas?

(3) What is the chemical potential μ?

(4) Does a positive b increase μ relative to the value for $b = 0$?

6.11 *Sign of chemical potential.* We have learned that a gas of N monatomic molecules in volume V has a chemical potential μ given as

$$\mu = k_B T \left(\ln \frac{N}{V} - \frac{3}{2} \ln \frac{2\pi m}{\beta h^2} \right)$$

Is this μ positive or negative?

6.12 *Two-part box with different temperatures.* Consider a two-part box connected by a porous, insulating plug. Box i of volume V_i is held at temperature T_i ($i = 1, 2$). The boxes hold monatomic molecules.

(1) Let N_i be the number of molecules in box i. Then, $N_1 + N_2 = $ constant. Write the Helmholtz free energy F of the two-part box using N_1 and N_2.

(2) Minimize F by changing N_1. What relationship holds between μ_1 and μ_2, where μ_i is the chemical potential of a molecule in box i?

7

Imperfect Gas

In this chapter, we employ the canonical-ensemble formulation to take into account interactions between molecules in a gas. We have found molecular-level expressions of the interaction in Section 3.4. We use them to derive the equation of state similar to that of van der Waals gas.

7.1 Virial Expansion

The equation of state for a perfect gas is $pV = Nk_B T$. We have proved in Section 4.6 that a canonical ensemble of noninteracting molecules gives this equation of state. In real gas, the interactions are always present. At low pressures, the effect of the interactions on the equation of state is negligible, since the interactions are much smaller compared with the other components of energy. With an increasing pressure, the interactions become a larger part of the total energy. We cannot neglect the interactions any more.

A deviation from the ideality is represented in the **virial expansion**:

$$pV = Nk_B T \left[1 + B_2 \frac{N}{V} + B_3 \left(\frac{N}{V} \right)^2 + \cdots \right] \tag{7.1}$$

It is an expansion of $pV/(Nk_B T)$ by the number density N/V (number of molecules in unit volume). The expansion coefficients are called virial coefficients; in increasing order, the **second virial coefficient** (B_2), the third virial coefficient (B_3), and so on. They are constants of N/V. In the low-density limit, the equation reduces to the equation of state of the perfect gas. The reason why the coefficients are called second, third, etc. – not first, second, etc. – will be clear, if you rewrite Eq. (7.1) into

$$p = k_B T \left[\frac{N}{V} + B_2 \left(\frac{N}{V} \right)^2 + B_3 \left(\frac{N}{V} \right)^3 + \cdots \right] \tag{7.2}$$

It is usually the case that, when $B_2 N/V \ll 1$, $B_2 N/V$ dominates in the expansion. That is, $B_2 N/V \gg B_3 (N/V)^2$. Then, at low densities, the correction by B_2 is sufficient.

Statistical Thermodynamics: Basics and Applications to Chemical Systems, First Edition. Iwao Teraoka.
© 2019 John Wiley & Sons, Inc. Published 2019 by John Wiley & Sons, Inc.
Companion website: www.wiley.com/go/Teraoka_StatsThermodynamics

The sign of B_2 can be positive or negative, and it may change with temperature, and the dependence is different for each gas. A positive B_2 makes the pressure higher compared with the one for the perfect gas. We will learn that repulsion between molecules leads to a positive B_2, and therefore increases the pressure. In contrast, a negative B_2 makes the pressure lower, which likely results from attractive interactions between molecules.

A slightly different expression is also used for the virial expansion:

$$\frac{pV}{nRT} = 1 + \widetilde{B}_2 \frac{n}{V} + \widetilde{B}_3 \left(\frac{n}{V}\right)^2 + \cdots \tag{7.3}$$

where n/V is the molar concentration. The dimension for $\widetilde{B}_2$ is m^3 mol^{-1}, as opposed to m^3 for B_2, in the SI unit. B_i and $\widetilde{B}_i$ are related to each other by

$$B_i N_A^{i-1} = \widetilde{B}_i \tag{7.4}$$

where $i = 2, 3, \ldots$.

It may be interesting to see what the virial coefficients are in the gas that follows van der Waals equation of state:

$$\left(p + \frac{n^2}{V^2}a\right)(V - nb) = nRT \tag{7.5}$$

We know that a positive b increases the pressure, and nb stands for a decrease in the volume accessible to the molecules. A positive a makes the pressure lower, which is due to the attractive interaction. As the pressure arises from the collision of molecules with the wall, a molecule being pulled back by nearby other molecules will decrease the speed of collision, thereby lowering the pressure.

Let us expand Eq. (7.5) into a power series of n/V:

$$p = \frac{nRT}{V\left(1 - \frac{n}{V}b\right)} - \frac{n^2}{V^2}a = \frac{nRT}{V}\left[1 + \frac{n}{V}b + \left(\frac{n}{V}b\right)^2 + \cdots\right] - \frac{n^2}{V^2}a \tag{7.6}$$

which is rewritten to

$$\frac{pV}{nRT} = 1 + \left(b - \frac{a}{RT}\right)\frac{n}{V} + b^2\left(\frac{n}{V}\right)^2 + \cdots \tag{7.7}$$

Comparison of Eq. (7.7) with Eq. (7.3) gives the virial coefficients in the van der Waals equation of state:

$$\widetilde{B}_2 = b - \frac{a}{RT} \tag{7.8}$$

$$\widetilde{B}_3 = b^2 \tag{7.9}$$

Since the values of a and b are listed for different gases, we can prepare a plot of $\widetilde{B}_2$ as a function of T. However, these two parameters are usually calculated from the critical temperature and pressure that are away from the ambient.

Therefore, $\widetilde{B}_2$ calculated with these parameters and Eq. (7.8) does not necessarily provide a good approximation to $\widetilde{B}_2$ observed in experiments conducted at relatively low pressures. It is more common to adopt a three-parameter empirical equation for $\widetilde{B}_2(T)$:

$$\widetilde{B}_2 = a - b\exp(c/(T/\text{K})) \tag{7.10}$$

where a, b, and c are the parameters that optimally fit experimental data for $\widetilde{B}_2(T)$. The a and b are different from those in the van der Waals equation of state. The values of a, b, and c are listed in Table 7.1 for some common gases.

Different parts of Figure 7.1 show $\widetilde{B}_2(T)$ for different gases, calculated with these parameters and Eq. (7.10) or from other sources. It is common to all the gases that $\widetilde{B}_2$ is negative at low temperatures and increases with an increasing temperature. Figure 7.1a compares $\widetilde{B}_2$ of noble gases and hydrogen. Heavy molecules have a stronger interaction compared with light molecules, and that is due to the dispersion force. For light molecules, especially for He, $\widetilde{B}_2$ turns to positive, and then starts to decrease at high temperatures; the decrease cannot be described by Eq. (7.10). Figure 7.1b compares four molecules with a molecular weight between 16 and 20 g mol^{-1}. The interaction due to the dispersion force should be similar. However, there is a huge difference in $\widetilde{B}_2$. While neon's $\widetilde{B}_2$ gets close to 0 and turns positive with an increasing temperature in the range shown, B_2 of water (steam) remains negative and retains a large absolute value. We can see a large contribution by the dipole–dipole interaction to water's $\widetilde{B}_2$. The dipole–dipole interaction is weaker for NH_3. Methane is a nonpolar molecule, and its weak interaction results in a small $|\widetilde{B}_2|$. Now we jump to Figure 7.1d. The molecules are linear alkanes, and the order of the curves can

Table 7.1 Parameters for the optimal fit of the second virial coefficient $\widetilde{B}_2(T)$ by $a - b\exp(c\text{K}/T)$.

Gas	a (cm^3 mol^{-1})	b (cm^3 mol^{-1})	c	Gas	a (cm^3 mol^{-1})	b (cm^3 mol^{-1})	c
He	114.1	98.7	3.215	CO_2	137.6	87.7	325.7
Ne	81.0	63.6	30.7	HCl	57.7	37.8	495.9
Ar	154.2	119.3	105.1	NH_3	44.3	23.6	766.6
Kr	189.6	148.0	145.3	H_2O	33.0	15.2	1300.7
Xe	245.6	190.9	200.2	CH_4	206.4	159.5	133
H_2	315.0	289.7	9.47	C_2H_6	267.3	191.5	256
N_2	185.4	141.8	88.7	C_3H_8	322.8	220.3	350
O_2	152.8	117.0	108.8	C_4H_{10}	165.4	135.4	555
CO	202.6	154.2	94.2				

Source: Data from Kaye and Laby [8].

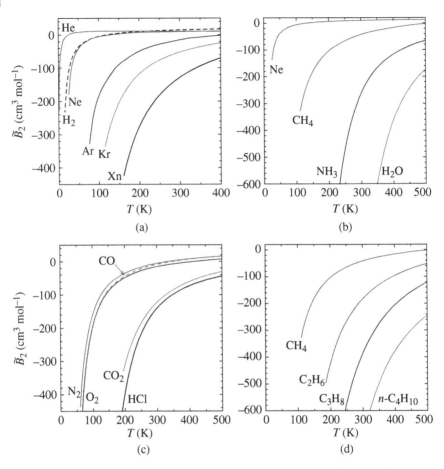

Figure 7.1 Second virial coefficients of different gases. (a) Noble gases, (b) gases of a molecular weight between 16 and 20 g mol^{-1}, (c) polar and nonpolar diatomic and polyatomic molecules, (d) linear alkanes. Note: the scales are different for each panel. Each curve starts at the normal boiling point. The plots were prepared from Ref. [9] for He and Ref. [8] for all the other gases. Source: Data from Kaye and Laby [8] and McCarty [9].

be understood by the dependence of the dispersion force on the number of C and H atoms. Figure 7.1c has a mixture of polar and nonpolar molecules. There are two groups of curves. The first group is for N_2, O_2, and CO. Their molecular weights are similar, and the three curves run close to each other. CO is the only polar molecule, but its weak dipole–dipole interaction does not make a large contribution to $\widetilde{B}_2$. HCl, in contrast, has a stronger interaction, as it is a lot more polar than CO is. The curve for HCl runs a lot lower compared with the first group, and is lower than is the curve for CO_2 that has a higher molecular weight but is nonpolar.

In Figure 7.1a, curves for some gases reach positive values at high temperatures within the range shown. The temperature at which B_2 becomes equal to 0 is called a **Boyle temperature** and is usually denoted by T_B. At T_B, the second virial coefficient vanishes; thus, the leading term in the virial expansion is the one with B_3. It means that the gas is close to ideal. Later in this chapter, we learn that the vanishing B_2 is a result of counteracting repulsive and attractive interactions that cancel each other. It is not that the interaction is absent.

Molecules with a strong attractive interaction (due to dipole–dipole interaction or dispersion force) has a higher T_B compared to molecules with a repulsive or weakly attractive interaction. If B_2's temperature dependence is given by Eq. (7.8), T_B is given as

$$T_B = \frac{a}{Rb} \tag{7.11}$$

The second virial coefficient of He peaks at around 200 K. A further increase in the temperature decreases $\widetilde{B}_2$. It is considered that other gases will also exhibit a peaking if the temperature can be raised sufficiently without decomposing the molecules. Both Eqs. (7.8) and (7.10) fail to describe this decrease. In the two equations, $\widetilde{B}_2$ is an increasing function of T in the whole range.

7.2 Molecular Expression of Interaction in the Canonical Ensemble

In this section, we first learn how to express the interaction in a molecular model. Then, we consider a canonical ensemble that consists of interacting molecules to obtain the partition function and calculate thermodynamic functions. We see how the interaction changes the pressure and heat capacity.

We treat the interaction as a correction to the energy. In the absence of the interaction, the energy of the system is given as a sum of the energies of individual molecules. The latter may contain rotational and vibrational components. When the interaction is present but is weak, the correction to the energy of the system arises predominantly from the interaction between a pair of molecules. We call this type of interaction **binary**. We can neglect the interaction that depends on three molecules (**tertiary**) and more molecules.

The energy of a system consisting of N interacting molecules can be written as

$$E = \sum_{i=1}^{N} \varepsilon_i + \sum_{i>j}^{N} \Phi(r_{ij}) \tag{7.12}$$

The first series represents the ideal part; the energy of the ith molecule, ε_i, consists of the center-of-mass kinetic energy, rotational energy, vibrational energy, and electronic energy:

$$\varepsilon_i = \frac{\mathbf{p}_i^2}{2m_i} + \varepsilon_{\text{rot},i} + \varepsilon_{\text{vib},i} + \varepsilon_{\text{elec},i} \tag{7.13}$$

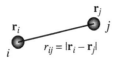

Figure 7.2 Two molecules i and j at $\mathbf{r}_i$ and $\mathbf{r}_j$, respectively, interact with each other. The interaction is a function of their distance r_{ij}.

In the second series, $\Phi(r_{ij})$ represents the interaction. It is natural to assume that the interaction between the ith and jth molecules is a function of r_{ij}, the distance between the two molecules (see Figure 7.2). The indices i and j run through all pairs of i and j that satisfy $1 \leq j < i \leq N$. There are a total $\frac{1}{2}N(N-1)$ pairs.

The goal of this section is to express the second virial coefficient B_2 using Φ. Once we express the partition function Z using Φ, use of $p = k_B T \, \partial \ln Z / \partial V$ will allow us to express B_2 using Φ.

The simplest model of gas is a system of monatomic molecules. In a system of N such molecules in volume V ($N \gg 1$), the ideal part of the energy is the center-of-mass kinetic energy only. The energy of the system is expressed as

$$E = \sum_{i=1}^{N} \frac{\mathbf{p}_i^{\,2}}{2m_i} + \sum_{i>j}^{N} \Phi(r_{ij}) \tag{7.14}$$

Then, the partition function is written as

$$Z = \frac{1}{N! h^{3N}} \int_V d\mathbf{r}_1 \cdots \int_V d\mathbf{r}_N \int_{-\infty}^{\infty} d\mathbf{p}_1 \cdots$$
$$\times \int_{-\infty}^{\infty} d\mathbf{p}_N \exp\left(-\beta \sum_{i=1}^{N} \frac{\mathbf{p}_i^{\,2}}{2m_i} - \beta \sum_{i>j}^{N} \Phi(r_{ij}) \right) \tag{7.15}$$

The integrals by $\mathbf{r}_1, \ldots, \mathbf{r}_N$ and the integrals by $\mathbf{p}_1, \ldots, \mathbf{p}_N$ can be separated:

$$Z = \frac{1}{N! h^{3N}} \int_V d\mathbf{r}_1 \cdots \int_V d\mathbf{r}_N \exp\left(-\beta \sum_{i>j}^{N} \Phi(r_{ij}) \right)$$
$$\times \int_{-\infty}^{\infty} d\mathbf{p}_1 \cdots \int_{-\infty}^{\infty} d\mathbf{p}_N \exp\left(-\beta \sum_{i=1}^{N} \frac{\mathbf{p}_i^2}{2m_i} \right) \tag{7.16}$$

To calculate the pressure p, we take the logarithm of Z and differentiate it by V. Obviously, the integrals by $\mathbf{p}_i$ is irrelevant to the pressure. Only the integrals by $\mathbf{r}_i$ count in calculating p.

Now we look at interacting diatomic molecules. In a crude approximation that preaverages the effect of molecular orientation, the interaction between a pair of molecules depends only on their distance. With this approximation, we can write the interaction between the diatomic molecules in the same way we do for the interaction between monatomic molecules. The system of diatomic molecules has the energy of rotation and the energy of vibration on top of the center-of-mass translation energy in the ideal part of the energy. The energy of the system is expressed as

$$E = \sum_{i=1}^{N} \frac{\mathbf{p}_i^2}{2m_i} + \sum_{i=1}^{N} \frac{h^2}{8\pi^2 I_i} J_i(J_i+1) + \sum_{i=1}^{N} h\nu_i \left(\frac{1}{2}+n_i\right) + \sum_{i>j}^{N} \Phi(r_{ij}) \quad (7.17)$$

Then,

$$Z = \frac{1}{N! h^{3N}} \int_V d\mathbf{r}_1 \cdots \int_V d\mathbf{r}_N \int_{-\infty}^{\infty} d\mathbf{p}_1 \int_{-\infty}^{\infty} d\mathbf{p}_N \cdots \sum_{J_1=0}^{\infty} (2J_1+1) \cdots$$

$$\times \sum_{J_N=0}^{\infty} (2J_N+1) \sum_{n_1=0}^{\infty} \cdots \sum_{n_N=0}^{\infty} \exp\left(-\beta \sum_{i=1}^{N} \frac{\mathbf{p}_i^2}{2m_i}\right.$$

$$\left. -\beta \sum_{i=1}^{N} \frac{h^2}{8\pi^2 I_i} J_i(J_i+1) - \beta \sum_{i=1}^{N} h\nu_i \left(\frac{1}{2}+n_i\right) - \beta \sum_{i>j}^{N} \Phi(r_{ij})\right) \quad (7.18)$$

which is rewritten to

$$Z = \frac{1}{N! h^{3N}} \int_V d\mathbf{r}_1 \cdots \int_V d\mathbf{r}_N \exp\left(-\beta \sum_{i>j}^{N} \Phi(r_{ij})\right) \int_{-\infty}^{\infty} d\mathbf{p}_1 \cdots$$

$$\times \int_{-\infty}^{\infty} d\mathbf{p}_N \exp\left(-\beta \sum_{i=1}^{N} \frac{\mathbf{p}_i^2}{2m_i}\right)$$

$$\times \sum_{J_1=0}^{\infty} (2J_1+1) \cdots \sum_{J_N=0}^{\infty} (2J_N+1) \exp\left(-\beta \sum_{i=1}^{N} \frac{h^2}{8\pi^2 I_i} J_i(J_i+1)\right)$$

$$\times \sum_{n_1=0}^{\infty} \cdots \sum_{n_N=0}^{\infty} \exp\left(-\beta \sum_{i=1}^{N} h\nu_i \left(\frac{1}{2}+n_i\right)\right) \quad (7.19)$$

As in the system of monatomic molecules, only the integrals by $\mathbf{r}_i$ count in calculating p. We can extend this statement to a system of polyatomic molecules.

The part in Z that depends on V is called the configurational part of the partition function:

$$Z_{\text{conf}} = \frac{1}{N!} \int_V d\mathbf{r}_1 \cdots \int_V d\mathbf{r}_N \exp\left(-\beta \sum_{i>j}^{N} \Phi(r_{ij})\right) \quad (7.20)$$

This Z_{conf} is common to monatomic molecules and diatomic molecules. In the absence of interactions ($\Phi = 0$), $Z_{\text{conf}} = V^N/N!$.

The overall partition function consists of two parts:

$$Z = Z_{\text{conf}} Z_{\text{non-conf}} \quad (7.21)$$

where $Z_{\text{non-conf}}$ does not involve V, and is different for the monatomic molecules and diatomic molecules. Then,

$$\ln Z = \ln Z_{\text{conf}} + \ln Z_{\text{non-conf}} \tag{7.22}$$

Differentiate $\ln Z$ by V and we obtain

$$p = k_{\text{B}} T \frac{\partial}{\partial V} \ln Z = k_{\text{B}} T \frac{\partial}{\partial V} \ln Z_{\text{conf}} \tag{7.23}$$

The pressure depends on Z_{conf} only. We calculate the N-fold integrals in Eq. (7.20).

We start with rewriting $e^{-\beta \Sigma \Phi}$ into a product:

$$\exp\left(-\beta \sum_{i>j} \Phi(r_{ij})\right) = \prod_{i>j} \exp(-\beta \Phi(r_{ij})) \tag{7.24}$$

We separate one from each exponential factor in the product:

$$\exp(-\beta \Phi(r_{ij})) = 1 + [\exp(-\beta \Phi(r_{ij})) - 1] \tag{7.25}$$

To simplify the notation, let us define u_{ij} as

$$u_{ij} \equiv \exp(-\beta \Phi(r_{ij})) - 1 \tag{7.26}$$

Thus defined, u_{ij} is a function of r_{ij}, although we do not write it explicitly. Figure 7.3 compares the potential $\beta \Phi$ and the newly introduced u for a Lennard-Jones potential. At $r_{ij} = 0$ where Φ is infinitely high, $u = -1$. As the Φ decreases to 0 with an increasing r_{ij}, u increases; and when $\Phi = 0$, $u_{ij} = 0$. In the range of $\Phi > 0$ (repulsive interaction), $u < 0$. As Φ drops below 0 (attractive interaction), u turns positive. At long distances, both Φ and u are close to 0. Since the reach of the Lennard-Jones potential is at most ~1 nm (for molecules in gas), u is almost 0 at $r_{ij} > 1$ nm. Since we consider weak interactions, $|u|$ is nearly 0 except at short distances.

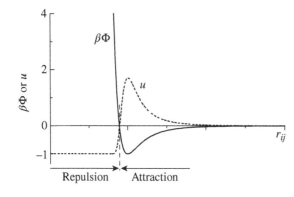

Figure 7.3 Plot of a potential $\beta \Phi$ (Lennard-Jones) and u as a function of intermolecular distance r_{ij}. For the range of r_{ij} where $\Phi > 0$ (repulsive), $u < 0$. For the range of r_{ij} where $\Phi < 0$ (attractive), $u > 0$.

With this u, Eq. (7.24) is rewritten to

$$\exp\left(-\beta \sum_{i>j} \Phi(r_{ij})\right) = \prod_{i>j}(1 + u_{ij}) = (1 + u_{21})(1 + u_{31}) \cdots (1 + u_{N1})$$
$$\times (1 + u_{32}) \cdots (1 + u_{N2})$$
$$\times \cdots \times$$
$$\times (1 + u_{N,N-1}) \tag{7.27}$$

The right-hand side is a $\tfrac{1}{2}N(N-1)$-fold product of $1 + u_{ij}$, $(1 + u_{21})(1 + u_{31}) \cdots$ $(1 + u_{N,N-1})$. For now, we assume that each u_{ij} is small compared with one. We may loosen this requirement later. When $(1 + u_{21})(1 + u_{31}) \cdots (1 + u_{N,N-1})$ is expanded, the leading term is 1. We get it by choosing 1 from every factor in the $\tfrac{1}{2}N(N-1)$-fold product. The second most dominant terms are obtained by choosing u_{ij} from one of $\tfrac{1}{2}N(N-1)$ factors and 1 from all the other factors. There are $\tfrac{1}{2}N(N-1)$ such terms. The third most dominant terms are obtained by choosing u_{ij} from two of $\tfrac{1}{2}N(N-1)$ factors and 1 from all the other factors. The expanded form of Eq. (7.27) is

$$\exp\left(-\beta \sum_{i>j} \Phi(r_{ij})\right) = 1 + (u_{21} + u_{31} + \cdots + u_{N,N-1}) + (u_{21}u_{31} + \cdots)$$
$$= 1 + \sum_{i>j} u_{ij} + \sum_{(i,j)>(k,l)} u_{ij}u_{kl} + \cdots \tag{7.28}$$

where the sum in the second series is taken for different (i, j) pair and (k, l) pair, avoiding double counting. Each term in the first series involves two molecules, and therefore the sum represents the binary interactions. Each element in $\Sigma u_{ij}u_{kl}$ involves three or four molecules, and therefore the sum represents the tertiary and quaternary interactions. Their magnitudes are compared here.

$$1 \gg \left|\sum_{i>j} u_{ij}\right| \gg \left|\sum_{(i,j)>(k,l)} u_{ij}u_{kl}\right| \gg \cdots \tag{7.29}$$

See Problem 7.1 as proof. Therefore, Eq. (7.28) can be approximated as

$$\exp\left(-\beta \sum_{i>j} \Phi(r_{ij})\right) \cong 1 + \sum_{i>j} u_{ij} \tag{7.30}$$

Then, Z_{conf} is approximated as

$$Z_{\text{conf}} = \frac{1}{N!} \int_V d\mathbf{r}_1 \cdots \int_V d\mathbf{r}_N \left(1 + \sum_{i>j} u_{ij}\right) = \frac{V^N}{N!} + Z_{\text{conf,non-ideal}} \tag{7.31}$$

where $Z_{\text{conf, nonideal}}$ is the nonideal part of Z_{conf}:

$$Z_{\text{conf,nonideal}} \equiv \frac{1}{N!} \int_V d\mathbf{r}_1 \cdots \int_V d\mathbf{r}_N \sum_{i>j} u_{ij}(r_{ij}) \tag{7.32}$$

We evaluate $Z_{\text{conf, nonideal}}$ as follows. Exchanging the order of integration and summation, and then explicitly writing the integrals by $\mathbf{r}_i$ and $\mathbf{r}_j$, we get

$$Z_{\text{conf,non-ideal}} = \frac{1}{N!} \sum_{i>j} \int_V d\mathbf{r}_1 \cdots \underline{\int_V d\mathbf{r}_j} \int_V d\mathbf{r}_i \cdots \int_V d\mathbf{r}_N u_{ij}(r_{ij})$$

$$= \frac{1}{N!} \sum_{i>j} \underline{\int_V d\mathbf{r}_1 \cdots \int_V d\mathbf{r}_N} \int_V d\mathbf{r}_j \int_V d\mathbf{r}_i u_{ij}(r_{ij}) \tag{7.33}$$

where the underlined integrals exclude those by $\mathbf{r}_i$ and $\mathbf{r}_j$. Then,

$$Z_{\text{conf,non-ideal}} = \frac{V^{N-2}}{N!} \sum_{i>j} \int_V d\mathbf{r}_j \int_V d\mathbf{r}_i u_{ij}(r_{ij}) \tag{7.34}$$

There are $\frac{1}{2}N(N-1)$ pairs of j and i. These integrals are identical, and therefore we let $j = 1$ and $i = 2$ represent them. Since $N \gg 1$,

$$Z_{\text{conf,non-ideal}} \equiv \frac{V^{N-2}N^2}{N!2} \int_V d\mathbf{r}_1 \int_V d\mathbf{r}_2 u_{21}(r_{21}) \tag{7.35}$$

We change the variables of integration from $(\mathbf{r}_1, \mathbf{r}_2)$ to $(\mathbf{r}_1, \mathbf{r})$, where $\mathbf{r} \equiv \mathbf{r}_1 - \mathbf{r}_2$:

$$\int_V d\mathbf{r}_1 \int_V d\mathbf{r}_2 u_{12}(r_{12}) = \int_V d\mathbf{r}_1 \int_V u(\mathbf{r}) d\mathbf{r} = V \int_V u(\mathbf{r}) d\mathbf{r} \tag{7.36}$$

By combining these two equations, we obtain

$$z_{\text{conf,non-ideal}} \cong \frac{V^{N-1}}{N!} N^2 \frac{1}{2} \int_V u(r) d\mathbf{r} \tag{7.37}$$

where $r = |\mathbf{r}|$. Recall that $u(r)$ is nonzero at $r < \sim 1$ nm. Therefore, the range of integration can be narrowed to $r < \sim 1$ nm, at most 10 nm. It means that the integral is independent of V.

Let us define $B(T)$ by

$$B(T) \equiv \frac{1}{2} \int_V [-u(r)] d\mathbf{r} = \frac{1}{2} \int_V [1 - e^{-\beta \Phi(r)}] d\mathbf{r} \tag{7.38}$$

Note that $B(T)$ is not a function of V. Remember the coefficient $\frac{1}{2}$ in the definition, and the negative sign in the integrand. Obviously, $B(T)$ has a dimension of volume. With $B(T)$,

$$Z_{\text{conf,non-ideal}} \cong -\frac{V^{N-1}}{N!} N^2 B(T) \tag{7.39}$$

and therefore

$$Z_{\text{conf}} \cong \frac{1}{N!}(V^N - V^{N-1}N^2 B(T)) = \frac{V^N}{N!} \left(1 - \frac{N^2}{V} B(T)\right) \tag{7.40}$$

Looking at this equation, we find that the requirement of $|u_{ij}| \ll 1$ for all r_{ij} should rather be $(N^2/V)|B| \ll 1$. $|B|$ can be large if the number density is low, and $|u_{ij}|$ can be large locally as long as its integral is sufficiently small. Equation (7.40) includes a major correction to Z_{conf} by the interaction.

We now proceed to calculate the corrections to the thermodynamic functions by the interaction. If $(N^2/V)B \ll 1$,

$$\ln Z_{conf} \cong -\ln N! + N \ln V + \ln \left(1 - \frac{N^2}{V} B(T) \right)$$

$$\cong -\ln N! + N \ln V - \frac{N^2}{V} B(T) \tag{7.41}$$

With Eq. (7.23), the pressure is calculated as

$$p = \frac{N k_B T}{V} \left(1 + \frac{N}{V} B(T) \right) \tag{7.42}$$

Comparison of this equation with Eq. (7.2) tells us that $B(T)$ is the second virial coefficient.

Now we consider U and C_V of the interacting gas. From Eqs. (7.22) and (7.41), we have

$$\ln Z \cong -\ln N! + N \ln V - \frac{N^2}{V} B(T) + \ln Z_{non-conf} \tag{7.43}$$

Only the last two terms on the right-hand side depend on T, and obviously $\ln Z_{non-conf}$ represents the ideal part. Therefore,

$$U = U_{ideal} + U_{non-ideal} \tag{7.44}$$

where

$$U_{ideal} = -\frac{\partial}{\partial \beta} \ln Z_{non-conf} \tag{7.45}$$

and

$$U_{non-ideal} = \frac{\partial}{\partial \beta} \frac{N^2}{V} B(T) = -k_B T^2 \frac{N^2}{V} B'(T) \tag{7.46}$$

The two-part construction is carried into C_V:

$$C_V = C_{V,ideal} + C_{V,non-ideal} \tag{7.47}$$

where

$$C_{V,non-ideal} = \frac{d}{dT} \left(-k_B T^2 \frac{N^2}{V} B'(T) \right) = -k_B \frac{N^2}{V} (2TB' + T^2 B'') \tag{7.48}$$

Likewise, we can consider the effect of the interaction on the chemical potential. Since $-(N^2/V)B(T)$ is the nonideal part of $\ln Z$, the nonideal part of the free energy is given as

$$F_{non-ideal} = k_B T \frac{N^2}{V} B(T) \tag{7.49}$$

Therefore, the nonideal part of the chemical potential is

$$\mu_{\text{non-ideal}} = \frac{\partial}{\partial N} F_{\text{non-ideal}} = 2k_B T \frac{N}{V} B(T) \tag{7.50}$$

A positive B (repulsive) increases μ, compared with the ideal gas, a reasonable result.

For a van der Waals gas with $B = N_A(b - a/T)$, $U_{\text{nonideal}} = -N_A k_B (N^2/V)a$ and $C_{V, \text{nonideal}} = 0$. A different B may give a different U_{nonideal} and a nonvanishing $C_{V, \text{nonideal}}$, but the effect of the interaction on C_V will be small.

7.3 Second Virial Coefficients in Different Models

In this section, we employ different models of the interaction potential Φ to calculate the second virial coefficient $B(T)$. Since Φ is a function of center-to-center distance r of two molecules, the spherical polar coordinates simplify the three-dimensional integral of a function $f(r)$ into a one-dimensional integral:

$$\int f(r)\mathrm{d}\mathbf{r} = \int_0^\infty f(r)r^2\mathrm{d}r \int_0^\pi \sin\theta\,\mathrm{d}\theta \int_0^{2\pi} \mathrm{d}\phi = 4\pi \int_0^\infty f(r)r^2\mathrm{d}r \tag{7.51}$$

In Eq. (7.38) that defines $B(T)$, $f(r) = \frac{1}{2}[1 - e^{-\beta\Phi(r)}]$, and

$$B(T) = 2\pi \int_0^\infty [1 - e^{-\beta\Phi(r)}]r^2\mathrm{d}r = 2\pi \int_0^\infty (-u(r))r^2\mathrm{d}r \tag{7.52}$$

Here, we consider three models of Φ and calculate $B(T)$ for each of them.

7.3.1 Hard-Core Repulsion Only

The simplest model for the interaction has an excluded-volume effect only. A molecule is modeled as a sphere of radius r_s. When the two spheres are not allowed to overlap, we call the sphere a hard sphere, and the resulting Φ a **hard-core potential**. We have seen this potential in Section 3.4.1. The fact that the center-to-center distance r of two spheres cannot be below $2r_s$ is translated into Φ given as

$$\Phi(r) = \begin{cases} \infty & (r < 2r_s) \\ 0 & (r > 2r_s) \end{cases} \tag{7.53}$$

For this $\Phi(r)$ shown in Figure 7.4, $B(T)$ is calculated as

$$B(T) = 2\pi \int_0^{2r_s} r^2\mathrm{d}r = \frac{16\pi}{3}r_s^3 = 4v_s \tag{7.54}$$

where $v_s = (4/3)\pi r_s^3$ is the volume of the sphere. For the hard-core potential, $B(T)$ is a positive constant of T.

Figure 7.4 Potential $\Phi(r)$ plotted as a function of center-to-center distance r between two spheres of radius r_s. The distance cannot be less than $2r_s$.

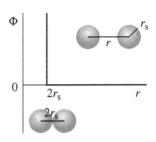

With $B_2 = B(T)$ given by Eq. (7.54), the equation of state may be written as

$$pV = Nk_{\rm B}T\left(1 + 4v_{\rm s}\frac{N}{V}\right) \tag{7.55}$$

The correction is $4v_{\rm s}N/V$, which is effective if $4v_{\rm s}N/V \ll 1$. Since B is constant, the hard-core potential has $U_{\rm nonideal} = 0$ and $C_{V,\,\rm nonideal} = 0$.

7.3.2 Square-well Potential

This model adds a short-range attractive interaction to the hard-core repulsion. If the center of the other sphere is within a layer of thickness $2(r_{\rm a} - r_{\rm s})$ around a given sphere, and the interaction is constant at $-\varepsilon_{\rm s}$, where $\varepsilon_{\rm s} > 0$. The potential curve $\Phi(r)$ and the corresponding $u(r)$ are sketched in Figure 7.5b; Figure 7.5a is a counterpart for the hard-core repulsion.

For this potential,

$$B(T) = 2\pi\left[\int_0^{2r_{\rm s}} r^2\,dr + \int_{2r_{\rm s}}^{2r_{\rm a}}[1 - \exp(\beta\varepsilon_{\rm s})]r^2\,dr\right] \tag{7.56}$$

which is calculated as

$$\frac{B(T)}{v_{\rm s}} = 4\left\{1 - \left[(r_{\rm a}/r_{\rm s})^3 - 1\right]\left[\exp(\beta\varepsilon_{\rm s}) - 1\right]\right\} \tag{7.57}$$

When the temperature is high, this expression of $B(T)$ can be simplified. Since $\beta\varepsilon_{\rm s} \ll 1$ at high temperatures, $\exp(\beta\varepsilon_{\rm s}) - 1 \cong \beta\varepsilon_{\rm s}$. Then,

$$\frac{B(T)}{v_{\rm s}} \cong 4\left\{1 - [(r_{\rm a}/r_{\rm s})^3 - 1]\frac{\varepsilon_{\rm s}}{k_{\rm B}T}\right\} \tag{7.58}$$

The right-hand side has the same T dependence as does Eq. (7.8) obtained for van der Waals gas. Comparison of the two equations allows us to express the two parameters of the van der Waals equation of state using the parameters of the square-well potential:

$$b = 4v_{\rm s}N_{\rm A} \tag{7.59}$$

$$a = 4v_{\rm s}[(r_{\rm a}/r_{\rm s})^3 - 1]\varepsilon_{\rm s}N_{\rm A}^2 \tag{7.60}$$

The two parameters, originally introduced to phenomenologically describe the state of real gases, now have acquired molecule-level interpretation.

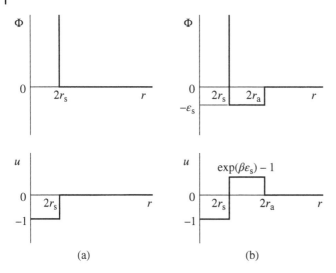

Figure 7.5 Potential Φ and the function $u = \exp(-\beta\Phi) - 1$, plotted as a function of the center-to-center distance r. (a) Excluded-volume interaction between hard spheres of radius r_s. (b) Square-well potential. Excluded-volume spheres of radius r_s is augmented with attractive interaction of $-\varepsilon_s$ at distances between $2r_s$ and $2r_a$.

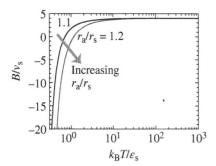

Figure 7.6 Second virial coefficient, reduced by the hard-sphere volume, plotted as a function of reduced temperature k_BT/ε_s, for a square-well potential for two values of attraction layer thickness. With an increasing thickness of the layer, the Boyle temperature increases.

Namely, b represents the excluded volume, and a is due to the short-range attraction. The b parameter is equal to four times the volume of molecules per mole.

Figure 7.6 shows a plot of the virial coefficient reduced by v_s, as a function of the reduced temperature $(\beta\varepsilon_s)^{-1} = k_BT/\varepsilon_s$. The figure compares the curve of B/v_s for two values of r_a/r_s. The high-temperature asymptote of B/v_s is 4, and with a decreasing temperature, B/v_s decreases and turns negative. The greater the r_a/r_s, the sooner (i.e. at a higher temperature) the B turns negative, i.e. the Boyle temperature is higher, which is a reasonable result since a greater r_a means a thicker layer of attraction. The temperature dependence is similar to the one in real gases (see Figure 7.1).

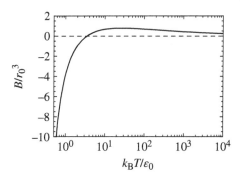

Figure 7.7 Second virial coefficient, reduced by the cube of r_0, plotted as a function of reduced temperature for the Lennard-Jones potential. The line is unique to the potential.

7.3.3 Lennard-Jones Potential

It is not easy to calculate $B(T)$ for the Lennard-Jones (L-J) potential:

$$\Phi(r) = \varepsilon_0 \left[\left(\frac{r_0}{r} \right)^{12} - 2 \left(\frac{r_0}{r} \right)^6 \right] \tag{3.47}$$

The best we can get is the following expression [10]:

$$B = -\frac{2\pi}{3} r_0^3 \sum_{n=0}^{\infty} \frac{2^n}{4n!} \Gamma\left(\frac{2n-1}{4} \right) (\beta \varepsilon_0)^{(2n+1)/4} \tag{7.61}$$

where $\Gamma(z)$ is a gamma function (see Appendix A.7). Problem 7.4 derives this Taylor expansion.

Figure 7.7 shows a plot of B/r_0^3, calculated using Eq. (7.61), as a function of $(\beta \varepsilon_0)^{-1} = k_B T/\varepsilon_0$. The curve is similar to the one for the square-well potential shown in Figure 7.6, but there is a difference: B decreases with an increasing temperature at high temperatures. The latter dependence agrees with the actual plot of B_2 for helium (see Figure 7.1a).

7.4   Joule–Thomson Effect

This section considers the Joule–Thomson effect as an application of the molecular expression of B. When a real gas is passed through a porous plug into a low-pressure zone (throttling) under thermal insulation, the temperature changes in general (see Figure 7.8).

If the gas were ideal, there would be no change in the temperature. It is the interaction between molecules that causes the temperature to change, as the mean distance between molecules changes when the pressure changes. This phenomenon is known as the **Joule–Thomson effect**.

In the absence of heat exchange, the change in the internal energy, $U_2 - U_1$, is equal to the work done to the gas system. The pressure is constant at p_1 to the

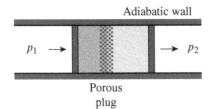

Adiabatic wall

$p_1 \rightarrow$ $\rightarrow p_2$

Porous
plug

Figure 7.8 Joule–Thomson effect. Throttling of gas in the left chamber at pressure p_1 through a porous plug into the right chamber at a lower pressure p_2. The wall is a thermal insulator. The temperature of the gas changes as a result of a change in the interaction between molecules.

left of the porous plug, and p_2 to the right of the plug ($p_1 > p_2$). The volume of the left chamber changes from V_1 to 0, and the one for the right chamber changes from 0 to V_2. Then, the Joule–Thomson process does the work of $p_1 V_1$ to the left chamber, and the gas does the work of $p_2 V_2$ to the surroundings. Therefore,

$$U_2 - U_1 = p_1 V_1 - p_2 V_2 \tag{7.62}$$

It means that the enthalpy $H = U + pV$ does not change. The Joule–Thomson coefficient μ_{JT} is defined as

$$\mu_{JT} \equiv \left(\frac{\partial T}{\partial p} \right)_H \tag{7.63}$$

Since p decreases in the process, a positive μ_{JT} indicates that the temperature drops.

Here, we use thermodynamics to convert this definition of μ_{JT} into one that does not involve H. We start with the definition of H:

$$dH = T\, dS + V\, dp \tag{7.64}$$

At a constant T,

$$\left(\frac{\partial H}{\partial p} \right)_T = T \left(\frac{\partial S}{\partial p} \right)_T + V \tag{7.65}$$

One of the Maxwell relationships is

$$\left(\frac{\partial S}{\partial p} \right)_T = -\left(\frac{\partial V}{\partial T} \right)_p \tag{7.66}$$

which we can derive from $dG = -S\, dT + V\, dp$. Recall the definition of the volume thermal expansion coefficient:

$$\alpha \equiv \frac{1}{V} \left(\frac{\partial V}{\partial T} \right)_p = \left(\frac{\partial \ln V}{\partial T} \right)_p \tag{7.67}$$

With Eqs. (7.66) and (7.67), Eq. (7.65) is now

$$\left(\frac{\partial H}{\partial p} \right)_T = V(1 - \alpha T) \tag{7.68}$$

Now, we write a chain rule for three variables, H, p, and T:

$$\left(\frac{\partial T}{\partial p}\right)_H \left(\frac{\partial p}{\partial H}\right)_T \left(\frac{\partial H}{\partial T}\right)_p = -1 \tag{7.69}$$

From Eqs. (7.63), (7.68) and (7.69), we obtain

$$\mu_{JT} = -\left(\frac{\partial H}{\partial p}\right)_T \left(\frac{\partial T}{\partial H}\right)_p = V(\alpha T - 1)\left(\frac{\partial T}{\partial H}\right)_p \tag{7.70}$$

We recall the definition of the heat capacity at constant pressure: $C_p = (\partial H/\partial T)_p$. Then, we arrive at

$$\mu_{JT} = \frac{V(\alpha T - 1)}{C_p} \tag{7.71}$$

In an ideal gas, $\alpha = 1/T$, and therefore $\mu_{JT} = 0$. Nonideality makes μ_{JT} nonzero. Since C_p of a real gas is not much different from that of an ideal gas, $\alpha T - 1$ is the main factor that determines μ_{JT}, especially its sign. Here, we find how $\alpha T - 1$ is related to the second virial coefficient $B(T)$.

Equation (7.1) expresses p as a function of V. We convert this expression into one that expresses V as a function of p. Follow the next transformation (B_2 is replaced by B).

$$V = \frac{Nk_B T}{p}\left(1 + \frac{N}{V}B(T)\right) \cong \frac{Nk_B T}{p}\left(1 + \frac{p}{k_B T}B(T)\right) \tag{7.72}$$

In the second equality, N/V was replaced with $p/(k_B T)$, since the zeroth-order approximation is sufficient. Differentiating Eq. (7.72) by T leads to

$$\left(\frac{\partial V}{\partial T}\right)_p = \frac{Nk_B}{p}\left(1 + \frac{p}{k_B}B'(T)\right) \tag{7.73}$$

With this equation, we can express α as

$$\alpha \cong \frac{1}{T}\frac{1 + \frac{p}{k_B}B'}{1 + \frac{p}{k_B T}B} \cong \frac{1}{T}\left[1 + \frac{p}{k_B}\left(B' - \frac{B}{T}\right)\right] \tag{7.74}$$

Therefore,

$$\alpha T - 1 = \frac{p}{k_B}\left(B' - \frac{B}{T}\right) \tag{7.75}$$

Now it is obvious that $\alpha T - 1$ is on the order of B. The C_p of an ideal gas is given by Eq. (5.16), and the correction to C_p by B is small. Therefore,

$$\mu_{JT} = \frac{V}{C_V + Nk_B} \times \frac{p}{k_B}\left(B' - \frac{B}{T}\right) \cong \frac{N(TB' - B)}{C_V + Nk_B} \tag{7.76}$$

Table 7.2 $TB' - B$ and the inversion temperature T_{inv} in the Joule–Thomson effect for four models of molecular interaction.

Model	$TB' - B$	T_{inv}
van der Walls equation of state	$\dfrac{1}{N_A}\left(\dfrac{2a}{RT} - b\right)$	$\dfrac{2a}{Rb}$
Hard-core repulsion	$-4v_s$	NA
Square-well potential	$4v_s\left\{\left[\left(\dfrac{r_a}{r_s}\right)^3 - 1\right]\left[\left(1 + \dfrac{\varepsilon_s}{k_BT}\right)\exp\left(\dfrac{\varepsilon}{k_BT}\right) - 1\right] - 1\right\}$	$\dfrac{2\varepsilon_s}{k_B}\left[\left(\dfrac{r_a}{r_s}\right)^3 - 1\right]$
Lennard-Jones potential	$\dfrac{2\pi}{3}r_0^3\displaystyle\sum_{n=0}^{\infty}\dfrac{2^n}{4n!}\dfrac{2n+5}{4}\Gamma\left(\dfrac{2n-1}{4}\right)\left(\dfrac{\varepsilon_0}{k_BT}\right)^{(2n+1)/4}$	$6.43\dfrac{\varepsilon_0}{k_B}$

The square-well potential employs the high-temperature approximation for T_{inv}.

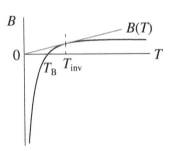

Figure 7.9 Graphical method to locate the inversion temperature T_{inv}. Draw a tangent to the curve of $B(T)$ from the origin, and the point of contact provides T_{inv}.

Here we look at how $TB' - B$ changes with T. Table 7.2 lists the dependence for four models of interacting gas molecules (van der Waals gas and the three models in Section 7.4). The temperature that vanishes μ_{JT} is called the inversion temperature T_{inv}. Except for the hard-core potential, each model has its own T_{inv}. At $T < T_{inv}$, $\mu_{JT} > 0$ in all of the models. The expansion into a lower pressure lowers the temperature. When the attractive interaction dominates, increasing the distances between molecules due to the expansion increases the interaction (attraction weakens). Table 7.2 also lists T_{inv} for the models except the hard-core repulsion model. The one for the square-well potential is a high-temperature approximation. The T_{inv} for the Lennard-Jones potential was obtained numerically.

It is possible to find T_{inv} graphically, if we have a plot of $B(T)$. Figure 7.9 illustrates the method: Draw a straight line through the origin that is tangent to the curve of $B(T)$. The point of contact is T_{inv}. Obviously, $T_{inv} > T_B$.

Figure 7.10 shows a plot of $(TB' - B)/(4v_s)$ as a function of temperature for the square-well potential with $r_a/r_s = 6^{1/3}$. For this potential, $k_BT_{inv}/\varepsilon_s \cong 10.6$. At $T < T_{inv}$, $\mu_{JT} > 0$. The expansion (p decreases) causes T to drop. At $T > T_{inv}$, the temperature rises.

Figure 7.10 $(TB' - B)/(4v_s)$ as a function of temperature, reduced by ε_s/k_B, in the square-well potential model with $(r_a/r_s)^3 - 1 = 5$. The inversion temperature T_{inv} is indicated by the arrow. Below $k_B T_{\text{inv}}/\varepsilon_s = 10.6$, $TB' - B > 0$ and $\mu_{JT} > 0$. The expansion causes a temperature drop.

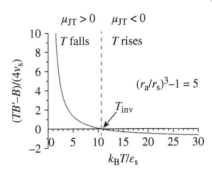

Problems

7.1 *Tertiary and quaternary interactions.* This problem considers the terms neglected in Eq. (7.32). The most significant term among them is

$$Z_{\text{conf,next}} \equiv \int_V d\mathbf{r}_1 \cdots \int_V d\mathbf{r}_N \sum_{(i,j)>(k,l)} u_{ij} u_{kl}$$

The summation consists of two type of products, $u_{ij} u_{ik}$ and $u_{ij} u_{kl}$, where none of i, j, k, and l are identical. We therefore divide $Z_{\text{conf, next}}$ into two parts:

$$Z_{\text{conf,next}} = Z_{\text{conf,3}} + Z_{\text{conf,4}}$$

where counting all possible combinations leads to

$$Z_{\text{conf,3}} \equiv \int_V d\mathbf{r}_1 \cdots \int_V d\mathbf{r}_N \sum_{i>j>k} (u_{ij} u_{ik} + u_{ik} u_{jk} + u_{ij} u_{jk})$$

$$Z_{\text{conf,4}} \equiv \int_V d\mathbf{r}_1 \cdots \int_V d\mathbf{r}_N \sum_{i>j>k>l} (u_{ij} u_{kl} + u_{ik} u_{jl} + u_{il} u_{jk})$$

(1) Show that $Z_{\text{conf,4}}/Z_{\text{conf,2}} \ll 1$, if $(N^2/V)|B| \ll 1$, where $Z_{\text{conf,2}} = N! Z_{\text{conf,nonideal}}$ in Eq. (7.39).

(2) Show that $Z_{\text{conf,3}}/Z_{\text{conf,2}} \ll 1$, if $(N^2/V)|B| \ll 1$. In evaluating the three-fold integral, assume, for simplicity, that u_{ij} is a constant at u in volume v and 0 elsewhere.

7.2 *Heat capacity and entropy.* For a system of N monatomic molecules in volume V with the second virial coefficient B,

(1) Calculate C_p and S up to the linear order of B.

(2) If $B = [b - a/(RT)]/N_A$, where a and b are constants of temperature, what are the nonideal parts of C_p and S? Also discuss the dependence of the corrections on a and b.

7.3 *Soft-core repulsion.* Let us consider a gas consisting of N indistinguishable monatomic molecules in volume V. They are "phantom" in the sense that overlap is allowed, although positive potential gives a penalty. The interaction Φ between a pair of molecules separated by r (center-to-center) has the potential shown here, where $\varepsilon > 0$.

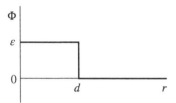

(1) Calculate the second virial coefficient B for the potential given in the figure and draw a sketch for a plot of B as a function of $\beta\varepsilon$.

(2) What is the internal energy? Answer up to the linear order of B.

(3) What is the heat capacity C_V? Note that ε is a constant of temperature. Draw a sketch for a plot of C_V as a function of temperature. Indicate the temperature that maximizes C_V.

(4) What is the state of the system at low temperatures? How about the system at high temperatures?

7.4 B_2 *of Lennard-Jones potential.* This problem derives Eq. (7.61) that analytically expresses the second virial coefficient of the Lennard-Jones potential (Eq. (3.47)) as a series.

(1) Use integral by parts to show

$$\frac{B}{8\pi\beta\varepsilon_0} = \int_0^\infty r^2 \left[\left(\frac{r_0}{r}\right)^{12} - \left(\frac{r_0}{r}\right)^6 \right] \exp\left(-\beta\varepsilon_0 \left[\left(\frac{r_0}{r}\right)^{12} - 2\left(\frac{r_0}{r}\right)^6 \right] \right) \mathrm{d}r$$

(2) Taylor expansion of $\exp(2\beta\varepsilon(r_0/r)^6)$ leads to

$$\frac{B}{8\pi\beta\varepsilon_0} = \sum_{n=0}^\infty \frac{(2\beta\varepsilon_0)^n}{n!} \int_0^\infty r^2 \left[\left(\frac{r_0}{r}\right)^{12} - \left(\frac{r_0}{r}\right)^6 \right] \left(\frac{r_0}{r}\right)^{6n}$$

$$\times \exp\left(-\beta\varepsilon_0 \left(\frac{r_0}{r}\right)^{12} \right) \mathrm{d}r$$

Replace the variable of integration from r to $t \equiv \beta\varepsilon(r_0/r)^{12}$ to express the integral using Gamma functions.

(3) Use the recurrence relationship of the Gamma function (Eq. (A.37)) to derive Eq. (7.61).

7.5 *Magnitude of nonideality.* We estimate how large or how small the nonideal part of U, U_{nonideal}, is compared with the ideal part, U_{ideal}. Let us choose water vapor at 300 K. For this vapor, $U_{\text{ideal}} \cong 3Nk_BT$.

(1) Express $U_{\text{nonideal}}/U_{\text{ideal}}$ using n/V (molar concentration), T, and a, b, and c in Eq. (7.10).

(2) Evaluate the ratio for the water vapor at 1 bar.

7.6 *Isothermal compressibility.* The isothermal compressibility β_T is defined as

$$\beta_T = -\frac{1}{V}\left(\frac{\partial V}{\partial p}\right)_T = -\left(\frac{\partial \ln V}{\partial p}\right)_T$$

(1) Equation (7.72) gives the volume as a function of pressure. For this gas, calculate β_T up to the linear order of B.

(2) Discuss the effect of the interaction on β_T.

7.7 *Open system of interacting particles.* The partition function of N indistinguishable and interacting molecules is approximately given as

$$Z_N = \frac{V^N}{N!}\left(1 - \frac{N^2}{V}B\right)Z_{1,\text{nc}}{}^N$$

where $B = B(T)$ is the second virial coefficient and $Z_{1,\text{nc}}$ is the nonconfiguration part of the single-molecule partition function.

(1) Show that the grand partition function is

$$\mathcal{Z} = \exp(e^{\beta\mu}VZ_{1,\text{nc}})[1 - Be^{\beta\mu}Z_{1,\text{nc}}(1 + e^{\beta\mu}VZ_{1,\text{nc}})]$$

up to the linear order of B.

(2) Find the coefficient of variation for the number of molecules up to the linear order of B.

(3) Obtain the equation of state up to the linear order of B.

7.8 *Temperature gradient.* Problem 6.12 shows that, across a porous plug separating two chambers at different temperatures, μ is shared. This conclusion can be extended to a system in which the temperature changes continuously.

Suppose we place a gas between a pair of parallel plates that are at temperatures T_1 and T_2 ($T_1 > T_2$). The temperature changes linearly with the distance x from the T_1 plate; see the given figure. It is expected that μ will be identical in the x direction.

(1) We know that the chemical potential for an ideal gas of monatomic molecules is given as

$$\mu = k_B T \left(\ln \frac{N}{V} - \frac{3}{2} \ln \frac{2\pi m}{\beta h^2} \right)$$

Draw a sketch for a plot of N/V as a function of x. Use a solid line.

(2) When interaction is present, μ is given as

$$\mu = k_B T \left(\ln \frac{N}{V} + 2B\frac{N}{V} - \frac{3}{2} \ln \frac{2\pi m}{\beta h^2} \right)$$

where B is the second virial coefficient. Assume that B is a positive constant of temperature to draw a sketch for a plot of N/V as a function of x in the same chart as the one you prepared in (1). Use a dashed line.

8

Rubber Elasticity

In this chapter, we learn how to model rubber elasticity that allows a large deformation. Before looking at a model for the rubber elasticity, we first learn how elastic a polymer chain is. The simplest model is a polymer chain in one dimension (Section 8.2). We apply the canonical-ensemble statistical mechanics to consider the 1D polymer chain and find the relationship of the chain's extension to the force applied. Then, we skip two dimensions, and proceed to consider a polymer chain in three dimensions in Section 8.3. We then briefly look at a model for the rubber elasticity.

8.1 Rubber

Rubber is a special class of an elastic material. All solids and glasses are elastic materials, but rubbers stand out, since they allow large deformations while still returning to the original dimension when the pulling force is released. Figure 8.1 shows a typical relationship between stress σ and strain ε for steel and rubber within the elasticity limit, i.e. in the absence of plastic deformation. There are different types of steel and different types of rubber (soft and hard). The figure shows a typical curve for steel and rubber, respectively.

The strain of rubber at its elasticity limit is large, often several hundred percent, but it can be reached with a weak stress. The slope of the stress–strain curve within the linearity limit is called the elastic modulus. Rubber has a small elastic modulus. In contract, the modulus is large for steel, and the strain at its elastic limit is rather small, typically around 0.1%.

Rubber, also called elastomer, is made usually by covalently cross-linking a polymer melt – a liquid state of the polymer. The cross-linking leads to a three-dimensional network.

Statistical Thermodynamics: Basics and Applications to Chemical Systems, First Edition. Iwao Teraoka.
© 2019 John Wiley & Sons, Inc. Published 2019 by John Wiley & Sons, Inc.
Companion website: www.wiley.com/go/Teraoka_StatsThermodynamics

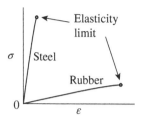

Figure 8.1 Typical relationship between stress σ and strain ϵ for steel and rubber within the elasticity limit.

8.2 Polymer Chain in One Dimension

Consider a one-dimensional (1D) chain of N bonds, each with length b. The direction of each bond is either vertically up or down. We use s_i to denote the state of the ith bond ($i = 1, 2, ..., N$): $s_i = +1$ when the bond is vertically down; -1 when vertically up (see Table 8.1).

Figure 8.2a shows an example for a 1D chain of 21 bonds. The state of the whole chain is specified by $\{s_i\}_{i=1, 2, ..., 21}$, as shown in Figure 8.2b.

Table 8.1 Variable s_i to specify the state of the ith bond.

Bond orientation	s_i
Vertically down	1
Vertically up	−1

There are two states.

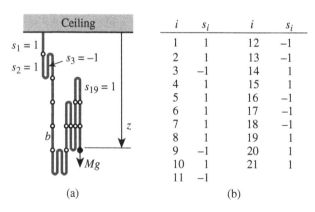

i	s_i	i	s_i
1	1	12	−1
2	1	13	−1
3	−1	14	1
4	1	15	1
5	1	16	−1
6	1	17	−1
7	1	18	−1
8	1	19	1
9	−1	20	1
10	1	21	1
11	−1		

(a)   (b)

Figure 8.2 (a) Model of a polymer chain in one dimension. N bonds of length b are concatenated. In the example shown, $N = 21$. The orientation of the ith bond is specified by s_i; 1 for downward and −1 for upward. The first bond is attached to the ceiling, and the last bond is to mass M. (b) List of i and s_i for the example of chain conformation shown in (a).

The system we consider here is a single chain of N bonds, and the state of the system is specified by $s_1, s_2, \ldots, s_N$. There are a total 2^N states. We call the model a **freely jointed chain**, as the probability for s_i to be 1 is equal to the one for -1 in the absence of force.

Now we attach one of the chain ends to the ceiling at $z = 0$, where positive z is taken to be vertically down. The ceiling is phantom, and a part of the chain can be above the ceiling without obstruction. The position z of the other end of the chain is given as

$$z = b \sum_{i=1}^{N} s_i \tag{8.1}$$

We want to find whether the chain behaves like a spring. To find a relationship between the force and the extension of the chain, we attach mass M at the other end of the chain. Then, the potential energy E of the load is

$$E = -Mgz = -Mgb \sum_{i=1}^{N} s_i \tag{8.2}$$

We assume, for simplicity, that the energy of the system is this potential energy only. Then, the partition function is given as

$$Z = \sum_{s_1=\pm 1} \cdots \sum_{s_N=\pm 1} \exp\left(\beta Mgb \sum_{i=1}^{N} s_i \right) \tag{8.3}$$

The N-fold sum can be rewritten to the product of N sums, and since all the sums are identical,

$$Z = \sum_{s_1=\pm 1} \exp(\beta Mgb s_1) \cdots \sum_{s_N=\pm 1} \exp(\beta Mgb s_N) = \left(\sum_{s=\pm 1} e^{\beta Mgb s} \right)^N$$
$$= [2 \cosh(\beta Mgb)]^N \tag{8.4}$$

Obviously, the partition function represents a system of N distinguishable particles, each capable of being at state 1 or -1. The probability of the two states are $e^{\beta Mgb}$ and $e^{-\beta Mgb}$, respectively (unnormalized). The distinguishability is due to the address of each bond: the first bond, the second bond, and so on.

As usual, we take the logarithm of Z:

$$\ln Z = N \ln 2 + N \ln[\cosh (\beta Mgb)] \tag{8.5}$$

Since $M > 0$, we expect that the mean of z to be positive. The mean is defined as

$$\langle z \rangle = \frac{1}{Z} \sum_{s_1=\pm 1} \cdots \sum_{s_N=\pm 1} \left(b \sum_{i=1}^{N} s_i \right) \exp\left(\beta Mgb \sum_{i=1}^{N} s_i \right) \tag{8.6}$$

which is calculated as

$$\langle z \rangle = \frac{1}{Mg} \frac{\partial \ln Z}{\partial \beta} \tag{8.7}$$

It can also be calculated by differentiating $\ln Z$ by β, followed by division by Mg. Yet another way is to use $\langle E \rangle = -\partial \ln Z / \partial \beta$ and $\langle E \rangle = -Mg\langle z \rangle$. With Eq. (8.5), $\langle z \rangle$ is calculated as

$$\langle z \rangle = Nb \tanh(\beta Mgb) \tag{8.8}$$

When $\beta Mgb \ll 1$,

$$\langle z \rangle = \frac{Nb^2 Mg}{k_B T} \tag{8.9}$$

which is rewritten to

$$Mg = \kappa \langle z \rangle \tag{8.10}$$

where

$$\kappa = \frac{k_B T}{Nb^2} \tag{8.11}$$

It means that the freely jointed chain of N bonds is equivalent to a spring of force constant κ given by this equation. The higher the temperature, the stronger the spring. Having more bonds or longer bonds makes the spring weak.

The origin of the elasticity is in the availability of different states specified by $s_1, s_2, \ldots, s_N$. The different states do not involve interactions, and therefore the elasticity is purely entropic (**entropy elasticity**).

In the absence of M, $\langle z \rangle = 0$. The probability of z being a specific value, say, z_1 (>0), is equal to the probability for $z = -z_1$, and therefore, $\langle z \rangle = 0$. In fact, z is distributed with a binomial distribution with mean $= 0$.

Although $\langle z \rangle$ is 0, $\langle z^2 \rangle$ is not. We can use Eq. (8.3) to evaluate $\langle z^2 \rangle$ as shown below. We start with deriving a formula for calculating $\langle z^2 \rangle$.

$$\langle z^2 \rangle = \frac{1}{Z} \sum_{s_1 = \pm 1} \cdots \sum_{s_N = \pm 1} \left(b \sum_{i=1}^{N} s_i \right)^2 \exp\left(\beta Mgb \sum_{i=1}^{N} s_i \right) = \frac{1}{Z} \frac{1}{(Mg)^2} \frac{\partial^2 Z}{\partial \beta^2} \tag{8.12}$$

Since $\partial Z / \partial \beta = MgZ\langle z \rangle$ (Eq. (8.7))

$$\frac{\partial^2 Z}{\partial \beta^2} = Mg \left(\frac{\partial Z}{\partial \beta} \langle z \rangle + Z \frac{\partial \langle z \rangle}{\partial \beta} \right) = MgZ \left(Mg\langle z \rangle^2 + \frac{\partial \langle z \rangle}{\partial \beta} \right) \tag{8.13}$$

Therefore,

$$\langle \Delta z^2 \rangle = \langle z^2 \rangle - \langle z \rangle^2 = \frac{1}{Mg} \frac{\partial \langle z \rangle}{\partial \beta} \tag{8.14}$$

Alternatively, we can derive this equation from $\langle \Delta E^2 \rangle = -\partial \langle E \rangle / \partial \beta$ and $E = -Mgz$.

The root-mean-square (rms) of z, $\langle z^2 \rangle^{1/2}$, which is equal to $\langle \Delta z^2 \rangle^{1/2}$ since $\langle z \rangle = 0$, is a typical distance between the two ends of the chain. It is also a good measure for the dimension of the 1D chain. Calculation of $\langle \Delta z^2 \rangle^{1/2}$ with Eq. (8.14) is easy. From Eq. (8.8), we get

$$\frac{\partial \langle z \rangle}{\partial \beta} = NMgb^2 \text{sech}^2(\beta Mgb) \tag{8.15}$$

Therefore,

$$\langle \Delta z^2 \rangle = Nb^2 \text{sech}^2(\beta Mgb) \tag{8.16}$$

Then,

$$\langle z^2 \rangle = Nb^2 \text{sech}^2(\beta Mgb) + N^2 b^2 \tanh^2(\beta Mgb) \tag{8.17}$$

When $\beta Mgb \ll 1$,

$$\langle z^2 \rangle_0 = Nb^2 \tag{8.18}$$

The subscript 0 indicates a mean in the unperturbed state, i.e. in the absence of applied force ($M = 0$). The rms of the end-to-end distance is $bN^{1/2}$ for a chain in one dimension.

We can rewrite Eq. (8.18) into

$$\frac{1}{2}\kappa \langle z^2 \rangle_0 = \frac{1}{2} k_B T \tag{8.19}$$

It means that the mean energy of the spring is equal to $\frac{1}{2} k_B T$. Recall that a 1D harmonic oscillator (classical) has the energy of $k_B T$, and the latter is equally divided into the elastic energy and the kinetic energy. The freely jointed chain is static and does not involve movement. There is no kinetic energy.

It is interesting to examine the linearity of the extension–force relationship in terms of a dimensionless extension X defined as

$$X \equiv \frac{\langle z \rangle}{bN^{1/2}} \tag{8.20}$$

We also introduce a dimensionless force F by

$$F \equiv \frac{MgbN^{1/2}}{k_B T} \tag{8.21}$$

The two dimensionless quantities are related to each other by (see Eq. 8.8)

$$F = \frac{1}{2} N^{1/2} \ln \frac{1 + N^{-1/2} X}{1 - N^{-1/2} X} \tag{8.22}$$

Three curves in Figure 8.3 show how the reduced force changes with the reduced extension for three chain lengths, $N = 10$, 30, and 100. When $X < 1$,

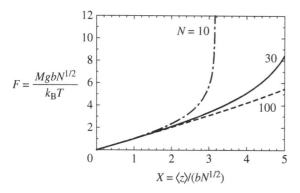

$$F = \frac{MgbN^{1/2}}{k_B T}$$

$$X = \langle z \rangle / (bN^{1/2})$$

Figure 8.3 Reduced force $F = MgbN^{1/2}/(k_B T)$, plotted as a function of reduced chain extension $X = \langle z \rangle/(bN^{1/2})$ for 1D chains of $N = 10$, 30, and 100.

the three curves are overlapped and run close to a straight line, $F = X$, which is equivalent to $Mg = \kappa \langle z \rangle$. For a chain with $N = 10$, the curve deviates upward at around $\langle z \rangle$ twice as large as $bN^{1/2}$. The deviation starts at a greater $\langle z \rangle$ for a chain with $N = 30$. The curve for $N = 100$ is nearly straight for the whole range of $\langle z \rangle/(bN^{1/2})$ shown. The greater the number of bonds, the broader the linear range.

The curve for $N = 10$ rises sharply at $\langle z \rangle/(bN^{1/2}) = 10^{1/2}$. At this chain length, the chain is fully stretched. The latter quantity is $N^{1/2}$ for a chain of length N. As long as the chain is not pulled beyond this length, the chain will return to the natural dimension, as soon as the pulling force vanishes. For a chain with $N = 100$, pulling the chain to five times its natural dimension still holds it in a nearly linear elasticity regime.

8.3 Polymer Chain in Three Dimensions

We now turn to a freely jointed chain of N bonds in three dimensions. Unlike the 1D chain, each bond can orient in any direction in the 3D space. The direction of the ith bond is specified by two variables, θ_i and ϕ_i. The angle the bond forms with the $+z$ direction is θ_i, and ϕ_i is the angle that the projection of the bond onto a horizontal plane forms with a given direction within the plane (see Figure 8.4). The system of the N bonds is described by (θ_1, ϕ_1), ..., and (θ_N, ϕ_N).

The range for θ_i is from 0 to π, and ϕ_i can be anywhere between 0 and 2π. As you may know, (θ_i, ϕ_i) is the angular part of the spherical polar coordinates. Unless something is attached to the bond, θ_i is distributed with a weight of sin θ_i, and there is no weight in the distribution of ϕ_i.

As we did for the 1D chain, we place one of the ends of the 3D chain on the ceiling at $z = 0$ and attach a mass of M at the other end. The ceiling is again phantom. We let all the bonds be oriented with distributions of (θ_i, ϕ_i) determined

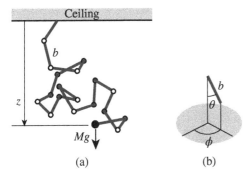

Figure 8.4 (a) Model of a polymer chain in three dimensions. N bonds of length b are concatenated. One of the chain ends is attached to the ceiling at $z = 0$, and mass M is attached to the other end at z. The bond angle is unrestricted, and the dihedral angle is also free, i.e. each bond can freely rotate relative to the preceding bond. (b) The orientation of each bond is specified by the polar angle θ (angle from the z direction) and the azimuthal angle ϕ.

by the canonical ensemble. The position of the other end is given as

$$z = b \sum_{i=1}^{N} \cos \theta_i \tag{8.23}$$

and therefore the potential energy of the system is

$$E = -Mgb \sum_{i=1}^{N} \cos \theta_i \tag{8.24}$$

The partition function of the system is

$$Z = \int_0^\pi \sin \theta_1 d\theta_1 \int_0^{2\pi} d\phi_1 \cdots \int_0^\pi \sin \theta_N d\theta_N \int_0^{2\pi} d\phi_N$$

$$\exp\left(\beta Mgb \sum_{i=1}^{N} \cos \theta_i \right) \tag{8.25}$$

which is calculated as follows:

$$Z = \int_0^\pi \sin \theta_1 d\theta_1 \int_0^{2\pi} d\phi_1 \exp(\beta Mgb \cos \theta_1)$$

$$\times \cdots \times \int_0^\pi \sin \theta_N d\theta_N \int_0^{2\pi} d\phi_N \exp(\beta Mgb \cos \theta_N)$$

$$= \left[\int_0^\pi \sin \theta d\theta \int_0^{2\pi} d\phi e^{\beta Mgb \cos \theta} \right]^N = \left[4\pi \frac{\sinh(\beta Mgb)}{\beta Mgb} \right]^N \tag{8.26}$$

We take the logarithm:

$$\ln Z = N \left[\ln 4\pi + \ln \frac{\sinh(\beta M g b)}{\beta M g b} \right] \tag{8.27}$$

The mean of z is calculated according to

$$\langle z \rangle = \frac{1}{Z} \int_0^\pi \sin \theta_1 d\theta_1 \int_0^{2\pi} d\phi_1 \cdots \int_0^\pi \sin \theta_N d\theta_N \int_0^{2\pi} d\phi_N$$

$$\times \left(b \sum_{i=1}^N \cos \theta_i \right) \exp \left(\beta M g b \sum_{i=1}^N \cos \theta_i \right) = \frac{1}{Mg} \frac{\partial \ln Z}{\partial \beta} \tag{8.28}$$

It is calculated as

$$\langle z \rangle = bN \left[\coth(\beta M g b) - \frac{1}{\beta M g b} \right] \tag{8.29}$$

The function that appears here is called a **Langevin function** (Eq. (A.19)):

$$\langle z \rangle = bNL(\beta M g b) \tag{8.30}$$

When $x \ll 1$, $L(x) = \frac{1}{3}x$ (Eq. (A.1.21)). Therefore, when $\beta M g b \ll 1$,

$$\langle z \rangle \cong \frac{1}{3} bN\beta M g b = \frac{Nb^2}{3k_B T} Mg \tag{8.31}$$

It means that the 3D chain of N bonds is, in practice, a spring with a force constant

$$\kappa = \frac{3k_B T}{Nb^2} \tag{8.32}$$

The coefficient is 3 in this 3D model, as opposed to 1 for the 1D chain; see Eq. (8.11). Obviously, the coefficient is equal to the dimensionality. We can show that the coefficient is 2 for a chain in two dimensions; see Problem 8.8.

Using the same method as the one adopted for the 1D chain, we can calculate $\langle \Delta z^2 \rangle$. Since

$$\frac{\partial \langle z \rangle}{\partial \beta} = bNMgbL'(\beta M g b) = Nb^2 Mg \left[\frac{1}{(\beta M g b)^2} - \frac{1}{\sinh^2(\beta M g b)} \right] \tag{8.33}$$

use of Eq. (8.14) (or its counterpart for the 3D chain) leads to

$$\langle \Delta z^2 \rangle = Nb^2 \left[\frac{1}{(\beta M g b)^2} - \frac{1}{\sinh^2(\beta M g b)} \right] \tag{8.34}$$

With Eq. (8.29),

$$\langle z^2 \rangle = Nb^2 \left[\frac{1}{(\beta M g b)^2} - \frac{1}{\sinh^2(\beta M g b)} \right] + N^2 b^2 \left[\coth(\beta M g b) - \frac{1}{\beta M g b} \right]^2 \tag{8.35}$$

When $\beta Mgb \ll 1$,

$$\langle z^2 \rangle_0 = \frac{1}{3} N b^2 \tag{8.36}$$

The square of the unperturbed dimension is a third of the one for a 1D chain. The square of the dimension is the same in x and y directions. Adding the three components, we obtain

$$\langle \mathbf{r}^2 \rangle_0 = \langle x^2 \rangle_0 + \langle y^2 \rangle_0 + \langle z^2 \rangle_0 = N b^2 \tag{8.37}$$

The rms of the end-to-end distance is $b N^{1/2}$. The mean energy of the spring is $\frac{1}{2}\kappa \langle \mathbf{r}^2 \rangle_0 = \frac{3}{2} k_B T$. The energy per degree of freedom is $\frac{1}{2} k_B T$, identical to the value for the 1D chain.

As we did for the 1D chain, we look at the force–extension relationship for reduced quantities. The dimensionless extension X is now

$$X \equiv \frac{\langle z \rangle}{b(N/3)^{1/2}} \tag{8.38}$$

Since $\langle z^2 \rangle_0$ is now different, the dimensionless force F is also different from the one in Eq. (8.21):

$$F \equiv \frac{Mgb(N/3)^{1/2}}{k_B T} \tag{8.39}$$

The relationship between X and F is

$$F = (N/3)^{1/2} L^{-1}(X/(3N)^{1/2}) \tag{8.40}$$

where $L^{-1}(x)$ is the inverse function of the Langevin function. Unfortunately, $L^{-1}(x)$ cannot be expressed with elementary functions. However, we can still display the relationship (see Figure 8.5).

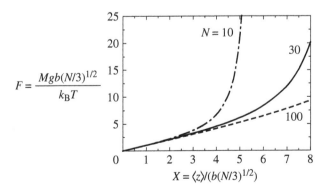

Figure 8.5 Reduced force $F = Mgb(N/3)^{1/2}/(k_B T)$, plotted as a function of reduced chain extension $X = \langle z \rangle/(b(N/3)^{1/2})$, for 3D chains of $N = 10$, 30, and 100.

8.4 Network of Springs

Rubber is a network of cross-linked polymer. The polymer before the cross-linking must have a series of functional groups that can cross-link. In poly(*cis* isoprene) and polybutadiene, it is a C=C bond in every repeat unit. These polymers are highly viscous liquid. Adding sulfur to the viscous liquid of polyisoprene results in natural rubber (see Figure 8.6). This process, called vulcanization, was discovered in the mid-nineteenth century and was one of the few instrumental developments in the industrialization.

In Figure 8.6, the part of a polymer chain between a pair of adjacent cross-links along the chain is not constrained otherwise. We call the part a partial chain. There are many other polymer chains nearby, but they are not covalently bonded to the partial chain. As we learned in Sections 8.2 and 8.3, the partial chain can be modeled as a spring.

Until the cross-linking reaction occurs, polymer chains can move through the matrix of other chains, although it is slow. In fact, the most likely place for one of the two ends of a given partial chain to be is where the other end is. The most probable distance between the two ends is not 0, but is close to the rms distance $\langle \mathbf{r}^2 \rangle_0^{1/2} = bN^{1/2}$. When vulcanized, this distance is frozen in, and many cross-links along any polymer chain lead to forming a network.

Recall that the chain remains elastic and does not undergo plastic deformation when it is pulled to extend over many times the rms distance. Furthermore, the partial chain will retain the same spring force constant when subjected to a large strain.

Estimating the elastic modulus of a rubber requires mechanical analysis of connected springs. Different models are possible, and one of them is a network of springs connected in a 3D grid, shown in Figure 8.7. We do not get into the details here.

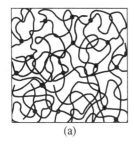
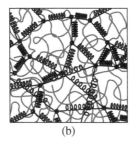

(a)   (b)

Figure 8.6 (a) Rubber network formed by cross-linking molten polymer chains. Each filled circle represents a cross-linking, for example, by S–S or S–S–S, where S is a sulfur atom. (b) A partial chain between a pair of adjacent cross-links along a chain can be modeled by a spring.

Figure 8.7 Two-dimensional representation of a network of springs.

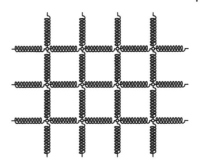

For the rubber to be elastic, it is necessary that the polymer is in a molten state. If the polymer is in a glassy state, i.e. below its glass transition temperature, the repeat units are not mobile, thus severely restricting the elasticity. In short, a rubber is a liquid state of cross-linked polymer.

The cross-links are usually covalent bonding, but there is a different type of cross-links, called physical cross-links. One of the physically cross-linked rubber consists of styrene–butadiene–styrene block copolymer. The styrene part is phase-separated from the butadiene part, and each styrene domain is a cross-link. Unlike chemically cross-linked rubber, this type of rubber may be molded by raising the temperature above the glass transition temperature of polystyrene (thermoplastic elastomer).

Gels are also cross-linked polymer chains. What makes the gel different from the rubber is the presence of solvent molecules. The gel does not dissolve in the solvent, but swells. In contrast, rubber does not have solvent molecules, and the space is filled with repeat units of the polymer.

Gels are also elastic, but their elastic moduli are orders of magnitude as small as the those for the rubbers. As for the rubber, there are chemically cross-linked gels and physically cross-linked gels. Gelatin is an example of the latter.

Problems

8.1 *Dimension of a 1D chain.* Both $\langle z \rangle$ and $\langle \Delta z^2 \rangle$ of a freely jointed chain in one dimension depend on Mg. Compare the dependence by drawing a sketch for the plots of $\langle z \rangle/(Nb)$ and $\langle \Delta z^2 \rangle^{1/2}/(N^{1/2}b)$ as a function of βMgb. Discuss the difference.

8.2 *Entropy of a 1D chain.* Calculate the entropy S of the 1D freely jointed chain. Draw a sketch of $S/(Nk_B)$ as a function of βMgb and discuss the result.

8.3 *1D chain of elongated and shrunk elements.* Consider a one-dimensional chain of N elements that can be either elongated (state e) or shrunk (state s); see the given figure. The length l_i of the ith element is l_e and l_s for the two states. When pulled by tension τ, the energy ε_i of the ith element is $-\tau l_i$, i.e., $-\tau l_e$ for state e and $-\tau l_s$ for state s. The illustration shows an example of such a chain.

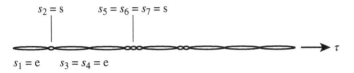

(1) Calculate the partition function Z.
(2) What is the mean of the length L of the chain?
(3) Show that Hooke's law holds when τ is weak. What is the spring force constant κ?
(4) Calculate $\langle \Delta L^2 \rangle$.
(5) Show that $\frac{1}{2}\kappa\langle \Delta L^2 \rangle = \frac{1}{2}k_B T$.

8.4 *1D chain of continuously variable length.* The 1D chain model we considered in Section 8.2 consists of bonds that can orient either upward or downward. We consider another model for a 1D chain that consists of N bonds with a continuously variable length. The ith bond is oriented always downward, but its length l_i can be anywhere between 0 and l. As we did earlier, we attach the first bond to the ceiling and mass M at the end.

(1) Calculate the partition function Z and express the mean for the position z of the chain end from the ceiling as a function of Mg.
(2) Draw a sketch for the plot of $\langle z \rangle/(Nl)$ as a function of βMgl.
(3) Show that the chain follows Hooke's law for a small mass (or at high temperatures) and evaluate the spring force constant κ.
(4) Calculate $\langle \Delta z^2 \rangle$.
(5) Show that $\frac{1}{2}\kappa\langle \Delta z^2 \rangle = \frac{1}{2}k_B T$.

8.5 *Mean and variance of $\cos\theta$ in a 3D chain.* Calculate $\langle \cos\theta_i \rangle$ and $\langle \cos^2\theta_i \rangle$ for a 3D freely jointed chain.

8.6 *Horizontal dimension of a 3D chain.* Calculate $\langle x \rangle$ and $\langle x^2 \rangle$ for a 3D freely jointed chain, where x is the x component of the distance between the two chain ends. The x direction is taken in the horizontal plane, and

$$x = b \sum_{i=1}^{N} \sin\theta_i \cos\phi_i$$

Also find how $\langle x^2 \rangle$ depend on βMgb when $\beta Mgb \ll 1$.

8.7 *Entropy of a 3D chain.* Calculate the entropy S of the 3D freely jointed chain. Draw a sketch of $S/(Nk_B)$ as a function of βMgb. What is the large-βMgb asymptote of $S/(Nk_B)$?

8.8 *2D chain.* In this chapter, we did not consider a freely jointed chain in two dimensions, but it is for a reason: The exact partition function Z cannot be expressed by elementary functions. However, we can obtain approximate expressions of Z in two asymptotes.

As we did for the 1D and 3D chains, we hang one of the chain ends from the ceiling and attach mass M at the end. The orientation of each bond is now restricted within a vertical plane that contains the chain end at the ceiling (note that the ceiling is just a line). The angle θ_i of the ith bond is defined in the same way.

(1) Show that the partition function of the chain is given as

$$Z = \left(\int_0^\pi e^{\beta Mgb \cos\theta} \, d\theta \right)^N$$

(2) We first consider a high-temperature approximation.
(2a) Calculate the mean of the energy of the chain up to $O((\beta Mgb)^2)$.
(2b) Find the force constant κ of the 2D chain.
(2c) What is $\frac{1}{2}\kappa\langle \Delta z^2 \rangle$?
(3) We then consider a low-temperature approximation. At low temperatures, $\beta Mgb \gg 1$, and the integral in (1) is contributed mostly from the part in which θ is close to 0. In that neighborhood, $\cos\theta = 1 - \frac{1}{2}\theta^2$, and the upper limit of the integral can be brought to $+\infty$:
(3a) Use this approximation to calculate the partition function of the 2D chain.
(3b) What is $\langle z \rangle$?
(4) Combine the results of the two approximations to draw a sketch for the plot of $\langle z \rangle$ as a function of βMgb.

9

Law of Mass Action

In the physical chemistry course, the law of mass action was derived by equating the free energy of the reactants with the one for the products. For a reaction of A and B to produce C and D with stoichiometric coefficients, a, b, c, and d, respectively,

$$a\text{A} + b\text{B} \rightleftarrows c\,\text{C} + d\,\text{D}$$

the law of mass action gives the relationship between the partial pressures of the four components, p_A, p_B, p_C, and p_D. The standard equilibrium constant K_p° is given as

$$K_p^\circ = \frac{(p_C/p^\circ)^c (p_D/p^\circ)^d}{(p_A/p^\circ)^a (p_B/p^\circ)^b}$$

where p° (typically, 1 bar or 1 atm) removes all the units.

In this chapter, we follow the method of Hückel to derive the law of mass action. The successful derivation is considered as one of the most triumphant accomplishments in the statistical thermodynamics applied to chemistry. In our derivation, we find what makes it possible for the reactants and products to follow the law of mass action, and what condition is required.

In Sections 9.3 and 9.4, we look at some simple reactions. In Section 9.3, we go back to the isomerization we looked at in Section 5.5 to uncover an implicit approximation employed in the method we learn in the first two sections. We prove the legitimacy of the method in Section 9.4 where we learn the method of the steepest descent.

In this chapter, N_A is not the Avogadro's number, but rather denotes the number of A molecules.

Statistical Thermodynamics: Basics and Applications to Chemical Systems, First Edition. Iwao Teraoka.
© 2019 John Wiley & Sons, Inc. Published 2019 by John Wiley & Sons, Inc.
Companion website: www.wiley.com/go/Teraoka_StatsThermodynamics

9.1 Reaction of Two Monatomic Molecules

Consider a reaction in which a monatomic molecule A and a monatomic molecule B react to form a diatomic molecule AB:

$$A + B \rightleftarrows AB \qquad (9.1)$$

The reaction occurs in a container of volume V. The numbers of the molecules and the single-molecule partition functions are listed for A, B, and AB in Table 9.1.

The system is a mixture of three gases, consisting of N_A molecules of A, N_B molecules of B, and N_{AB} molecules of AB. Its partition function Z is given as

$$Z = \frac{Z_{1A}{}^{N_A} Z_{1B}{}^{N_B} Z_{1AB}{}^{N_{AB}}}{N_A! N_B! N_{AB}!} \qquad (9.2)$$

In this expression, reactants and products are not distinguished; the system is a mixture of three gases. Do not write Z_{1A} in the denominator. The division by $N_A! N_B! N_{AB}!$ is because the molecules of A are indistinguishable within themselves, and so are the molecules of B and the molecules of AB. However, an A molecule is distinguishable from a B molecule and an AB molecule.

Note that the stoichiometric coefficients do not appear in the expression of Z. Since all of the stoichiometric coefficients are equal to one in this reaction, it may not be obvious; but later when we look at other examples of reaction with stoichiometric coefficients other than one, the absence of stoichiometric coefficients in the expression of Z will become apparent.

We take the logarithm of Eq. (9.2):

$$\ln Z = N_A(\ln Z_{1A} - \ln N_A + 1) + N_B(\ln Z_{1B} - \ln N_B + 1)$$
$$+ N_{AB}(\ln Z_{1AB} - \ln N_{AB} + 1) \qquad (9.3)$$

The stoichiometry enters as the condition that the total number of A atoms not change in the reaction, and so does the total number of B atoms. Since

$$\text{Number of A atoms} = N_A + N_{AB} \qquad (9.4)$$
$$\text{Number of B atoms} = N_B + N_{AB} \qquad (9.5)$$

Table 9.1 Number of molecules and the single-molecule partition function for each of the three components in the reaction, $A + B \leftrightarrow AB$.

Component	Number of molecules	Single-molecule partition function
A	N_A	Z_{1A}
B	N_B	Z_{1B}
AB	N_{AB}	Z_{1AB}

N_A, N_B, and N_{AB} are not independent. Rather, they change according to

$$\frac{\partial N_A}{\partial N_{AB}} = -1, \frac{\partial N_B}{\partial N_{AB}} = -1 \tag{9.6}$$

As one AB molecule is produced, one molecule of A and one molecule of B are consumed.

The equilibrium in the system is reached when the Helmholtz free energy F minimizes. The necessary condition is given as

$$\frac{\partial}{\partial N_{AB}}\left(-\frac{F}{k_B T}\right) = \frac{\partial \ln Z}{\partial N_{AB}} = 0 \tag{9.7}$$

With Eq. (9.3),

$$\frac{\partial}{\partial N_{AB}}\left(-\frac{F}{k_B T}\right) = \frac{\partial N_A}{\partial N_{AB}} \frac{\partial}{\partial N_A} N_A (\ln Z_{1A} - \ln N_A + 1)$$

$$+ \frac{\partial N_B}{\partial N_{AB}} \frac{\partial}{\partial N_B} N_B (\ln Z_{1B} - \ln N_B + 1) + \frac{\partial}{\partial N_{AB}} N_{AB}(\ln Z_{1AB} - \ln N_{AB} + 1) \tag{9.8}$$

which is calculated as

$$\frac{\partial}{\partial N_{AB}}\left(-\frac{F}{k_B T}\right) = -(\ln Z_{1A} - \ln N_A) - (\ln Z_{1B} - \ln N_B)$$

$$+ (\ln Z_{1AB} - \ln N_{AB}) \tag{9.9}$$

The minimization condition is obtained by equating Eq. (9.9) to 0:

$$\frac{N_{AB}}{N_A N_B} = \frac{Z_{1AB}}{Z_{1A} Z_{1B}} \tag{9.10}$$

Problem 9.2 confirms that Eq. (9.10) is sufficient for the minimization of F.

The left-hand side of Eq. (9.10) defines the equilibrium constant for the number of molecules in the system, as dictated by the law of mass action. We denote the ratio K_N. The equilibrium constant is equal to the corresponding ratio of single-molecule partition functions:

$$K_N \equiv \frac{N_{AB}}{N_A N_B} = \frac{Z_{1AB}}{Z_{1A} Z_{1B}} \tag{9.11}$$

To derive Eq. (9.11), we assumed indistinguishability. If we assume that all of the molecules are distinguishable, minimizing the Helmholtz free energy leads to a wrong conclusion (see Problem 9.1).

It is easy to convert Eq. (9.11) into an equation that relates the partial pressures of A, B, and AB. The equilibrium constant K_p can be calculated from K_N as

$$K_p = \frac{p_{AB}}{p_A p_B} = \frac{N_{AB}}{N_A N_B} \frac{V}{k_B T} = \frac{Z_{1AB}}{Z_{1A} Z_{1B}} \frac{V}{k_B T} \tag{9.12}$$

Now, we look at the chemical potentials of A, B, and AB. The chemical potential μ_A of an A molecule is calculated from $\ln Z$ as

$$\mu_A = -k_B T \frac{\partial \ln Z}{\partial N_A} = -k_B T(\ln Z_{1A} - \ln N_A) \tag{9.13}$$

Thus,

$$Z_{1A} = N_A \exp(-\mu_A/k_B T) \tag{9.14}$$

Similar expressions are obtained for Z_{1B} and Z_{1AB}. Then, Eq. (9.11) transforms into

$$\mu_A + \mu_B = \mu_{AB} \tag{9.15}$$

The sum of the chemical potentials remains unchanged in the forward and reverse reactions. You may have learned it already in a thermodynamics course.

In Chapters 4–6, we derived the partition functions for monatomic molecules and heteronuclear diatomic molecules. We can use them to have an explicit expression for K_N. Given here is such an attempt.

A and B are monatomic, and AB is diatomic. Therefore,

$$Z_{1A} = V \left(\frac{2\pi m_A k_B T}{h^2} \right)^{3/2} \tag{9.16}$$

$$Z_{1B} = V \left(\frac{2\pi m_B k_B T}{h^2} \right)^{3/2} \tag{9.17}$$

$$Z_{1AB} = V \left[\frac{2\pi(m_A + m_B) k_B T}{h^2} \right]^{3/2} Z_{1AB,\text{rot}} Z_{1AB,\text{vib}} \exp(-\beta \varepsilon_{AB}) \tag{9.18}$$

where m_i is the mass of atom i (i = A, B), and ε_{AB} is the electronic energy of the AB diatomic molecule relative to the energy of the dissociated state (an A atom and a B atom).

Substituting Z_{1A}, Z_{1B}, and Z_{1AB} in Eq. (9.11) with Eqs. (9.16)–(9.18), we obtain

$$K_N = \frac{1}{V} \left(\frac{h^2}{2\pi \mu_{\text{red}} k_B T} \right)^{3/2} Z_{1AB,\text{rot}} Z_{1AB,\text{vib}} \exp(-\beta \varepsilon_{AB}) \tag{9.19}$$

where μ_{red} is the reduced mass of the diatomic molecule. The pressure equilibrium constant K_p is

$$K_p = \frac{1}{k_B T} \left(\frac{h^2}{2\pi \mu_{\text{red}} k_B T} \right)^{3/2} Z_{1AB,\text{rot}} Z_{1AB,\text{vib}} \exp \left(-\frac{\varepsilon_{AB}}{k_B T} \right) \tag{9.20}$$

At temperatures around room temperature, $(\Theta_{\text{rot,AB}} \ll T \ll \Theta_{\text{vib,AB}})$, Eq. (9.20) can be further simplified to

$$K_p \cong \frac{1}{k_B T} \left(\frac{h^2}{2\pi \mu_{\text{red}} k_B T} \right)^{3/2} \frac{T}{\Theta_{\text{rot,AB}}} \exp \left(-\frac{\Theta_{\text{vib,AB}}}{2T} - \frac{\varepsilon_{AB}}{k_B T} \right) \tag{9.21}$$

Figure 9.1 Plot of $\ln Z$ as a function of N_{AB} for hypothetical systems: $m_A = 40\,\text{g mol}^{-1}$, $m_B = 60\,\text{g mol}^{-1}$, $T = 300\,\text{K}$, $V = 1\,\text{L}$, $\Theta_{rot,AB} = 10\,\text{K}$, and $\Theta_{vib,AB} = 1000\,\text{K}$. Three values of ε_{AB} were employed. The total number of A atoms is 10^{24}, and so is the total number of B atoms.

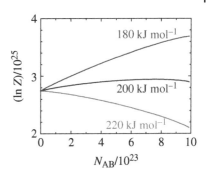

The temperature dependence of K_p is simple:

$$\frac{\partial \ln K_p}{\partial T} = -\frac{3}{2T} + \left(\frac{\Theta_{vib,AB}}{2} + \frac{\varepsilon_{AB}}{k_B}\right)\frac{1}{T^2} \tag{9.22}$$

The $\ln Z$ in Eq. (9.3) is defined for nonnegative values of N_A, N_B, and N_{AB}. Maximization of $\ln Z$ may occur at one of the two limits, either $N_{AB} = 0$ or $N_{AB,max}$, the greater of the number of A atoms and the number of B atoms in the system. If $\ln Z$ maximizes at one of the two limits, $\partial \ln Z/\partial N_{AB}$ will not be 0 at the maximum. Figure 9.1 shows a plot of $\ln Z$ as a function of N_{AB} for three values of ε_{AB} in a hypothetical system. For the values of parameters selected here (see Figure 9.1 caption), $\ln Z$ does not maximize in the range of $0 < N_{AB} < N_{AB,max}$ when $\varepsilon_{AB} = 180$ or $220\,\text{kJ mol}^{-1}$. When $\varepsilon_{AB} = 180\,\text{kJ mol}^{-1}$, the positive ε_{AB} does not place a sufficiently high enthalpic penalty against A and B forming a covalent bond. When $\varepsilon_{AB} = 220\,\text{kJ mol}^{-1}$, the increase in the enthalpy and a decrease in the entropy cause the dissociated state being preferred.

The single-molecule partition function of the diatomic molecule neglects restrictions on the rotational states by the nuclear spin statistics and quantum statistics.

9.2 Decomposition of Homonuclear Diatomic Molecules

We consider decomposition of a homonuclear diatomic molecule A_2 into two atoms of A:

$$A_2 \rightleftarrows 2A \tag{9.23}$$

As we did in the preceding section, we consider a system consisting of N_A atoms of A and N_{A_2} molecules of A_2 in volume V. We can obtain a similar expression of the equilibrium constant K_N for the numbers of molecules or atoms as

$$K_N = \frac{N_A^{\,2}}{N_{A_2}} = \frac{Z_{1A}^{\,2}}{Z_{1A_2}} \tag{9.24}$$

where Z_{1A} and Z_{1A_2} are the single-atom and single-molecule partition function for A and A_2, respectively. See Problem 9.3. As we did in Section 9.1, we can derive the following relationship between the chemical potentials of A_2 and A:

$$\mu_{A_2} = 2\mu_A \tag{9.25}$$

The pressure equilibrium constant K_p is

$$K_p = \frac{p_A^2}{p_{A_2}} = \frac{N_A^2}{N_{A_2}}\frac{k_B T}{V} = \frac{Z_{1A}^2}{Z_{1A_2}}\frac{k_B T}{V} \tag{9.26}$$

As we did in the preceding section, we can express K_N and K_p using molecular parameters. The single-molecule partition function of A_2 is expressed as

$$Z_{1A_2} = V\left(\frac{2\pi \cdot 2m_A k_B T}{h^2}\right)^{3/2} Z_{1A_2,rot}Z_{1A_2,vib}\exp(-\beta\varepsilon_{A_2}) \tag{9.27}$$

where ε_{A_2} is the electronic energy of the homonuclear diatomic molecule A_2 relative to the energy of the dissociated state. The single-atom partition function of A is given by Eq. (9.16). Then,

$$K_N = V\left(\frac{\pi m_A k_B T}{h^2}\right)^{3/2}\frac{1}{Z_{1A_2,rot}Z_{1A_2,vib}}\exp\left(\frac{\varepsilon_{A_2}}{k_B T}\right) \tag{9.28}$$

and

$$K_p = k_B T\left(\frac{\pi m_A k_B T}{h^2}\right)^{3/2}\frac{1}{Z_{1A_2,rot}Z_{1A_2,vib}}\exp\left(\frac{\varepsilon_{A_2}}{k_B T}\right) \tag{9.29}$$

As we noted earlier, we are neglecting nuclear spin statistics that may place a restriction on the rotational states of the homonuclear diatomic molecule.

At high temperatures ($\Theta_{rot,A_2} \ll T \ll \Theta_{vib,A_2}$), Eq. (9.29) can be further simplified to

$$K_p = \left(\frac{\pi m_A k_B T}{h^2}\right)^{3/2} k_B\Theta_{rot,A_2} 2\sinh\left(\frac{\Theta_{vib,A_2}}{2T}\right)\exp\left(\frac{\varepsilon_{A_2}}{k_B T}\right) \tag{9.30}$$

Since Θ_{rot,A_2} and Θ_{vib,A_2} are either known or calculated, Eq. (9.30) expresses K_p as a function of T with one unknown parameter ε_{A_2}. The latter can be estimated using a tabulated data of standard free energy of formation, $\Delta_f G_{298}^\circ$, of atom A from A_2:

$$K_p(T^\circ) = p^\circ \exp(-2\Delta_f G_{298}^\circ/RT^\circ) \tag{9.31}$$

where typically $T^\circ = 298$ K and $p^\circ = 1$ bar.

Figure 9.2 shows K_p as a function of temperature for dissociation of O_2. The data used are $\Delta_f G_{298}^\circ = 231.73$ kJ mol^{-1}, $\Theta_{rot,O_2} = 2.08$ K, and $\Theta_{vib,O_2} = 2256$ K. Again, the nuclear spin statistics was neglected for the evaluation of K_p.

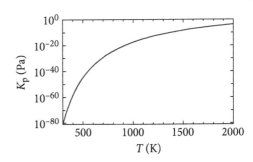

Figure 9.2 Pressure equilibrium constant for dissociation of oxygen molecules, plotted as a function of temperature T.

9.3 Isomerization

By now, you may have noticed a trick we have employed in the preceding two sections: we did not calculate the partition function of the system. In Section 9.1, for example, $Z(N_A, N_B, N_{AB}, V, T)$ given by Eq. (9.2) is just one of the terms that constitute a series for the partition function of the closed system. The system is specified by V, T, the number of atoms A, and the number of atoms B, not by N_A, N_B, and N_{AB}. The latter three numbers are variables; actually, only one of them is independent. For example, N_{AB} is a running index in the series. We maximized $\ln Z(N_A, N_B, N_{AB}, V, T)$ by changing N_{AB} to evaluate N_A, N_B, and N_{AB} at equilibrium. Rather, we should evaluate $\langle N_A \rangle$, $\langle N_B \rangle$, and $\langle N_{AB} \rangle$ of the system, and define K_N as

$$K_N \equiv \frac{\langle N_{AB} \rangle}{\langle N_A \rangle \langle N_B \rangle} \tag{9.32}$$

Before working on correct partition functions for the systems we considered Sections 9.1 and 9.2, we consider a simpler system here. That is isomerization of a molecule:

$$A \rightleftarrows B \tag{9.33}$$

For example, states A and B are boat and chair conformations of cyclohexane (see Section 5.5).

First, we treat this system in the same way as we did in Sections 9.1 and 9.2. Consider a system consisting of N_A molecules in state A and N_B molecules in state B. The single-molecule partition functions are Z_{1A} and Z_{1B} for the two states. The partition function for a given N_A and N_B is

$$Z(N_A, N_B) = \frac{Z_{1A}{}^{N_A} Z_{1B}{}^{N_B}}{N_A! N_B!} \tag{9.34}$$

We take a logarithm:

$$\ln Z = N_A(\ln Z_{1A} - \ln N_A + 1) + N_B(\ln Z_{1B} - \ln N_B + 1) \tag{9.35}$$

Differentiating $\ln Z$ by N_B gives

$$\frac{\partial}{\partial N_B}\left(-\frac{F}{k_B T}\right) = \frac{\partial N_A}{\partial N_B}\frac{\partial}{\partial N_A}N_A(\ln Z_{1A} - \ln N_A + 1)$$

$$+ \frac{\partial}{\partial N_B}N_B(\ln Z_{1B} - \ln N_B + 1)$$

$$= -(\ln Z_{1A} - \ln N_A) + (\ln Z_{1B} - \ln N_B) \qquad (9.36)$$

Therefore, minimizing the free energy F leads to

$$-(\ln Z_{1A} - \ln N_A) + (\ln Z_{1B} - \ln N_B) = 0 \qquad (9.37)$$

That is

$$K_N = \frac{N_B}{N_A} = \frac{Z_{1B}}{Z_{1A}} \qquad (9.38)$$

We now derive the correct partition function. In a closed system of N molecules, N_A ranges from 0 to N. Once N_A is specified, $N_B = N - N_A$. The correct partition function Z_c is

$$Z_c = \sum_{N_A=0}^{N} \frac{Z_{1A}^{N_A} Z_{1B}^{N-N_A}}{N_A!(N-N_A)!} \qquad (9.39)$$

With the binomial theorem, it is calculated as

$$Z_c = \frac{1}{N!}\sum_{N_A=0}^{N} \frac{N!}{N_A!(N-N_A)!}Z_{1A}^{N_A} Z_{1B}^{N-N_A} = \frac{(Z_{1A} + Z_{1B})^N}{N!} \qquad (9.40)$$

This equation can also be obtained as the Nth power of single-molecule partition function $Z_{1A} + Z_{1B}$, divided by $N!$, to account for indistinguishability.

In the canonical ensemble, we can find mean values of the variables. For N_A,

$$\langle N_A \rangle = \frac{1}{Z_c}\sum_{N_A=0}^{N} N_A \frac{Z_{1A}^{N_A} Z_{1B}^{N-N_A}}{N_A!(N-N_A)!} \qquad (9.41)$$

which is transformed as follows:

$$\langle N_A \rangle = \frac{1}{Z_c}Z_{1A}\frac{\partial Z_c}{\partial Z_{1A}} = Z_{1A}\frac{\partial \ln Z_c}{\partial Z_{1A}} = N\frac{Z_{1A}}{Z_{1A} + Z_{1B}} \qquad (9.42)$$

Likewise,

$$\langle N_B \rangle = N\frac{Z_{1B}}{Z_{1A} + Z_{1B}} \qquad (9.43)$$

Therefore,

$$K_N = \frac{\langle N_B \rangle}{\langle N_A \rangle} = \frac{Z_{1B}}{Z_{1A}} \qquad (9.44)$$

identical to Eq. (9.38).

Is the equality of the two expressions just by chance? The next section gives a theoretical ground for this equality.

9.4 Method of the Steepest Descent

The partition function given by Eq. (9.40) is expressed as

$$Z_c = \sum_{N_B=0}^{N} Z(N - N_B, N_B) \tag{9.45}$$

where $Z(N - N_B, N_B)$ is given by Eq. (9.34). We further rewrite Eq. (9.45) into

$$Z_c = \sum_{N_B=0}^{N} \exp(\ln Z(N - N_B, N_B)) \tag{9.46}$$

If $N \gg 1$, we can approximate the sum by an integral:

$$Z_c \cong \int_0^{\infty} \exp(\ln Z(N - N_B, N_B)) dN_B \tag{9.47}$$

If $\ln Z(N - N_B, N_B)$ maximizes (F minimizes in Eq. (9.36)) at $N_B = N_{Bm}$ (see Figure 9.3), we can approximate $\ln Z(N - N_B, N_B)$ around $N_B = N_{Bm}$ by a parabola peaked at $N_B = N_{Bm}$:

$$\ln Z(N - N_B, N_B) \cong \ln Z(N - N_{Bm}, N_{Bm}) - \frac{1}{2}\sigma^{-2}(N_B - N_{Bm})^2 \tag{9.48}$$

where

$$\sigma^{-2} = -\left.\frac{\partial^2 \ln Z(N - N_B, N_B)}{\partial N_B^2}\right|_{N_{Bm}} \tag{9.49}$$

Figure 9.3 Plots of $\ln Z(N - N_B, N_B)$ and $Z(N - N_B, N_B)$ around their peaks at N_{Bm}.

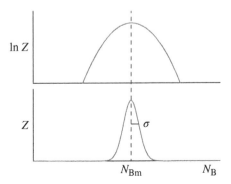

Then, the integrand in Eq. (9.47) is just a normal distribution of N_B centered at N_{Bm} with a variance equal to σ^2. The integral is approximated as

$$Z_c \cong Z(N - N_{Bm}, N_{Bm}) \int_0^\infty \exp\left(-\frac{(N_B - N_{Bm})^2}{2\sigma^2}\right) dN_B \tag{9.50}$$

which is calculated as

$$Z_c \cong Z(N - N_{Bm}, N_{Bm})(2\pi\sigma^2)^{1/2} \tag{9.51}$$

For $\ln Z(N - N_B, N_B)$ given by Eq. (9.35), σ^2 is calculated as

$$\sigma^2 = \frac{1}{N - N_{Bm}} + \frac{1}{N_{Bm}} \tag{9.52}$$

Now, we calculate $\langle N_B \rangle$, as shown here:

$$\begin{aligned}
\langle N_B \rangle &= \frac{1}{Z_c} Z(N - N_{Bm}, N_{Bm}) \int_0^\infty N_B \exp\left(-\frac{(N_B - N_{Bm})^2}{2\sigma^2}\right) dN_B \\
&= \int_0^\infty N_B (2\pi\sigma^2)^{-1/2} \exp\left(-\frac{(N_B - N_{Bm})^2}{2\sigma^2}\right) dN_B \\
&= N_{Bm}
\end{aligned} \tag{9.53}$$

This result legitimizes the minimization of F in Sections 9.1 and 9.2.

The approximation of the integral (Eq. (9.47)) by Eq. (9.50) is called the **method of the steepest descent** or a **saddle-point method**, as the method was first applied to an integral on a complex plane.

The partition function for the isomerization can be calculated exactly, but it is an exception. For example, in the reaction $A + B \leftrightarrow AB$ we considered in Section 9.1, the partition function for the system with given numbers of A atoms and B atoms ($N_{A\,at}$ and $N_{B\,at}$, respectively),

$$Z_c = \sum_{N_{AB}=0}^{N_{smaller}} \frac{Z_{1A}^{N_{A\,at}-N_{AB}} Z_{1B}^{N_{B\,at}-N_{AB}} Z_{1AB}^{N_{AB}}}{(N_{A\,at} - N_{AB})!(N_{B\,at} - N_{AB})!N_{AB}!} \tag{9.54}$$

where $N_{smaller}$ is the smaller of $N_{A\,at}$ and $N_{B\,at}$. The summation cannot be calculated easily. However, we can apply the method of the steepest descent to obtain an approximate expression of Z. The N_{AB} we obtained by maximizing $\ln Z$ in Eq. (9.8) is identical to $\langle N_{AB} \rangle$ when N_{AB} is allowed to vary according to the canonical distribution.

Problems

9.1 *Distinguishable molecules.* For the reaction given by Eq. (9.1), obtain the partition function assuming that all of the molecules are distinguishable.

Then, show that minimizing the Helmholtz free energy leads to inconsistency.

9.2 *Sufficient condition for minimization of F.* Equation (9.7) gives a condition necessary for F to minimize, but it is not sufficient since maximization of F satisfies the same Eq. (9.7). Show that F actually minimizes for (9.10).

9.3 *Decomposition of a diatomic molecule.* Develop a discussion similar to the one in Section 9.1 to derive K_N given by Eq. (9.24).

9.4 *Variance.* Calculate the variance of N_A for the isomerization reaction in Section 9.3.

9.5 *Formation of HCl.* The equilibrium constant K_N for the chemical reaction,

$$H_2 + Cl_2 \rightleftarrows 2HCl$$

is expressed by the single-molecule partition functions:

$$K_N = \frac{N_{HCl}^2}{N_{H_2} N_{Cl_2}} = \frac{Z_{1HCl}^2}{Z_{1H_2} Z_{1Cl_2}}$$

All of the three components are in a gas phase.
(1) Express the pressure equilibrium constant K_p using molecular parameters such as m_H, m_{Cl}, Θ_{rot,H_2}, Θ_{rot,Cl_2}, $\Theta_{rot, HCl}$, Θ_{vib,H_2}, Θ_{vib,Cl_2}, $\Theta_{vib, HCl}$, and ε_{HCl} (with reference to H_2 and Cl_2).
(2) Use $\Delta_f G^\circ_{298}(HCl) = -95.30 \, kJ \, mol^{-1}$ and the data below to find ε_{HCl}.

Gas	Θ_{rot} (K)	Θ_{vib} (K)
H_2	87.6	6331
Cl_2	0.351	805.5
HCl	15.2	4302

9.6 *Haber process.* We consider the Haber process

$$3H_2 + N_2 \rightleftarrows 2NH_3$$

in a container of volume V and temperature T. Let A, B, and C represent H_2, N_2, and NH_3, respectively. The system has N_A molecules of

H_2, N_B molecules of N_2, and N_C molecules of NH_3. The single-molecule partition functions are Z_{1A}, Z_{1B}, and Z_{1C}, respectively.

(1) Express the equilibrium constant K_N using Z_{1A}, Z_{1B}, and Z_{1C}.

(2) Express K_N using microscopic parameters of the molecules. How does K_N depend on V? Does it agree with Le Chatelier's principle?

(3) Convert your answer in (1) into the pressure equilibrium constant K_p for partial pressures, p_A, p_B, and p_C. How does K_p depend on V?

10

Adsorption

In this chapter, we apply the grand canonical-ensemble formulation to adsorption phenomena. The latter are widely observed on a solid surface, on a liquid surface, and at the interface between two liquid phases. After reviewing the adsorption phenomena in Section 10.1, we learn two models of adsorption in the next two sections. The first is a Langmuir isotherm, and the second a Brenauer, Emmett, and Teller (BET) isotherm. In Section 10.4, we look at a more complicated situation in which diatomic molecules dissociate into atoms upon adsorption.

10.1 Adsorption Phenomena

When a gas is in contact with a surface, some of the gas molecules may adsorb onto the surface. The surface is called adsorbent, and the adsorbed molecules are adsorbate. If the surface is flat, the amount of the adsorbate is usually too small to change the pressure of the gas. However, if a finely divided material is used as an adsorbent, the surface can adsorb a sufficient number of molecules to allow measurement of a pressure drop or a mass increase of the adsorbent.

Finely divided materials include fine powders of metal, activated carbon powders, and porous silica particles (see Figure 10.1a–c). Fine powders are usually also porous. The smaller the pore diameter, the greater the surface area. For example, 1 g of silica gel of pore size less than 10 nm has typically 200–500 m^2 of surface area, ready to adsorb water; silica gels are widely used as desiccant. A 1 g of nickel or its alloy in powder form has more than 10 m^2 of surface and can store H_2 in fuel-cell automobiles.

If molecules that constitute the air such as N_2, O_2, Ar, CO_2 do not adsorb, and if the gas of interest adsorbs rather strongly onto the surface, the experiment can be done at ambient. The pressure is now the partial pressure of the gas molecules of interest. For example, when we consider adsorption of water onto porous silica, it is the partial pressure of water in the air. The partial pressure of water can be anywhere from 0 to the vapor pressure of water, and the latter

Statistical Thermodynamics: Basics and Applications to Chemical Systems, First Edition. Iwao Teraoka.
© 2019 John Wiley & Sons, Inc. Published 2019 by John Wiley & Sons, Inc.
Companion website: www.wiley.com/go/Teraoka_StatsThermodynamics

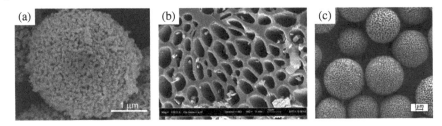

Figure 10.1 Scanning electron micrographs of finely divided particles. (a) Nickel particles, (b) activated charcoal, and (c) silica gel particle. Particles of different particle diameters and surface areas (per mass of the particles) are available [11–13].

increases with an increasing temperature. The pressure of other molecules present in the air does not affect the adsorbate–vapor equilibrium of water, as the interaction is weak between molecules in the vapor phase.

The relationship between the pressure of the gas and the amount of the adsorbate at a given temperature is called an adsorption isotherm. In this chapter, we learn two molecular models for the adsorption, a Langmuir isotherm and a BET isotherm. In both models, the surface has identical sites to adsorb molecules.

Quantitative studies of adsorption date back to the late nineteenth century. In 1909, Freundlich introduced an empirical formula: the surface coverage is proportional to the power of the pressure. The isotherms we learn here are different from this empirical equation.

A solution in contact with a surface also exhibits adsorption. The formulas we have obtained for gas molecules apply to molecules in solution just by reinterpreting the pressure as the osmotic pressure. Likewise, the formulas for adsorption we are deriving in this chapter apply as they are to the adsorption from a solution.

10.2 Langmuir Isotherm

We consider a surface adjacent to a gas of monatomic molecules. As illustrated in Figure 10.2, the surface has N identical sites. In the simplest model, each site can accommodate up to one molecule; each site is either unoccupied or occupied by one molecule. When the site is occupied, it lowers the energy by ε (>0) and acquires chemical potential μ.

The state of a site and its energy are listed in Table 10.1 for the two states.

A grand canonical ensemble is convenient for treating the adsorption phenomenon. The system is not the gas molecules, but rather the surface of N sites in contact with a gas (see Figure 10.3). The system is free to adsorb gas molecules or release them. In this open system, the space adjacent to the surface is a reservoir of molecules, each having a chemical potential of μ.

Figure 10.2 Model of an adsorbing surface (adsorbent). The surface consists of N sites, and each site can adsorb one monatomic molecule.

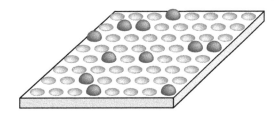

Table 10.1 States of sites on the surface, their energies and chemical potentials, and numbers.

State	Energy	Chemical potential	Number of sites
Unoccupied	0	0	$N - n$
Occupied	$-\varepsilon$	μ	n

Figure 10.3 Adsorption sites on the surface in equilibrium with a vapor. The vapor phase is a reservoir of heat and particles with the temperature and the chemical potential specified.

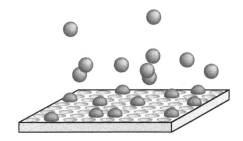

Let n be the number of occupied sites, where $n = 0, 1, ..., N$. For a given n, there are $_NC_n$ ways to have the occupied sites. Each of these states has an energy of $-n\varepsilon$. Since each state has n molecules of chemical potential μ, the chemical potential of the state is $n\mu$, relative to the chemical potential in the absence of adsorbed molecules. Then, the grand partition function $\mathcal{Z}$ is expressed as

$$\mathcal{Z} = \sum_{n=0}^{N} e^{\beta n \mu} \binom{N}{n} e^{\beta n \varepsilon} = \sum_{n=0}^{N} \binom{N}{n} e^{\beta(\mu+\varepsilon)n} \tag{10.1}$$

We can calculate the sum of the series using the binomial theorem. In Eq. (A.25), $a = e^{\beta(\mu+\varepsilon)}$ and $b = 1$:

$$\mathcal{Z} = [1 + e^{\beta(\mu+\varepsilon)}]^N \tag{10.2}$$

Another way to obtain the grand partition function is to consider all the sites independently. Let n_i be the number of molecules at the ith site, where $i = 1, 2, ..., N$. In our simple model, $n_i = 0$ or 1 for all i. The adsorption of each site is independent of the states of its neighboring sites. The state of the surface is specified by $n_1, n_2, ..., n_N$. The total energy of the surface is

$-\varepsilon(n_1 + n_2 + \cdots + n_N)$ relative to the adsorbate-free surface, and the chemical potential is $\mu(n_1 + n_2 + \cdots + n_N)$. Then, the grand partition function $\mathcal{Z}$ is expressed as

$$\mathcal{Z} = \sum_{n_1=0,1} \sum_{n_2=0,1} \cdots \sum_{n_N=0,1} \exp\left(\beta\mu \sum_{i=1}^{N} n_i + \beta\varepsilon \sum_{i=1}^{N} n_i \right)$$

$$= \left[\sum_{n_1=0,1} \exp(\beta(\mu + \varepsilon)n_1) \right]^N \tag{10.3}$$

which gives the same result as Eq. (10.2).

Now that we have found $\mathcal{Z}$, we are ready to calculate the mean number of occupied sites, $\langle n \rangle$, according to

$$\langle n \rangle = \frac{1}{\mathcal{Z}} \sum_{n=0}^{N} n \binom{N}{n} e^{\beta(\mu+\varepsilon)n} = \frac{1}{\mathcal{Z}} \frac{1}{\beta} \frac{\partial \mathcal{Z}}{\partial \mu} = \frac{1}{\beta} \frac{\partial \ln \mathcal{Z}}{\partial \mu} \tag{10.4}$$

With Eq. (10.2), $\langle n \rangle$ is calculated as

$$\langle n \rangle = \frac{1}{\beta} \frac{\partial}{\partial \mu} N \ln(1 + e^{\beta(\mu+\varepsilon)}) = \frac{N}{1 + e^{-\beta(\mu+\varepsilon)}} \tag{10.5}$$

We define the surface coverage θ as

$$\theta \equiv \langle n \rangle / N \tag{10.6}$$

Then,

$$\theta = \frac{1}{1 + e^{-\beta(\mu+\varepsilon)}} \tag{10.7}$$

Now we want to convert this equation into a function of pressure p. Earlier we found that μ is expressed as a function of p for a gas of monatomic molecules:

$$\frac{\mu}{k_B T} = \ln p - \frac{5}{2} \ln(k_B T) - \frac{3}{2} \ln \frac{2\pi m}{h^2} \tag{6.40}$$

We rewrite it to

$$e^{-\beta\mu} = \frac{k_B T}{p} \left(\frac{2\pi m k_B T}{h^2} \right)^{3/2} \tag{10.8}$$

Now, we introduce q as

$$q \equiv k_B T \left(\frac{2\pi m k_B T}{h^2} \right)^{3/2} \tag{10.9}$$

that is a function of T and has a dimension of the pressure. Earlier, we defined the thermal wavelength, $\lambda_T = h/(2\pi m k_B T)^{1/2}$, see Eq. (5.59). Then,

$$q = \frac{k_B T}{\lambda_T^3} \tag{10.10}$$

and

$$\frac{p}{q} = \frac{p\lambda_T^3}{k_B T} \tag{10.11}$$

It means that p/q represents the number of molecules in volume λ_T^3. In other words, q is equal to the pressure when the gas has, on the average, one molecule in volume λ_T^3. That pressure is a lot higher compared with 1 atm at accessible temperatures. Therefore, $p/q \ll 1$, unless the pressure is excessively high.

Equation (10.8) is now

$$e^{-\beta\mu} = \frac{q}{p} \tag{10.12}$$

and the surface coverage is expressed as

$$\theta = \frac{p}{p + qe^{-\beta\varepsilon}} \tag{10.13}$$

This relationship between θ and p is known as a **Langmuir isotherm**. At low pressures, $p \ll qe^{-\beta\varepsilon}$, and the coverage is proportional to p: $\theta \cong (p/q)e^{\beta\varepsilon}$. With an increasing pressure, θ approaches one. Figure 10.4 shows the curve.

In an experiment that involves adsorption, the amount g of the adsorbate (typically, mass) is measured as the pressure is changed. To confirm that the results follow the Langmuir isotherm, a linearized form of the Langmuir isotherm is employed; only a linear relationship can provide unbiased assessment of the fitting theoretical equation. There are two such forms for the Langmuir isotherm as shown here.

The surface coverage should be evaluated as $\theta = g/g_0$, where g_0 is the amount of the adsorbate at saturation. Since g_0 is not known before the analysis, θ cannot be evaluated. Rather, $1/g$ is plotted as a function of $1/p$ (see Figure 10.5a). Experimental data will be scattered along a straight line. From the intercept and the slope of a linear fit, we can estimate g_0 and $qe^{-\beta\varepsilon}$. Another way is to plot p/g as a function of p (see Figure 10.5b).

The Langmuir isotherm can also be observed in adsorption from a solution. Figure 10.6 is an example that exhibits applicability of the isotherm model. In this experiment, peat was immersed into a solution of divalent cation.

Figure 10.4 Langmuir isotherm. Surface coverage θ is plotted as a function of pressure p. The dash-dotted line is a tangent to the curve at $p = 0$.

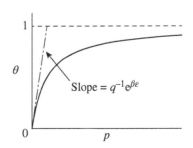

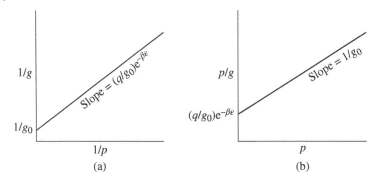

Figure 10.5 Linearized Langmuir isotherm. (a) Reciprocal of the amount g adsorbed is plotted as a function of the reciprocal of the pressure p. (b) The ratio of the pressure to the amount adsorbed is plotted as a function of the pressure.

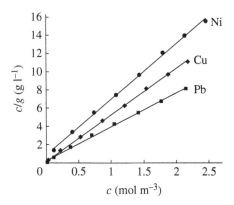

Figure 10.6 Linearized Langmuir isotherm demonstrated for adsorption of divalent cations onto peat. Peat was immersed into an aqueous solution of salt with a divalent cation. The ratio of the cation concentration to the amount of mass take up by the peat is plotted as a function of the concentration. Source: Data from Ho et al. 2002 [14].

It is not easy to use a canonical ensemble for a combined system of molecules on the surface and the vapor phase to derive the isotherm. This example shows that not all the problems can be easily formulated using the two ensembles; when we solve a problem, one of the two ensembles can offer a far easier solution. A canonical ensemble is more convenient for some problems, and a grand canonical ensemble is the method of choice for others. The adsorption is an example of the latter.

The Langmuir isotherm was originally derived using a kinetic model of adsorption and desorption (see Problem 10.1).

10.3 BET Isotherm

The Langmuir isotherm inherently assumes that a fully covered surface is a monolayer. In this section, we allow each site of the surface to adsorb more than

Figure 10.7 Model of a surface for a BET isotherm. Each site can adsorb one molecule or more.

one molecule. Such a model was introduced by Brenauer, Emmett, and Teller in 1938, and the adsorption isotherm of their model is called a **BET isotherm**. The multilayer adsorbate can be thick and, therefore, the amount of adsorption can easily exceed the limit set by the monolayer adsorption we considered in the preceding section.

The BET isotherm is built upon the Langmuir model and its isotherm. The BET theory assumes that each site can adsorb any number of molecules, independent of the states of the adjacent sites. We denote by θ_i the fraction of sites, each having i molecules ($i = 0, 1, \ldots$). An example is shown in Figure 10.7 for a total 12 sites; $\theta_0 = 2/12$, $\theta_1 = 5/12$, $\theta_2 = 1/12$, $\theta_3 = 2/12$, $\theta_4 = 1/12$, $\theta_5 = 1/12$, and $\theta_6 = \theta_7 = \cdots = 0$.

We begin with rewriting the relationship in the Langmuir isotherm (Eq. (10.13)) into

$$q \exp(-\beta \varepsilon_1)\theta_1 = p\theta_0 \tag{10.14}$$

where $\theta_0 = 1 - \theta_1$ is the fraction of unoccupied sites, and ε_1 is the energy released by the binding of the first molecule. Note that θ_1 and ε_1 were θ and ε, respectively, in Section 10.2. For the second-layer adsorption, a slightly different relationship holds between θ_1 and θ_2:

$$q \exp(-\beta \varepsilon_2)\theta_2 = p\theta_1 \tag{10.15}$$

where ε_2 represents the energy released by adding one molecule to the site that has already a molecule. The latter is equal to ε_L, the energy of liquefaction (enthalpy of vaporization per molecule). In general, $\varepsilon_L \neq \varepsilon_1$. Therefore,

$$q \exp(-\beta \varepsilon_L)\theta_2 = p\theta_1 \tag{10.16}$$

The same relationship holds for all of subsequent adsorption:

$$q \exp(-\beta \varepsilon_L)\theta_i = p\theta_{i-1} \quad (i = 2, 3, \ldots) \tag{10.17}$$

Recapping, we write

$$\frac{\theta_i}{\theta_{i-1}} = \begin{cases} \dfrac{p}{q}\exp(\beta \varepsilon_1) \equiv r_1 & (i = 1) \\[2ex] \dfrac{p}{q}\exp(\beta \varepsilon_L) \equiv r_L & (i = 2, 3, \ldots) \end{cases} \tag{10.18}$$

State   Fraction of sites

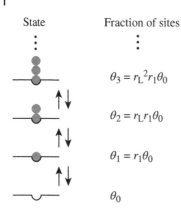

$$\theta_3 = r_{\mathrm{L}}^2 r_1 \theta_0$$

$$\theta_2 = r_{\mathrm{L}} r_1 \theta_0$$

$$\theta_1 = r_1 \theta_0$$

$$\theta_0$$

Figure 10.8 Layer-by-layer adsorption equilibrium in the BET adsorption model. Each layer consists of sites having a fixed number of molecules (occupation numbers). An equilibrium is established between adjacent layers with occupation numbers different by 1.

where r_1 and r_{L} were introduced. By cascading the ratios, each θ_i is expressed as

$$\theta_i = \frac{\theta_i}{\theta_{i-1}} \times \frac{\theta_{i-1}}{\theta_{i-2}} \times \cdots \times \frac{\theta_2}{\theta_1} \times \frac{\theta_1}{\theta_0} \times \theta_0 = r_{\mathrm{L}}^{i-1} r_1 \theta_0 \quad (i = 1, 2, \ldots) \quad (10.19)$$

Figure 10.8 explains the calculation for the first four states of sites.

We impose the condition that the sum of θ_i be equal to unity:

$$1 = \sum_{i=0}^{\infty} \theta_i = \theta_0 + \sum_{i=1}^{\infty} r_{\mathrm{L}}^{i-1} r_1 \theta_0 = \theta_0 + \frac{r_1 \theta_0}{1 - r_{\mathrm{L}}} \quad (10.20)$$

which is rewritten to

$$\frac{1}{\theta_0} = 1 + \frac{r_1}{1 - r_{\mathrm{L}}} \quad (10.21)$$

Note that $r_{\mathrm{L}} < 1$ is required to avoid divergence of the series in Eq. (10.20).

Let n_{ads} be the total number of adsorbed molecules. Its mean $\langle n_{\mathrm{ads}} \rangle$ is calculated as

$$\frac{\langle n_{\mathrm{ads}} \rangle}{N} \equiv \sum_{i=1}^{\infty} i\theta_i = r_1 \theta_0 \sum_{i=1}^{\infty} i r_{\mathrm{L}}^{i-1} = \frac{r_1 \theta_0}{(1 - r_{\mathrm{L}})^2} \quad (10.22)$$

See the formula, Eq. (A.24).

From Eqs. (10.21) and (10.22), we finally obtain

$$\frac{\langle n_{\mathrm{ads}} \rangle}{N} = \frac{r_1}{(1 - r_{\mathrm{L}} + r_1)(1 - r_{\mathrm{L}})} = \frac{1}{1 - r_{\mathrm{L}}} - \frac{1}{1 - r_{\mathrm{L}} + r_1} \quad (10.23)$$

In terms of p, q, ε_1, and ε_{L},

$$\frac{\langle n_{\mathrm{ads}} \rangle}{N} = \frac{1}{1 - \dfrac{p}{q} \exp(\beta\varepsilon_{\mathrm{L}})} - \frac{1}{1 - \dfrac{p}{q} \exp(\beta\varepsilon_{\mathrm{L}}) + \dfrac{p}{q} \exp(\beta\varepsilon_1)} \quad (10.24)$$

We introduce p^* and c by

$$p^* = q \exp(-\beta\varepsilon_{\mathrm{L}}) \quad (10.25)$$

$$c = \frac{p^*}{p} r_1 = \exp[\beta(\varepsilon_1 - \varepsilon_L)] \tag{10.26}$$

where $c > 0$. With p^* and c, Eq. (10.24) can be rewritten to a standard BET equation:

$$\frac{1}{\langle n_{ads}\rangle} \frac{p}{p^* - p} = \frac{1}{Nc} + \frac{c-1}{Nc} \frac{p}{p^*} \tag{10.27}$$

Here, we examine ramifications of the BET isotherm.

A. The requirement of $r_L < 1$ is now $p < p^*$. We find that, in Eq. (10.27), as p increases, $\langle n_{ads}\rangle$ increases as well (see Problem 10.6), and as p approaches p^*, $\langle n_{ads}\rangle \to \infty$. The surface would be wet with the adsorbate. When $p > p^*$, condensation occurs. Obviously, p^* is the vapor pressure.
B. If the second-layer adsorption is not allowed, i.e. $\varepsilon_L = -\infty$, Eq. (10.24) reduces to the Langmuir isotherm, Eq. (10.13).
C. At low pressures, $r_1 \ll 1$ and $r_L \ll 1$. Therefore, Eq. (10.23) reduces to

$$\frac{\langle n_{ads}\rangle}{N} \cong 1 + r_L - (1 + r_L - r_1) = r_1 = \frac{p}{q} \exp(\beta\varepsilon_1) = c\frac{p}{p^*} \tag{10.28}$$

Equation (10.28) is identical to the low-pressure asymptote of the Langmuir isotherm.

Figure 10.9 shows a plot of $\langle n_{ads}\rangle/N$ as a function of cp/p^* for $1/c = 1, 0.6, 0.3$, and 0. The curve for $1/c = 0$ is the Langmuir isotherm. With an increasing $1/c$, thus facilitating second-layer adsorption and beyond, the number of molecules adsorbed deviates upward. The deviation is greater at high pressures.

As noted in A, $\langle n_{ads}\rangle/N$ diverges when cp/p^* increases to approach c. We see a hint of the divergence in the curve of $1/c = 1$, but not in the others.

D. The parameter c indicates which adsorption releases more energy, the first-layer adsorption or the adsorption onto the subsequent layers (see Table 10.2).

Figure 10.9 Number of molecules adsorbed per adsorption site, $\langle n_{ads}\rangle/N$, is plotted as a function of $cp/p^* = r_1 = (p/q)\exp(\beta\varepsilon_1)$ for various values of c. The number adjacent to each curve indicates the value of $1/c$. At low pressures, $\langle n_{ads}\rangle/N \cong cp/p^*$, common to different values of $1/c$.

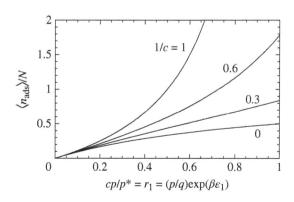

Table 10.2 Meaning of parameter c.

c	Energy	Energy release by adsorption
<1	$\varepsilon_L > \varepsilon_1$	The second (and beyond) layer adsorption releases more energy compared with the first layer.
$=1$	$\varepsilon_L = \varepsilon_1$	The energy release is identical for all layers.
>1	$\varepsilon_L < \varepsilon_1$	The second (and beyond) layer adsorption releases less energy compared with the first layer.

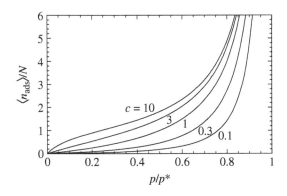

Figure 10.10 Number of molecules adsorbed per adsorption site, $\langle n_{ads} \rangle / N$, is plotted as a function of pressure reduced by p^* for $c = 0.1, 0.3, 1, 3,$ and 10.

Figure 10.10 shows the plot of $\langle n_{ads} \rangle / N$ as a function of p/p^* for $c = 0.1$, $0.3, 1, 3,$ and 10. The curve for $c = 1$ is a reference: $\langle n_{ads} \rangle / N = p/(p^* - p)$. With an increasing p, $\langle n_{ads} \rangle / N$ increases in proportion to p, and when p gets close to p^*, $\langle n_{ads} \rangle / N$ diverges to $+\infty$. When $c > 1$, $\langle n_{ads} \rangle / N$ is pushed up at low pressures, but not as much at high pressures. When $c < 1$, in contrast, $\langle n_{ads} \rangle / N$ is pushed down at low pressures, but not as much when p gets close to p^*.

Equation (10.27) shows a standard BET equation. It is a linearized form that allows us to examine whether the experiment follows the BET isotherm or not and to estimate the values of the parameters of the BET model. As we did for the Langmuir isotherm, we replace $\langle n_{ads} \rangle / N$ by g/g_0, where g is the amount of the adsorbate, and g_0 is the amount for monolayer coverage. Unlike the Langmuir isotherm, g/g_0 can exceed one. Figure 10.11 shows a plot of $p/[g(p^* - p)]$ as a function of p/p^*. If the adsorption data follow the BET isotherm, they will scatter along a straight line in this kind of plot. From the intercept and the slope of the linear fit, we can estimate g_0 and c. This method is widely used in characterization of surfaces.

The BET isotherm can be derived directly in a grand canonical ensemble (see Problem 10.5).

Figure 10.11 Linearized plot of the BET isotherm. $p/[g(p^* - p)]$ is plotted as a function of p/p^*. The intercept is $1/(g_0 c)$, and the slope is $(1 - 1/c)/g_0$.

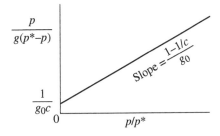

10.4 Dissociative Adsorption

In Section 10.2, we considered adsorption of monatomic molecules. The treatment was straightforward, since the molecule does not change upon adsorption except that it loses the center-of-mass movement. When the adsorbate is a diatomic molecule, the situation can be different. In this section, we consider adsorption that splits the diatomic molecule into two atoms (see Figure 10.12). The atoms are movable from site to site, and therefore the adsorbed molecules are indistinguishable. We can apply the same procedure as the one used in Section 10.2, but the resultant expression for the isotherm will be slightly different.

We consider an equilibrium between a gas of diatomic molecules with chemical potential μ_m and a surface consisting of N sites, each capable of adsorbing up to one atom. Let μ_a be the chemical potential of the atom on the surface. The equilibrium leads to

$$\mu_m = 2\mu_a \tag{10.29}$$

as we have learned in Section 9.2.

The surface can have $2n$ atoms, where $n = 0, 1, \ldots, N/2$. The grand partition function $\mathcal{Z}$ is

$$\mathcal{Z} = \sum_{n=0}^{N/2} \binom{N}{2n} \exp(2\beta(\mu_a + \varepsilon)n) \tag{10.30}$$

where the surface lowers its energy by -2ε when a molecule adsorbs.

Figure 10.12 Surface in equilibrium with a vapor of diatomic molecules. Upon adsorption, each molecule splits into atoms.

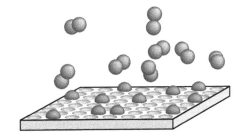

We can calculate the sum of the series using the following formula that can be derived from the binomial theorem:

$$\frac{(1+x)^N + (1-x)^N}{2} = \sum_{n=0}^{N/2} \binom{N}{2n} x^{2n} \tag{10.31}$$

With this formula, Eq. (10.30) is calculated as

$$\mathcal{Z} = \frac{1}{2}\{[1 + \exp(\beta(\mu_a + \varepsilon))]^N + [1 - \exp(\beta(\mu_a + \varepsilon))]^N\} \tag{10.32}$$

Since

$$\left|\frac{1 - \exp(\beta(\mu_a + \varepsilon))}{1 + \exp(\beta(\mu_a + \varepsilon))}\right| < 1 \tag{10.33}$$

and $N \gg 1$, we can neglect $[1 - \exp(\beta(\mu_a + \varepsilon))]^N$ relative to $[1 + \exp(\beta(\mu_a + \varepsilon))]^N$. Then, Eq. (10.32) is simplified to

$$\mathcal{Z} \cong \frac{1}{2}[1 + \exp(\beta(\mu_a + \varepsilon))]^N \tag{10.34}$$

The mean number of atoms on the surface is obtained from

$$\langle 2n \rangle = \frac{1}{\beta} \frac{\partial}{\partial \mu_a} \ln \mathcal{Z} \tag{10.35}$$

Therefore, the surface coverage is

$$\theta = \frac{\langle 2n \rangle}{N} = \frac{\exp(\beta(\mu_a + \varepsilon))}{1 + \exp(\beta(\mu_a + \varepsilon))} \tag{10.36}$$

When expressed in terms of site properties (μ_a and ε), this surface coverage is identical to the one in Eq. (10.7).

At around room temperature, the chemical potential of a diatomic molecule is approximated as

$$\exp(-\beta\mu_m) = \frac{k_B T}{p}\left(\frac{2\pi m k_B T}{h^2}\right)^{3/2} \frac{T}{\Theta_r} \tag{10.37}$$

We define q for the diatomic molecule as

$$\frac{q}{p} \equiv \exp(-\beta\mu_m) = \exp(-2\beta\mu_a) \tag{10.38}$$

Again, q is a function of T. That is,

$$(p/q)^{1/2} = \exp(\beta\mu_a) \tag{10.39}$$

Then, the isotherm is

$$\theta = \frac{(p/q)^{1/2}e^{\beta\varepsilon}}{1 + (p/q)^{1/2}e^{\beta\varepsilon}} \tag{10.40}$$

Figure 10.13 compares the isotherm of the dissociative diatomic molecules with the one for the monatomic molecules, both of Langmuir type. The dissociation of diatomic molecules causes the square-root dependence at low pressures and a slow approach to saturation.

Figure 10.13 Isotherm of dissociative diatomic molecules is compared with the one for the Langmuir isotherm of monatomic molecules. The abscissa is $(p/q)e^{-\beta\varepsilon}$ for the diatomic molecules, and $(p/q)e^{-2\beta\varepsilon}$ for the diatomic molecules.

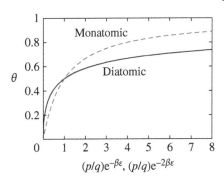

10.5 Interaction Between Adsorbed Molecules

In the preceding sections, we assumed that adsorption at a specific site on a given surface does not affect adsorption at adjacent sites, that is, the sites are independent. That is why we could write the grand partition function of the system as the Nth power of the single-site grand partition function. We could obtain the exact isotherm in each of Sections 10.2–10.4.

Some adsorption phenomena may exhibit interaction between an adjacent pair of adsorption sites. For example, when one of the sites has already a molecule adsorbed, that may facilitate adsorption at adjacent sites. In other words, there may be an interaction between s_i and s_j, if i and j are a nearest-neighbor pair. If the interaction is sufficiently strong, it may lead to formation of patches of adsorption on the surface. We consider this situation in Section 11.1 (B) and Problem 11.18.

Problems

10.1 *Adsorption kinetics.* Let us consider the Langmuir isotherm within reaction kinetics. The reaction is

$$[\text{empty site}] + [\text{molecule in vapor phase}] \underset{k_d}{\overset{k_a}{\rightleftharpoons}} [\text{occupied site}]$$

Let k_a and k_d be the rate constants of the forward and reverse reactions. The surface has N sites, and we define θ as we did in Eq. (10.6).

(1) We denote the concentration of molecules in the vapor phase by [M]. The "concentrations" of the empty sites and occupied sites can be expressed as $N(1 - \theta)$ and $N\theta$, respectively. Balance the forward and reverse rates.

(2) Express θ with k_a and k_d.

(3) Assume an ideal gas law for the vapor phase to express θ as a function the pressure p.

(4) How is k_d/k_a related to q defined by Eq. (10.9)?

10.2 *Microcalorimetry.* Section 10.2 considered a surface with N adsorption sites in contact with vapor-phase molecules of chemical potential μ. The surface lowers the energy per site by ε when it captures a molecule.
(1) Calculate the mean of the energy of the surface, $\langle E \rangle$.
(2) The chemical potential of a molecule in the vapor of pressure p is given by Eq. (6.40). Express $\langle E \rangle$ as a function of p and T.
(3) What is the high-temperature asymptote of $\langle E \rangle$?
(4) Draw a sketch for a plot of $\langle E \rangle / (N\varepsilon)$ as a function of T.
(5) Calculate the heat capacity C of the surface. Draw a sketch for a plot of C as a function of T.

10.3 *Maximizing fluctuation.* Show that the fluctuations in the surface coverage in the Langmuir isotherm maximizes when $\theta = 1/2$.

10.4 *Maximizing entropy.* This problem continues on what we learned in the Langmuir isotherm.
(1) Find the entropy S of the surface in the Langmuir adsorption model. Express $S/(Nk_B)$ as a function of $x = e^{\beta(\mu + \varepsilon)}$.
(2) Draw a sketch for a plot of $S/(Nk_B)$ as a function of x. Find when S maximizes.

10.5 *BET adsorption.* Section 10.3 used the result of the Langmuir isotherm in layer-by-layer adsorption to derive the Brunauer–Emmett–Teller (BET) isotherm. Here, we treat it directly in the grand canonical ensemble. Each of the N sites of the surface adsorbs molecules independently. We denote by i_j the number of molecules on the jth site ($j = 1, \ldots, N$). The energy $\varepsilon(i_j)$ of the site follows the following table.

i_j	Energy, $\varepsilon(i_j)$
0	0
1	$-\varepsilon_1$
2	$-\varepsilon_1 - \varepsilon_L$
3	$-\varepsilon_1 - 2\varepsilon_L$
4	$-\varepsilon_1 - 3\varepsilon_L$
$\vdots$	$\vdots$

The surface is in contact with a reservoir of molecules of chemical potential μ at temperature T.

(1) The state of the N sites on the surface is specified by $i_1, i_2, \ldots, i_N$. Write the grand partition function $\mathcal{Z}$ for the system and calculate $\mathcal{Z}$.

(2) Derive a formula to calculate the mean number of molecules adsorbed per site, $\langle i \rangle$, and calculate it. Confirm that it is identical to the result we obtained in Section 10.3.

(3) Derive a formula to calculate the mean number of liquefied molecules per site, $\langle i \rangle_L$, and calculate it. Liquefied molecules are those in the second, third, ..., layers.

(4) What is the fraction of empty sites, θ_{empty}?

(5) Use $e^{-\beta\mu} = q/p$ to rewrite your answer in (4). What is the low-pressure asymptote of θ_{empty}?

10.6 *Pressure dependence in the BET isotherm.* Show that $\langle n_{\text{ads}} \rangle$ increases when p increases ($p < p^*$) in Eq. (10.27).

10.7 *Competitive adsorption.* A surface of n sites is in contact with a mixture of gas A and B. Each site can adsorb up to one molecule, either A or B. When the site adsorbs molecule i ($i = $ A, B), its energy is lowered by ε_i ($\varepsilon_i > 0$). The chemical potential of molecule i is μ_i. Let n_A and n_B be the numbers of sites that have A molecules and B molecules, respectively.

(1) What is the grand partition function?

(2) Find $\langle n_A \rangle$ and $\langle n_B \rangle$.

(3) As we did for the Langmuir isotherm of pure monatomic gas, we introduce q_i by

$$p_i/q_i = \exp(\beta\mu_i)$$

where p_i is the partial pressure of component i. Rewrite your answers in (2) using $y_A = (p_A/q_A)\exp(\beta\varepsilon_A)$ and $y_B = (p_B/q_B)\exp(\beta\varepsilon_B)$.

10.8 *Diatomic molecules.* This problem considers adsorption of diatomic molecules of mass m onto a surface that consists of N adsorption sites. The molecules in the reservoir (single-molecule partition function is Z_{1F}) are in contact with the surface. The molecule occupies one of the sites when adsorbed, and the single-molecule partition function changes to Z_{1A}.

(1) First, we consider M molecules in the reservoir of volume V. The Z_{1F} is given as

$$Z_{1F} = V\left(\frac{2\pi m}{\beta h^2}\right)^{3/2} Z_{1\text{Frv}}$$

where $Z_{1\text{Frv}}$ is the rotovibrational part of Z_{1F}. What is the chemical potential μ?

(2) Now, we consider the system of adsorbed molecules. Express the grand partition function Z of the system using μ. What is the surface coverage θ?

(3) We can express Z_{1Frv} as $Z_{1Frv} = (T/\Theta_r)Z_{1Fv}$, where Z_{1Fv} is the vibrational part of Z_{1F}. Likewise, we can write Z_{1A} as $Z_{1A} = Z_{1Av}e^{\beta\varepsilon}$, where Z_{1Av} is the vibrational part of Z_{1A}, and adsorption lowers the energy by ε. We assume, for simplicity, that the adsorbed molecule retains the same vibrational degree of freedom as the molecule in free space but cannot rotate. Use these relationships to express θ.

10.9 *Adsorption of two particles.* Consider a system that has N adsorption sites. Each site can adsorb one particle or more. The site acquires energy ε each time it adsorbs a particle. The particles are indistinguishable.

(1) Two particles are on the N sites. If they do not interact, what is the partition function of the system?

(2) If the two particles interact and the interaction Φ exists for the two particles sharing a site, what is the partition function of the system? Also calculate the mean energy.

11

Ising Model

The systems we have considered so far are either ideal or close to ideal. The interaction between particles is either absent or sufficiently weak to change the thermodynamic properties only slightly. In the several chapters that follow, we extend our statistical methods to systems consisting of strongly interacting particles. Usually, the strong interaction is between a pair of nearest neighbors or between a pair in close proximity.

In this chapter, we learn an Ising model as the simplest model for strongly interacting particles. The Ising model and its variations are usually for magnets. Spontaneous magnetization in a ferromagnet is due to strong interactions between nearby electron spins that favor aligned spin orientations. The Ising model originates in a magnetic system, and the relevance of the model to the magnetic systems still remains. Here we learn the Ising model, because there are many systems in chemistry and biochemistry that can be recast into the Ising model.

In Section 11.1, we introduce the model, and look at how some systems in chemistry and biochemistry are recast into the model. In Section 11.2, we calculate the partition function for small systems in one dimension. Approximation methods we learn in Section 11.3 allow us to obtain the partition function of the system with many particles and calculate some average quantities. Section 11.4 covers a method of a transfer matrix for the exact calculation of the partition function. In Section 11.5, we learn variations of the Ising model.

11.1 Model

The Ising model dates back to the 1920s, when Wilhelm Lenz at the University of Hamburg gave a problem to his student Ernst Ising. The system of the Ising model consists of many spins, and each spin is quantized to adopt either an up or a down orientation. We assign discrete values of $+1$ and -1 to the two orientations.

Statistical Thermodynamics: Basics and Applications to Chemical Systems, First Edition. Iwao Teraoka.
© 2019 John Wiley & Sons, Inc. Published 2019 by John Wiley & Sons, Inc.
Companion website: www.wiley.com/go/Teraoka_StatsThermodynamics

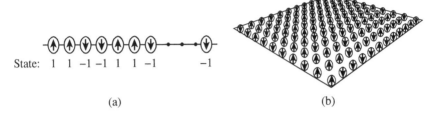

State: 1 1 −1 −1 1 1 −1 −1

(a) (b)

Figure 11.1 Ising model in (a) one dimension and (b) two dimensions. Spins are in a regular arrangement. Each spin can be either up (state 1) or down (state −1).

When the spins are arranged along a string, the model is called a one-dimensional (1D) **Ising model** (Figure 11.1a). A planar arrangement of spins is described by a two-dimensional Ising model (Figure 11.1b). A three-dimensional model or models in higher dimensions can be considered in the same way.

Here are examples of systems in chemistry and biochemistry that can be recast into the Ising model.

A. Consider a binary mixture of two liquids A and B. Unlike a gas, molecules adjacent to each other interact strongly. We can assign +1 if a given space of the liquid is occupied by a molecule of liquid A, and −1 if it belongs to liquid B. A (+1, +1) pair and a (−1, −1) pair represent a contact between A molecules and a contact between B molecules, respectively, and a (+1, −1) pair represents an A–B contact. The energy of the contact is different for the three pairs, and that can lead to phase separation. A three-dimensional Ising model is appropriate for describing the binary mixture. Figure 11.2 is a two-dimensional rendering of a lattice model for the liquid mixture. The site on each lattice point is occupied by either an A or a B molecule.

B. Adsorption onto the surface, as we learned in Chapter 10, can be treated in the Ising model. The state of each site on the surface (adsorbent) is either occupied or empty, i.e. binary. We can assign 1 and 0 (or 1 and −1) to the

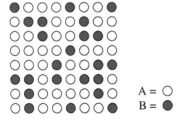

A = ○
B = ●

Figure 11.2 Two-dimensional rendering of a cubic lattice model for an A–B liquid mixture. Each site is either A or B; no vacancies.

two states. In Chapter 10, we assumed that there is no interaction between adsorbed molecules. In a real system, there can be interaction. For example, cohesion between adsorbed molecules may be strong. In other words, presence of an adsorbed molecule facilitates adsorption of molecules to the neighboring sites. Then, the adsorption may lead to patches being formed on the surface (see Figure 11.3).

C. A third example is a helix-forming polymer. There is a class of polymers that form a helix of a reversible sense (see Figure 11.4). Each repeat unit of the polymer can be either right-handed or left-handed relative to the preceding unit. We can assign +1 to the right-handed state and −1 to the left-handed state. A whole chain will have a single sense, if continuing the same sense of helix has a lot lower energy than reversing the helical sense or helped by an environment that prefers one sense over the other. Chapter 12 covers the helical polymer.

D. Helix–coil transition and helix unfolding of helix-forming bio-macromolecules such as polypeptides can also be included as an example. In these molecules, the state of each repeat unit is either a helix (of a specific sense) or a random coil, again binary (see Figure 11.5). We learn about helix–coil transition in Chapter 13. A tertiary structure of a protein consists of helices, random coils, and other structural motifs, and their spatial arrangement is frozen.

Figure 11.3 Surface with adsorbed molecules (dark balls). If cohesion is strong between adsorbed molecules, they will form patches or isolated islands.

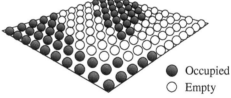

● Occupied
○ Empty

Figure 11.4 Helical polymer with a reversible helical sense.

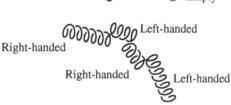

Left-handed
Right-handed
Right-handed
Left-handed

Figure 11.5 A polypeptide chain consisting of segments of helix and those of a random coil.

11.2 Partition Function

11.2.1 One-Dimensional Ising Model

Figure 11.6 illustrates a one-dimensional (1D) Ising model. A site is represented by an ellipse, and each ellipse has a spin. A total of N spins are arranged along a line with an equal spacing. The spin of the ith site, σ_i, is either $+1$ or -1:

$$\sigma_i = \begin{cases} 1 & \text{(up)} \\ -1 & \text{(down)} \end{cases} \tag{11.1}$$

where $i = 1, 2, \ldots, N$. The state of the N-spin system is specified by $\sigma_1, \sigma_2, \ldots, \sigma_N$. If all the spins are down, the state is $-1, -1, \ldots, -1$. If all the spins are up, it is $1, 1, \ldots, 1$. Since each spin can be either $+1$ or -1, the system has a total 2^N states.

When an external magnetic field H is applied, the spin acquires the energy of $-H\sigma_i$. The energy is low when the spin is in the same direction as the field, and high when the spin is in the opposite direction.

A potentially strong, short-range interaction is included in the 1D Ising model as $-J\sigma_i\sigma_{i+1}$, where J is usually taken to be positive or zero. The interaction is between an adjacent pair of spins. A pair of spins in the same direction (parallel) has a lower energy compared with a pair of the opposite spins (antiparallel). Their energy difference is $2J$.

Since σ_i is dimensionless, H and J have the dimensions of the energy. We can regard H as the product of magnetic field and the magnetic moment of the spin.

In the 1D Ising model, the total energy E of the N spins is expressed as

$$E = -H \sum_{i=1}^{N} \sigma_i - J \sum_{i=1}^{N-1} \sigma_i\sigma_{i+1} \tag{11.2}$$

Let us consider which state is stable among a total of 2^N states. When $H \gg J$, on the one hand, $E \cong -H\Sigma\sigma_i$. Then, the more up spins, the more stable they are. Two states, ↑↑↑↓↑↓↑↑ and ↑↑↑↓↓↑↑↑, for $N = 8$ are nearly equally probable, because they have the same number of up spins and the same number of down

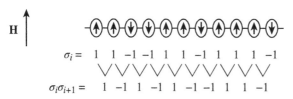

Figure 11.6 One-dimensional Ising model in a linear arrangement. The spin σ_i of the ith spin and the product of adjacent spin pairs, $\sigma_i\sigma_{i+1}$, are also shown. External magnetic field **H** may be applied.

Figure 11.7 One-dimensional Ising model in a ring arrangement.

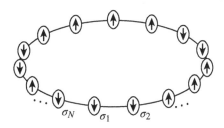

spins. What counts in the likelihood of a state is the overall fraction of up spins in that state. When $H \ll J$, on the other hand, $E \cong -J\Sigma\sigma_i\sigma_{i+1}$. Then, alteration of spins is discouraged. Two states, ↑↑↑↑↑↑↓↓ and ↑↑↑↓↓↓↓↓ are nearly equally probable. What matters is the fraction of alternating spin pairs.

Since the state of the system is specified by $\sigma_1, \ldots, \sigma_N$, the partition function of the N-spin system is written as

$$Z = \sum_{\sigma_1=\pm1} \sum_{\sigma_2=\pm1} \cdots \sum_{\sigma_N=\pm1} \exp\left(\beta H \sum_{i=1}^{N} \sigma_i + \beta J \sum_{i=1}^{N-1} \sigma_i\sigma_{i+1} \right) \qquad (11.3)$$

Note that index i in the series of $\sigma_i\sigma_{i+1}$ ends at $N-1$. We call this system a linear Ising model. When $N \gg 1$, we can include $i = N$ in the series with a nonexistent σ_{N+1}. Adding an additional term will not appreciably change the energy of the system.

To eliminate the end effect, a ring is sometimes formed, as illustrated in Figure 11.7. The partition function of the N-spin system is written as

$$Z = \sum_{\sigma_1=\pm1} \sum_{\sigma_2=\pm1} \cdots \sum_{\sigma_N=\pm1} \exp\left(\beta H \sum_{i=1}^{N} \sigma_i + \beta J \sum_{i=1}^{N} \sigma_i\sigma_{i+1} \right) \qquad (11.4)$$

where $\sigma_{N+1} = \sigma_1$. We call this system a ring Ising model.

11.2.2 Calculating Statistical Averages

Once Z is obtained as a function of β, H, and J, we can calculate $\langle\sigma\rangle$ and $\langle\sigma_i\sigma_{i+1}\rangle$. For a linear Ising model,

$$\langle\sigma\rangle = \frac{1}{Z} \sum_{\sigma_1=\pm1} \cdots \sum_{\sigma_N=\pm1} \left(\frac{1}{N}\sum_{i=1}^{N}\sigma_i \right) \exp\left(\beta H \sum_{i=1}^{N} \sigma_i + \beta J \sum_{i=1}^{N-1} \sigma_i\sigma_{i+1} \right)$$

$$= \frac{1}{N\beta} \frac{\partial \ln Z}{\partial H} \qquad (11.5)$$

and

$$\langle\sigma_j\sigma_{i+1}\rangle = \frac{1}{Z} \sum_{\sigma_1=\pm1} \cdots \sum_{\sigma_N=\pm1} \left(\frac{1}{N-1}\sum_{i=1}^{N-1}\sigma_i\sigma_{i+1} \right)$$

$$\times \exp\left(\beta H \sum_{i=1}^{N} \sigma_i + \beta J \sum_{i=1}^{N-1} \sigma_i \sigma_{i+1} \right)$$

$$= \frac{1}{(N-1)\beta} \frac{\partial \ln Z}{\partial J} \tag{11.6}$$

For a ring Ising model, $\langle \sigma \rangle$ is calculated using the same formula, Eq. (11.5). The other mean is calculated as

$$\langle \sigma_i \sigma_{i+1} \rangle = \frac{1}{Z} \sum_{\sigma_1 = \pm 1} \cdots \sum_{\sigma_N = \pm 1} \left(\frac{1}{N} \sum_{i=1}^{N} \sigma_i \sigma_{i+1} \right)$$

$$\times \exp\left(\beta H \sum_{i=1}^{N} \sigma_i + \beta J \sum_{i=1}^{N} \sigma_i \sigma_{i+1} \right)$$

$$= \frac{1}{N\beta} \frac{\partial \ln Z}{\partial J} \tag{11.7}$$

When $N \gg 1$, Eqs. (11.6) and (11.7) are identical.

We can calculate other averages either from $\langle \sigma \rangle$ or $\langle \sigma_i \sigma_{i+1} \rangle$, or directly from Z. We look at two examples here.

11.2.2.1 Average Number of Up Spins

Let N_+ and N_- be the numbers of up spins and down spins, respectively. They add up to N:

$$N_+ + N_- = N \tag{11.8}$$

By definition, the mean of $\Sigma \sigma_i$ is equal to $\langle N_+ \rangle - \langle N_- \rangle$:

$$\langle N_+ \rangle - \langle N_- \rangle = N \langle \sigma \rangle \tag{11.9}$$

These two equations lead to

$$\langle N_+ \rangle = \frac{N}{2}(1 + \langle \sigma \rangle) \tag{11.10}$$

Since we know how to find $\langle \sigma \rangle$, we can find $\langle N_+ \rangle$ as well. When $\langle \sigma \rangle = 1$, $\langle N_+ \rangle = N$; when $\langle \sigma \rangle = -1$, $\langle N_+ \rangle = 0$, as required.

11.2.2.2 Average of the Number of Spin Alterations (Number of Domains – 1)

Let us define a domain as a sequence of spins in the same orientation. Within each domain, the product of σ at adjacent sites is $+1$. Across the boundary of adjacent domains, the product is equal to -1 (see Figure 11.8). Therefore, counting the pairs whose product is equal to -1 gives the number of spin alterations, N_{alt}, that is equal to the number of domains – 1 in the linear Ising model. The mean of N_{alt} is

$$\langle N_{\text{alt}} \rangle = \left\langle \sum_{i=1}^{N} \frac{1 - \sigma_i \sigma_{i+1}}{2} \right\rangle = \frac{N}{2}(1 - \langle \sigma_i \sigma_{i+1} \rangle) \tag{11.11}$$

Figure 11.8 A sequence of identical spins defines a domain. The example shows three domains. There is a spin reversal across the domain boundary.

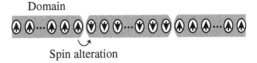

Domain

Spin alteration

where $N \gg 1$ was assumed. The mean of $\sigma_i \sigma_{i+1}$ is calculated from Eq. (11.6). For a ring Ising model, $\langle N_{alt} \rangle$ is equal to the number of domains.

11.2.2.3 Domain Size

The size of the domain is estimated from $\langle N_{alt} \rangle$. We denote the size by L_{th}, as the domain size is determined by the competition between the energy of spin alteration $(2J)$ and $k_B T$. Since a chain of N spins is divided into $\langle N_{alt} \rangle$ domains (to be exact, $\langle N_{alt} \rangle + 1$ domains), the domain size may be estimated as

$$L_{th} = \frac{N}{\langle N_{alt} \rangle} \tag{11.12}$$

11.2.2.4 Size of a Domain of Uniform Spins

The estimate of the domain size by Eq. (11.12) is appropriate for a long chain of spins. When the chain is short, it may not have any spin alterations. The whole chain is in a single domain.

The domain is terminated when (1) the next spin has the opposite direction; or (2) the chain end is reached. (1) occurs at a frequency of L_{th}^{-1} (at every L_{th}-th site on the average), and (2) occurs at a frequency of N^{-1}. These two phenomena are mutually independent. When the two domain-ending mechanisms coexist, the domain ends at a frequency of $L_{th}^{-1} + N^{-1}$. The resultant domain size L may be given as the reciprocal of the combined frequency:

$$\frac{1}{L} = \frac{1}{L_{th}} + \frac{1}{N} \tag{11.13}$$

The L defined in this way gives the average domain size for short and long chains of spins. Two extreme situations are worth noting: When $N \gg L_{th}$ ($\langle N_{alt} \rangle \gg 1$; long chain), on the one hand, $L \cong L_{th}$. The system consists of many domains. When $N \ll L_{th}$ ($\langle N_{alt} \rangle \ll 1$; short chain), on the other hand, $L \cong N$. The whole system is a monodomain.

11.2.3 A Few Examples of 1D Ising Model

Here we calculate partition functions and some statistical averages for a few simple systems.

11.2.3.1 Linear Ising Model, $N = 3$

There are a total eight states in the system of the linear Ising model with $N = 3$. Table 11.1 lists all possible configurations of σ_1, σ_2, and σ_3, and the energy E:

Table 11.1 States and energy in the 1D linear Ising model of $N = 3$.

σ_1	σ_2	σ_3	E
1	1	1	$-3H - 2J$
1	1	-1	$-H$
1	-1	1	$-H + 2J$
1	-1	-1	H
-1	1	1	$-H$
-1	1	-1	$H + 2J$
-1	-1	1	H
-1	-1	-1	$3H - 2J$

From the table, we calculate Z as

$$Z = e^{3\beta H + 2\beta J} + e^{-3\beta H + 2\beta J} + e^{\beta H - 2\beta J} + e^{-\beta H - 2\beta J} + 2e^{\beta H} + 2e^{-\beta H} \quad (11.14)$$

It may be convenient to rewrite it to

$$Z = 2e^{2\beta J} \cosh 3\beta H + 2(2 + e^{-2\beta J}) \cosh \beta H \quad (11.15)$$

Using Eq. (11.5), $\langle \sigma \rangle$ is calculated as

$$\langle \sigma \rangle = \frac{1}{3\beta} \frac{\partial \ln Z}{\partial H} = \frac{1}{3} \frac{3e^{2\beta J} \sinh 3\beta H + (2 + e^{-2\beta J}) \sinh \beta H}{e^{2\beta J} \cosh 3\beta H + (2 + e^{-2\beta J}) \cosh \beta H} \quad (11.16)$$

When $H = 0$, $\langle \sigma \rangle = 0$, as required. When $\beta H \gg 1$ (strong field), $\langle \sigma \rangle = 1$; the three spins are in the direction of the field. When $-\beta H \gg 1$, $\langle \sigma \rangle = -1$; the spins reverse to align to the opposite field.

It is interesting to see how $\langle \sigma \rangle$ behaves when the interaction is weak and when the interaction is strong. When the interaction is weak, i.e. $\beta J \ll 1$, Eq. (11.16) reduces to

$$\langle \sigma \rangle \cong \frac{\sinh 3\beta H + \sinh \beta H}{\cosh 3\beta H + 3 \cosh \beta H} = \tanh \beta H \quad (11.17)$$

identical to $\langle \sigma \rangle$ of a single spin. The result is reasonable, since the state of each spin will be determined independently by H.

When the interaction is strong, i.e. $\beta J \gg 1$,

$$\langle \sigma \rangle \cong \tanh 3\beta H \quad (11.18)$$

The three spins of the system collectively respond to H.

We can consider other limiting situations such as high temperatures, $\beta H \ll 1$ and $\beta J \ll 1$, and low temperatures, $\beta H \gg 1$ and $\beta J \gg 1$.

Table 11.2 States and energy for the 1D ring Ising model of $N = 3$.

σ_1	σ_2	σ_3	E
1	1	1	$-3H - 3J$
1	1	-1	$-H + J$
1	-1	1	$-H + J$
1	-1	-1	$H + J$
-1	1	1	$-H + J$
-1	1	-1	$H + J$
-1	-1	1	$H + J$
-1	-1	-1	$3H - 3J$

Using Eq. (11.6), we calculate $\langle \sigma_i \sigma_{i+1} \rangle$ as

$$\langle \sigma_i \sigma_{i+1} \rangle = \frac{1}{2\beta} \frac{\partial \ln Z}{\partial J} = \frac{e^{2\beta J} \cosh 3\beta H - e^{-2\beta J} \cosh \beta H}{e^{2\beta J} \cosh 3\beta H + (2 + e^{-2\beta J}) \cosh \beta H} \tag{11.19}$$

When $\beta J \ll 1$,

$$\langle \sigma_i \sigma_{i+1} \rangle \cong \frac{\cosh 3\beta H - \cosh \beta H}{\cosh 3\beta H + 3 \cosh \beta H} = \tanh^2 \beta H \tag{11.20}$$

The weak interaction leads to independence for σ_i and σ_{i+1}. Therefore, $\langle \sigma_i \sigma_{i+1} \rangle \cong \langle \sigma_i \rangle \langle \sigma_{i+1} \rangle \cong \tanh^2 \beta H$. When $\beta J \gg 1$,

$$\langle \sigma_i \sigma_{i+1} \rangle \cong 1 \tag{11.21}$$

a reasonable result.

11.2.3.2 Ring Ising Model, $N = 3$

In the ring Ising model with $N = 3$, the table for the system's states and energies is slightly different (Table 11.2):

From the table, we calculate Z as

$$Z = e^{3\beta H + 3\beta J} + e^{-3\beta H + 3\beta J} + 3e^{\beta H - \beta J} + 3e^{-\beta H - \beta J} \tag{11.22}$$

which is further rewritten to

$$Z = 2e^{3\beta J} \cosh 3\beta H + 6e^{-\beta J} \cosh \beta H \tag{11.23}$$

11.2.3.3 Ring Ising Model, $N = 4$

For a ring Ising model with $N = 4$, the table is longer with 16 states (see Table 11.3).

Table 11.3 States and energy for the 1D ring Ising model of $N = 4$.

σ_1	σ_2	σ_3	σ_4	E
1	1	1	1	$-4H - 4J$
1	1	1	-1	$-2H$
1	1	-1	1	$-2H$
1	1	-1	-1	0
1	-1	1	1	$-2H$
1	-1	1	-1	$4J$
1	-1	-1	1	0
1	-1	-1	-1	$2H$
-1	1	1	1	$-2H$
-1	1	1	-1	0
-1	1	-1	1	$4J$
-1	1	-1	-1	$2H$
-1	-1	1	1	0
-1	-1	1	-1	$2H$
-1	-1	-1	1	$2H$
-1	-1	-1	-1	$4H - 4J$

The partition function is

$$Z = e^{4\beta H + 4\beta J} + e^{-4\beta H + 4\beta J} + 4e^{2\beta H} + 4e^{-2\beta H} + 2e^{-4\beta J} + 4$$
$$= 2e^{4\beta J} \cosh 4\beta H + 8 \cosh 2\beta H + 2e^{-4\beta J} + 4 \tag{11.24}$$

We calculated Z for the ring model with $N = 3$ and 4 and for the linear model with $N = 3$. With a further increase in N, it becomes increasingly difficult to compute Z, if we need to follow the table. However, there is a method to obtain an exact Z for arbitrary N; that is a method of transfer matrix that Ernst Ising came up with. We learn it in Section 11.4. In the next section, we learn a few rather easy methods to approximate the partition function. The solution is not correct; but if we know its limits, the method is still useful for gaining insights into the system.

11.3 Mean-Field Theories

Mean-field (MF) approximation is widely used in statistical mechanics, when exactly solving the problem is impossible or difficult. Even when an exact solution is available, the MF theory may help us better understand the nature of the system. Here, we learn a few variations of the MF theory.

11.3.1 Bragg–Williams (B–W) Approximation

Here we learn one of the MF approximation methods, called **Bragg–Williams** (B–W) **approximation**, applied to the 1D Ising model. In the B–W approximation, σ_{i+1} in Eq. (11.2) is replaced by its average, $\langle\sigma\rangle$:

$$E = -H\sum_{i=1}^{N}\sigma_i - J\sum_{i=1}^{N}\sigma_i\langle\sigma\rangle = -(H + J\langle\sigma\rangle)\sum_{i=1}^{N}\sigma_i \tag{11.25}$$

The approximation eliminates interactions and places all the spins in a uniform field $H + J\langle\sigma\rangle$. Recall the requirement of an "ideal" system – E consists of components, each of which is determined by the state variable of an individual particle. Equation (11.25) satisfies this condition. Then, Z is expressed as the Nth power of a single-particle partition function:

$$Z = \left[\sum_{\sigma=\pm1} e^{\beta(H+J\langle\sigma\rangle)\sigma}\right]^{N} \tag{11.26}$$

which is calculated as

$$Z = [2\cosh(\beta(H + J\langle\sigma\rangle))]^{N} \tag{11.27}$$

The partition function has $\langle\sigma\rangle$ as a parameter.

Applying Eqs. (11.5) and (11.27) leads to

$$\langle\sigma\rangle = \frac{1}{\beta}\frac{\partial}{\partial H}\ln(2\cosh\beta(H + J\langle\sigma\rangle)) = \tanh(\beta(H + J\langle\sigma\rangle)) \tag{11.28}$$

Note that, although $\langle\sigma\rangle$ varies with H, we need to fix $\langle\sigma\rangle$ in taking the partial derivative.

Equation (11.28) has $\langle\sigma\rangle$ on both sides. Therefore, $\langle\sigma\rangle$ must be determined self-consistently. Here, we obtain self-consistent solutions.

First, we consider the solutions when $H = 0$. Equation (11.28) is now

$$\langle\sigma\rangle = \tanh(\beta J\langle\sigma\rangle) \tag{11.29}$$

Figure 11.9a shows a curve of $y = \tanh \beta Jx$ for different values of βJ. Note that the relevant range of $x = \langle\sigma\rangle$ is $-1 \le x \le 1$, as each σ_i can be either -1 or 1. The slope of the tangent to the curve at the origin is βJ (Figure 11.9b).

Since $\langle\sigma\rangle$ is a solution of $x = \tanh \beta Jx$, $\langle\sigma\rangle$ is the x coordinate of an intersection between a straight line, $y = x$, and the curve, $y = \tanh \beta Jx$. Obviously, the line and the curve intersect at the origin for different values of βJ. Whether there are other intersections or not depends on whether βJ is greater than 1 or not.

When $\beta J < 1$, the straight line and the curve intersect only at the origin (Figure 11.10a). Therefore, $\langle\sigma\rangle = 0$ is the only solution. Weak interaction between adjacent spins (high temperatures) leads to equal probabilities for up spins and down spins.

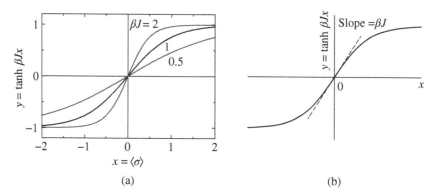

Figure 11.9 (a) Plot of $y = \tanh \beta J x$ for $\beta J = 0.5$, 1, and 2. Only $-1 \le x \le 1$ is relevant to the Ising model. (b) The tangent to the curve $y = \tanh \beta J x$ at the origin has a slope of βJ.

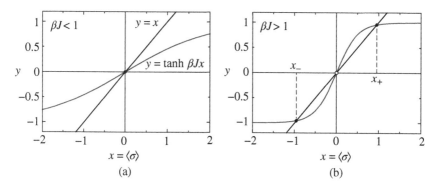

Figure 11.10 (a) When $\beta J < 1$, the straight line, $y = x$, and the curve, $y = \tanh \beta J x$, intersect at the origin only. The origin is a stable solution of the equation, $x = \tanh \beta J x$. (b) When $\beta J > 1$, the straight line and the curve intersect at the origin and two points at x_+ and $x_- = -x_+$. The origin is an unstable solution. The other two solutions are stable.

When $\beta J > 1$, the straight line and the curve intersect at three points. Their x-coordinates are x_-, 0, and x_+, all between -1 and 1 (Figure 11.10b). Therefore,

$$\langle \sigma \rangle = \begin{cases} x_- & \text{(stable)} \\ 0 & \text{(unstable)} \\ x_+ & \text{(stable)} \end{cases} \tag{11.30}$$

Figure 11.11 shows how to estimate x_+ graphically. Start at any value (x_0) between 0 and 1. Calculate $\tanh \beta J x_0$. Then, feed this value into x in $\tanh \beta J x$ and evaluate the function. In effect, $\tanh(\beta J(\tanh \beta J x_0))$ will be calculated. The up and right arrows in the figure represent these steps. Repeating this procedure brings x close to x_+ at the intersection. In other words, x and $\tanh \beta J x$ become nearly equal to each other. You can also start at x_0 greater than 1, and

Figure 11.11 Numerical method to solve $x = \tanh \beta Jx$ for $\beta J > 1$ by iteration. The value of $\tanh \beta Jx$ is fed into x in $\tanh \beta Jx$ repeatedly until the difference between x and $\tanh \beta Jx$ becomes sufficiently small, typically the magnitude of errors in the computer. Starting at $x = x_0$, this procedure brings x to x_+ where the curve meets the straight line. The starting point can be to the right of x_+.

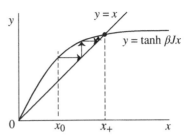

Figure 11.12 Solution of $\langle \sigma \rangle = \tanh \beta J \langle \sigma \rangle$, plotted as a function of βJ. For $\beta J < 1$, $\langle \sigma \rangle = 0$; for $\beta J > 1$, $\langle \sigma \rangle = x_+$. There is a sharp switch from 0 to x_+ as βJ exceeds 1.

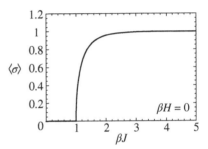

you will get to the same point of convergence. A computer program for this procedure is not difficult for those who know how to program in numerical computation. Likewise, starting at any negative value leads to x_-. The solution of $x = 0$ cannot be reached from any starting value; this solution is unstable. If x_0 is slightly greater than 0, it is sufficient to bring x to x_+.

Since the interaction between adjacent spins is strong, it is reasonable that an aligned spin direction, either $+1$ or -1, is preferred.

Now we will find an approximate expression for x_+. Since $\exp(\beta Jx_+) \gg 1$,

$$\tanh \beta Jx_+ = \frac{1 - \exp(-2\beta Jx_+)}{1 + \exp(-2\beta Jx_+)} \cong 1 - 2\exp(-2\beta Jx_+) \tag{11.31}$$

Since $x_+ = \tanh \beta Jx_+$,

$$x_+ \cong 1 - 2\exp(-2\beta Jx_+) \cong 1 - 2e^{-2\beta J} \tag{11.32}$$

In the last near-equality, $x_+ \cong 1$ was used. Likewise,

$$x_- \cong -1 + 2e^{-2\beta J} \tag{11.33}$$

Figure 11.12 shows a plot of x_+ as a function of βJ.

It may appear that the B–W approximation does a good job, but it is easy to prove that its results are wrong. Consider three sequences of spins in Table 11.4. Each sequence has only one spin alteration, and therefore the energy is identical in the absence of H. Therefore, the probability is identical for the three states. Any other sequence of spins that has only one alteration is equally probable.

Table 11.4 Three equally probable states in the absence of H.

State	$\langle \sigma \rangle$
↑↑↑↑↑↑↑↑↑↓	0.8
↑↑↑↑↑↓↓↓↓↓	0
↑↑↓↓↓↓↓↓↓↓	−0.6

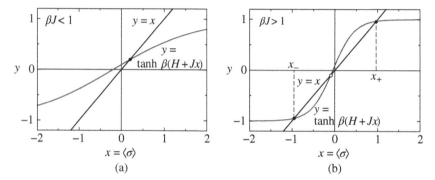

Figure 11.13 A positive H moves the curve to the left. (a) When $\beta J < 1$, the straight line, $y = x$, and the curve, $y = \tanh \beta(H + Jx)$ intersect at a point to the right of the origin. (b) When $\beta J > 1$, the straight line and the curve intersect at a point close to the origin and two points at x_+ and $x_- = -x_+$.

Therefore, the mean of σ must be zero, not x_+ or x_- when $H = 0$. Obviously, the B–W approximation fails to provide an estimate of $\langle \sigma \rangle$ that is close to reality.

Now we look at the situation when $H > 0$, but H is only slightly positive ($\beta H \ll 1$). The curve of $y = \tanh(\beta(H + Jx))$ does not pass the origin. The curve is shifted to the left compared with the one in the absence of H. Figure 11.13a,b shows the curve together with the line $y = x$ for $\beta J < 1$ and $\beta J > 1$, respectively.

When $\beta J < 1$, the line and the curve intersect at a point with positive x. When $\beta J > 1$, they intersect at three points, and x_+ and x_- are stable solutions. This situation is the same as the one we saw for $H = 0$. The x_+ has moved to the right compared with the point for $H = 0$.

We can find an explicit expression for $x = \langle \sigma \rangle$ when $\beta(H + J\langle \sigma \rangle) \ll 1$. Since $\tanh(\beta(H + J\langle \sigma \rangle)) \cong \beta(H + J\langle \sigma \rangle)$, we obtain

$$\langle \sigma \rangle \cong \frac{\beta H}{1 - \beta J} \tag{11.34}$$

When $\beta(H + J\langle \sigma \rangle) \gg 1$, $\langle \sigma \rangle = x_+$ is

$$\langle \sigma \rangle \equiv 1 - 2e^{-2\beta(H + J\langle \sigma \rangle)} \equiv 1 - 2e^{-2\beta(H + J)} \tag{11.35}$$

Figure 11.14 Numerical solution of $\langle\sigma\rangle = \tanh(\beta(H + J\langle\sigma\rangle))$, plotted as a function of βJ for different values of βH indicated adjacent to the curves. As βH decreases to 0, the curve approaches the one shown in Figure 11.12.

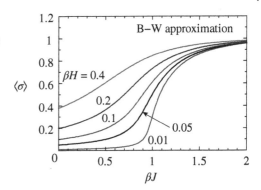

Figure 11.14 shows how $\langle\sigma\rangle$ changes with βJ for a fixed value of βH. Lines are shown for several values of βH.

Now we calculate $\langle\sigma_i\sigma_{i+1}\rangle$ using Eq. (11.6) ($N \gg 1$). We need to fix $\langle\sigma\rangle$ in taking the partial derivative. The result is

$$\langle\sigma_i\sigma_{i+1}\rangle = \frac{1}{N\beta}\frac{\partial \ln Z}{\partial J} = \langle\sigma\rangle\frac{\sinh(\beta(H+J\langle\sigma\rangle))}{\cosh(\beta(H+J\langle\sigma\rangle))} = \langle\sigma\rangle^2 \tag{11.36}$$

The result is reasonable, since σ_i and σ_{i+1} are independent of each other in the B–W approximation.

Other properties can be calculated as follows. The mean number of spin alterations is

$$\langle N_{\text{alt}}\rangle = \frac{N}{2}(1 - \langle\sigma\rangle^2) \tag{11.37}$$

Therefore, the domain size by thermal activation is

$$L_{\text{th}} = \frac{2}{1 - \langle\sigma\rangle^2} \tag{11.38}$$

if $N \gg 1$.

11.3.2 Flory–Huggins (F–H) Approximation

This subsection introduces another approximation method, called **Flory–Huggins** (F–H) **mean-field theory**. The theory starts with dividing the sum of states $\sum_{\sigma_1 = \pm 1}\sum_{\sigma_2 = \pm 1}\cdots\sum_{\sigma_N = \pm 1}$ in Eq. (11.3) according to total magnetization M:

$$M = \sum_{i=1}^{N}\sigma_i \tag{11.39}$$

where $-N \leq M \leq N$.

Each state with M has $(N + M)/2$ up spins and $(N - M)/2$ down spins. Therefore, the number of states in that group is equal to $_N C_{(N+M)/2}$.

Table 11.5 Numbers of nearest-neighbor pairs assuming random mixing.

Spin pair	Probability
Up–up	$[(N + M)/2N]^2$
Down–down	$[(N - M)/2N]^2$
Up–down (and down–up)	$2[(N + M)/2N][(N - M)/2N]$
Total	1

The F–H approximation calculates the interaction between adjacent pairs of spins, assuming random mixing of up and down spins. The probability for a nearest-neighbor pair of $+1$ and $+1$ is listed in Table 11.5, and so are the probability for a pair of -1 and -1, and $+1$ and -1.

Then,

$$\frac{1}{N}\sum_{i=1}^{N}\sigma_i\sigma_{i+1} = \left(\frac{N+M}{2N}\right)^2(+1)^2 + \left(\frac{N-M}{2N}\right)^2(-1)^2$$

$$+ 2\frac{N + MN - M}{2N2N}(+1)(-1) = \frac{M^2}{N^2} \qquad (11.40)$$

Therefore, the mean energy of the states with M is $-HM - JM^2/N$. The partition function is calculated as the sum of $\exp(-\beta \times \text{Energy})$ times the number of states with M:

$$Z = \sum_{M=-N}^{N}\binom{N}{(N+M)/2}\exp\left(\beta HM + \beta J\frac{M^2}{N}\right) \qquad (11.41)$$

where the running index M in the sum takes every other integer ($M = -N$, $-N + 2, \ldots N - 2, N$).

The sum in Eq. (11.41) cannot be expressed by elementary functions. Here, we apply the method of steepest descent we learned in Section 9.4 to approximately calculate Z.

For convenience, we rewrite the partition function into

$$Z = \sum_{M=-N}^{N} f(M) \qquad (11.42)$$

where

$$f(M) \equiv \binom{N}{(N+M)/2}\exp\left(\beta HM + \beta J\frac{M^2}{N}\right) \qquad (11.43)$$

Figure 11.15 shows a plot of $f(M)$ for a few different values of βH and βJ. When N is large, the curve looks like a normal distribution. Then, we should be able to approximate the sum with an integral of the curve of the normal distribution.

Figure 11.15 Distribution function of M for $N = 1000$ and different values of βH and βJ. The values of βH and βJ are indicated adjacent to each curve.

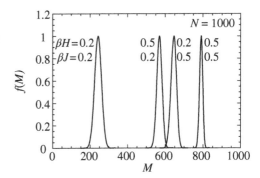

Now, we approximate $f(M)$ by a normal distribution of M:

$$f(M) \cong f(M_{max}) \exp\left(-\frac{A_2}{2}(M - M_{max})^2\right) \tag{11.44}$$

where $f(M)$ maximizes at $M = M_{max}$, and a parameter A_2 is optimized for the approximation.

To find M_{max}, we work on the logarithm of $f(M)$:

$$\ln f(M) = \ln N! - \ln \left(\frac{N+M}{2}\right)! - \ln \left(\frac{N-M}{2}\right)! + \beta HM + \beta J\frac{M^2}{N} \tag{11.45}$$

Using Stirling's formula, $\ln f(M)$ is approximated as

$$\ln f(M) = N \ln N - \frac{N+M}{2} \ln \frac{N+M}{2} - \frac{N-M}{2} \ln \frac{N-M}{2}$$
$$+ \beta HM + \beta J\frac{M^2}{N} \tag{11.46}$$

We then differentiate $f(M)$ by M, regarding M as a continuous variable:

$$\frac{\partial}{\partial M} \ln f(M) = -\frac{1}{2} \ln \frac{N+M}{2} + \frac{1}{2} \ln \frac{N-M}{2} + \beta H + 2\beta J\frac{M}{N} \tag{11.47}$$

The derivative vanishes at $M = M_{max}$ given by

$$\ln \frac{N+M_{max}}{N-M_{max}} = 2\beta H + 4\beta J\frac{M_{max}}{N} \tag{11.48}$$

Later, we show that $\ln Z$ is approximated by its peak value, $\ln f(M_{max})$; see Eq. (11.58). Since $\partial \ln f(M)/\partial H = \beta M$ in Eq. (11.45), we have, with Eq. (11.5),

$$\langle \sigma \rangle = \frac{M_{max}}{N} \tag{11.49}$$

Then, Eq. (11.48) is rewritten to

$$\ln \frac{1 + \langle \sigma \rangle}{1 - \langle \sigma \rangle} = 2\beta H + 4\beta J\langle \sigma \rangle \tag{11.50}$$

which is also written as

$$\langle\sigma\rangle = \tanh(\beta H + 2\beta J\langle\sigma\rangle) \tag{11.51}$$

This equation is similar to the one derived in the B–W approximation (Eq. (11.28)). The difference is a factor of 2 for the MF interaction within the hyperbolic tangent function. The difference is due to the squared-M dependence of the interaction in the F–H approximation, as opposed to the linear M dependence in the B–W approximation.

Now we find an expression of $\ln f(M_{max})$ using $\langle\sigma\rangle$. With $M_{max} = N\langle\sigma\rangle$, Eq. (11.46) evaluated at $M = M_{max}$ is

$$\ln f(M_{max}) = N\left[\ln 2 - \frac{1}{2}\ln(1-\langle\sigma\rangle^2) - \frac{1}{2}\langle\sigma\rangle\ln\frac{1+\langle\sigma\rangle}{1-\langle\sigma\rangle}\right.$$
$$\left. + \beta H\langle\sigma\rangle + \beta J\langle\sigma\rangle^2\right] \tag{11.52}$$

With Eq. (11.50), we arrive at

$$\ln f(M_{max}) = N\left[\ln 2 - \frac{1}{2}\ln(1-\langle\sigma\rangle^2) - \beta J\langle\sigma\rangle^2\right] \tag{11.53}$$

We thus find that $\ln f(M_{max})$ scales with N.

The parameter A_2 in Eq. (11.44) is found from

$$A_2 = -\frac{\partial^2 \ln f(M)}{\partial M^2}\bigg|_{M=M_{max}} \tag{11.54}$$

which is calculated as

$$A_2 = \frac{N}{N^2 - M_{max}^2} - \frac{2\beta J}{N} \tag{11.55}$$

Since M takes every other integral value,

$$Z \cong \frac{1}{2}f(M_{max})\int_{-\infty}^{\infty}\exp\left[-\frac{A_2}{2}(M-M_{max})^2\right]dM = \left(\frac{\pi}{2A_2}\right)^{1/2}f(M_{max}) \tag{11.56}$$

Obviously, $\ln Z$ consists of two parts:

$$\ln Z \cong \frac{1}{2}\ln\frac{\pi}{2A_2} + \ln f(M_{max}) \tag{11.57}$$

Equation (11.55) shows that $\ln(\pi/2A_2) \sim \ln N$, as opposed to $\ln f(M_{max})$ that scales with N. Therefore, we can neglect the first term in Eq. (11.57):

$$\ln Z \cong N\left[\ln 2 - \frac{1}{2}\ln(1-\langle\sigma\rangle^2) - \beta J\langle\sigma\rangle^2\right] \tag{11.58}$$

It may appear that Eq. (11.53) shows that $f(M_{max})$ is independent of H or J. It is not correct: $\langle\sigma\rangle$ is an implicit function of β, H, and J, and therefore $\ln Z \cong \ln f(M_{max})$ depends on β, H, and J.

Let us confirm that $\ln Z \cong \ln f(M_{max})$ with $\ln f(M_{max})$ given by Eq. (11.52) gives a correct pathway to calculating statistical averages. First, we calculate $\langle \sigma \rangle$. Using Eq. (11.5), we can easily confirm that it works (see Problem 11.5). Next, we calculate $\langle \sigma_i \sigma_{i+1} \rangle$ using the formula, Eq. (11.6) (see Problem 11.6):

$$\langle \sigma_i \sigma_{i+1} \rangle = \frac{1}{N\beta} \frac{\partial \ln f(M_{max})}{\partial J} = \langle \sigma \rangle^2 \tag{11.59}$$

The result is reasonable, since the F–H approximation neglects details of the spin sequence.

11.3.3 Approximation by a Mean-Field (MF) Theory

There is yet another method within the MF approximation that can be convenient in many systems including Ising models. That method is simply called a **mean-field theory**, and we will apply the method to one-dimensional Ising model. It starts with dividing σ_i into its statistical mean $\langle \sigma \rangle$ and deviation $\delta\sigma_i$:

$$\sigma_i = \langle \sigma \rangle + \delta\sigma_i \tag{11.60}$$

Note that $\langle \sigma \rangle$ is not $N^{-1}\Sigma\sigma_i$: The latter can be different from replica to replica for a given system. Then,

$$\sigma_i\sigma_{i+1} = (\langle \sigma \rangle + \delta\sigma_i)(\langle \sigma \rangle + \delta\sigma_{i+1}) = \langle \sigma \rangle^2 + \langle \sigma \rangle(\delta\sigma_i + \delta\sigma_{i+1}) + \delta\sigma_i\delta\sigma_{i+1} \tag{11.61}$$

The MF theory neglects terms of $O(\delta\sigma_i^2)$ and higher. Then,

$$\sum_{i=1}^{N} \sigma_i\sigma_{i+1} \cong N\langle \sigma \rangle^2 + 2\langle \sigma \rangle \sum_{i=1}^{N} \delta\sigma_i \tag{11.62}$$

We now replace $\delta\sigma_i$ with $\sigma_i - \langle \sigma \rangle$:

$$\sum_{i=1}^{N} \sigma_i\sigma_{i+1} \cong N\langle \sigma \rangle^2 + 2\langle \sigma \rangle \sum_{i=1}^{N} (\sigma_i - \langle \sigma \rangle) = 2\langle \sigma \rangle \sum_{i=1}^{N} \sigma_i - N\langle \sigma \rangle^2 \tag{11.63}$$

The energy is now expressed as

$$E = -H \sum_{i=1}^{N} \sigma_i - J \sum_{i=1}^{N} \sigma_i\sigma_{i+1} \cong JN\langle \sigma \rangle^2 - \sum_{i=1}^{N} \sigma_i(H + 2J\langle \sigma \rangle) \tag{11.64}$$

This expression of E simplifies the calculation of the partition function.

$$Z = \exp(-\beta NJ\langle \sigma \rangle^2) \left[\sum_{\sigma=\pm 1} e^{\beta(H+2J\langle \sigma \rangle)\sigma} \right]^N$$

$$= \exp(-\beta NJ\langle \sigma \rangle^2)[2\cosh(\beta(H + 2J\langle \sigma \rangle))]^N \tag{11.65}$$

We now express $\langle \sigma \rangle$ using this expression of Z:

$$\langle \sigma \rangle = \frac{1}{N\beta} \frac{\partial \ln Z}{\partial H} = \tanh(\beta(H + 2J\langle \sigma \rangle)) \tag{11.66}$$

Note that $\langle \sigma \rangle$ is fixed in taking the partial derivative. This result is identical to the one obtained in the F–H approximation; see Eq. (11.51).

Likewise, $\langle \sigma_i \sigma_{i+1} \rangle$ is calculated as

$$\langle \sigma_i \sigma_{i+1} \rangle = \frac{1}{N\beta} \frac{\partial \ln Z}{\partial J} = \frac{1}{N\beta} \left[-\beta N \langle \sigma \rangle^2 + N \frac{\sinh(\beta(H + 2J\langle \sigma \rangle))}{\cosh(\beta(H + 2J\langle \sigma \rangle))} 2\beta \langle \sigma \rangle \right]$$

$$= -\langle \sigma \rangle^2 + 2\langle \sigma \rangle \tanh(\beta(H + 2J\langle \sigma \rangle)) = \langle \sigma \rangle^2 \tag{11.67}$$

Again, $\langle \sigma \rangle$ is fixed when differentiating $\ln Z$ by J, although $\langle \sigma \rangle$ varies with J. The relationship between $\langle \sigma_i \sigma_{i+1} \rangle$ and $\langle \sigma \rangle$ is identical to the one in the earlier two approximations.

We can prove that Eq. (11.65) is identical to Eq. (11.58). Therefore, the F–H approximation is equivalent to the MF theory. Neglecting the product of deviations from the statistical mean in the interaction is equivalent to calculating the interaction according to the probability of forming an adjacent pair.

Note that none of these approximation methods provide asymptotic expressions of $\langle \sigma \rangle$ in large N. They provide the large-N limit only. To find the large-N asymptotes, we need to rely on the exact solution you learn in the following section.

11.4 Exact Solution of 1D Ising Model

11.4.1 General Formula

The 1D linear Ising model can be solved exactly. In fact, Ernst Ising came up with the exact solution several years after he was given the problem. We will follow his method here.

We start with explicitly writing the partition function Z_n for a linear array of the first n spins out of a total N spins:

$$Z_n = \sum_{\sigma_1} \sum_{\sigma_2} \cdots \sum_{\sigma_n} \exp\left(\beta H \sum_{i=1}^{n} \sigma_i + \beta J \sum_{i=1}^{n-1} \sigma_i \sigma_{i+1} \right) \tag{11.68}$$

where $n = 1, 2, \ldots, N$. We then separate the n-fold sum into the $(n-1)$-fold sum and the sum by σ_n:

$$Z_n = \sum_{\sigma_1} \sum_{\sigma_2} \cdots \sum_{\sigma_{n-1}} \exp\left(\beta H \sum_{i=1}^{n-1} \sigma_i + \beta J \sum_{i=1}^{n-2} \sigma_i \sigma_{i+1} \right)$$

$$\times \sum_{\sigma_n} \exp(\beta H \sigma_n + \beta J \sigma_{n-1} \sigma_n) \tag{11.69}$$

Let us define

$$Z_n(\sigma_n) \equiv \sum_{\sigma_1} \sum_{\sigma_2} \cdots$$

$$\times \sum_{\sigma_{n-1}} \exp\left(\beta H \sum_{i=1}^{n-1} \sigma_i + \beta J \sum_{i=1}^{n-2} \sigma_i \sigma_{i+1} + \beta H \sigma_n + \beta J \sigma_{n-1} \sigma_n \right)$$

(11.70)

This function accounts for the partition function for the n-spin system in which the nth spin is σ_n. The overall partition function Z_n for the n spins consists of $Z_n(1)$ for $\sigma_n = 1$ and $Z_n(-1)$ for $\sigma_n = -1$:

$$Z_n = Z_n(1) + Z_n(-1)$$

(11.71)

In Eq. (11.70), $\beta H \sigma_n$ and $\beta J \sigma_{n-1} \sigma_n$ are due to the nth spin. $Z_n(1)$ can be constructed as a sum of the product of $Z_{n-1}(1)$ and an additional factor, $e^{\beta(H+J)}$, and the product of $Z_{n-1}(-1)$ and an additional factor, $e^{\beta(H-J)}$. The first product is for $\sigma_{n-1} = \sigma_n = 1$ and the second product for $\sigma_{n-1} = -1$ and $\sigma_n = 1$. This relationship is conveniently expressed by a matrix:

$$Z_n(1) = [Z_{n-1}(1) \ Z_{n-1}(-1)] \begin{bmatrix} e^{\beta(H+J)} \\ e^{\beta(H-J)} \end{bmatrix}$$

(11.72)

Likewise, $Z_n(-1)$ can be constructed as a sum of the product of $Z_{n-1}(1)$ and an additional factor, $e^{\beta(-H-J)}$, and the product of $Z_{n-1}(-1)$ and an additional factor, $e^{\beta(-H+J)}$. The first product is for $\sigma_{n-1} = 1$ and $\sigma_n = -1$ and the second product for $\sigma_{n-1} = \sigma_n = -1$. In the matrix form,

$$Z_n(-1) = [Z_{n-1}(1) \ Z_{n-1}(-1)] \begin{bmatrix} e^{\beta(-H-J)} \\ e^{\beta(-H+J)} \end{bmatrix}$$

(11.73)

The two matrix expressions can be combined to

$$[Z_n(1) \ Z_n(-1)] = [Z_{n-1}(1) \ Z_{n-1}(-1)] \begin{bmatrix} e^{\beta(H+J)} & e^{\beta(-H-J)} \\ e^{\beta(H-J)} & e^{\beta(-H+J)} \end{bmatrix}$$

(11.74)

By multiplying a 2×2 matrix from the right, $[Z_{n-1}(1) \ Z_{n-1}(-1)]$ is converted to $[Z_n(1) \ Z_n(-1)]$. The matrix may be called a transfer matrix.

To simplify the notation, let us introduce h and j as

$$h \equiv e^{\beta H}, \ j \equiv e^{\beta J}$$

(11.75)

We denote the transfer matrix by $\mathbf{A}$:

$$\mathbf{A} \equiv \begin{bmatrix} hj & h^{-1}j^{-1} \\ hj^{-1} & h^{-1}j \end{bmatrix}$$

(11.76)

Then, Eq. (11.74) is

$$[Z_n(1) \ Z_n(-1)] = [Z_{n-1}(1) \ Z_{n-1}(-1)]\mathbf{A}$$

(11.77)

Repeatedly applying **A**, we arrive at

$$[Z_n(1) \ Z_n(-1)] = [Z_{n-2}(1) \ Z_{n-2}(-1)]\mathbf{A}^2 = \cdots = [Z_1(1) \ Z_1(-1)]\mathbf{A}^{n-1}$$

$$(11.78)$$

The first 1×2 matrix is

$$[Z_1(1) \ Z_1(-1)] = [h \ h^{-1}] \tag{11.79}$$

Then,

$$[Z_n(1) \ Z_n(-1)] = [h \ h^{-1}]\mathbf{A}^{n-1} \tag{11.80}$$

An explicit expression of Z_n is

$$Z_n = Z_n(1) + Z_n(-1) = [h \ h^{-1}]\mathbf{A}^{n-1} \begin{bmatrix} 1 \\ 1 \end{bmatrix} \tag{11.81}$$

$\mathbf{A}^n$ can be calculated by diagonalizing $\mathbf{A}$ by multiplying a matrix $\mathbf{B}$ from the right and its inverse matrix from the left:

$$\mathbf{B}^{-1}\mathbf{A}\mathbf{B} = \Lambda \tag{11.82}$$

The diagonal matrix

$$\Lambda = \begin{bmatrix} \lambda_1 & 0 \\ 0 & \lambda_2 \end{bmatrix} \tag{11.83}$$

has two elements λ_1 and λ_2 as the two eigenvalues of $\mathbf{A}$:

$$\mathbf{A}\mathbf{b}_i = \lambda_i \mathbf{b}_i \tag{11.84}$$

The 2×2 matrix $\mathbf{B}$ is constructed from eigenvectors $\mathbf{b}_1$ and $\mathbf{b}_2$ of the two eigenvalues as

$$\mathbf{B} = [\mathbf{b}_1 \ \mathbf{b}_2] \tag{11.85}$$

Multiplying Eq. (11.82) n times gives

$$\underbrace{(\mathbf{B}^{-1}\mathbf{A}\mathbf{B})(\mathbf{B}^{-1}\mathbf{A}\mathbf{B}) \cdots (\mathbf{B}^{-1}\mathbf{A}\mathbf{B})}_{n \text{ factors}} = \Lambda^n \tag{11.86}$$

The left-hand side is simplified to

$$\mathbf{B}^{-1}\mathbf{A}^n\mathbf{B} = \Lambda^n \tag{11.87}$$

Multiplying $\mathbf{B}$ from left and $\mathbf{B}^{-1}$ from right changes the above equation to

$$\mathbf{A}^n = \mathbf{B}\Lambda^n\mathbf{B}^{-1} \tag{11.88}$$

Our next job is to find λ_1 and λ_2. The two eigenvalues vanish the determinant:

$$|\mathbf{A} - \lambda\mathbf{I}| = \begin{vmatrix} hj - \lambda & h^{-1}j^{-1} \\ hj^{-1} & h^{-1}j - \lambda \end{vmatrix} = 0 \tag{11.89}$$

Therefore, λ_1 and λ_2 are the zeros of the quadratic equation:

$$\lambda^2 - j(h + h^{-1})\lambda + j^2 - j^{-2} = 0 \qquad (11.90)$$

Using the quadratic formula, we find that the two eigenvalues λ_1 and λ_2 $(\lambda_1 > \lambda_2)$ are

$$\lambda_1 = \frac{1}{2}\{j(h + h^{-1}) + [j^2(h - h^{-1})^2 + 4j^{-2}]^{1/2}\} \qquad (11.91)$$

$$\lambda_2 = \frac{1}{2}\{j(h + h^{-1}) - [j^2(h - h^{-1})^2 + 4j^{-2}]^{1/2}\} \qquad (11.92)$$

The eigenvalues are listed here for two simple situations. When $h = 1$ $(H = 0)$,

$$\lambda_1 = j + j^{-1} = 2\cosh\beta J, \quad \lambda_2 = j - j^{-1} = 2\sinh\beta J \qquad (11.93)$$

When $j = 1$ $(J = 0)$,

$$\lambda_1 = h + h^{-1} = 2\cosh\beta H, \quad \lambda_2 = 0 \qquad (11.94)$$

Once the eigenvalues are found, we could proceed to find the matrix $\mathbf{B}$. However, when $N \gg 1$, we can show that it is not necessary, and $\ln Z_N$ is approximated as

$$\ln Z_N \cong N \ln \lambda_1 \qquad (11.95)$$

Here, we first use this approximate expression of Z_N to find $\langle\sigma\rangle$ and $\langle\sigma_i\sigma_{i+1}\rangle$. Subsequently, we learn how to obtain the exact expression of Z_N.

11.4.2 Large-N Approximation

Since we are using h and j in place of H and J, we rewrite Eq. (11.5) into

$$\langle\sigma\rangle = \frac{1}{N\beta}\frac{\partial\ln Z_N}{\partial H} = \frac{1}{N}\frac{\partial\ln Z_N}{\partial\ln h} = \frac{h}{N}\frac{\partial\ln Z_N}{\partial h} \qquad (11.96)$$

Applying this formula to Eq. (11.95) with Eq. (11.91) gives

$$\langle\sigma\rangle \cong \frac{j(h - h^{-1})}{[j^2(h - h^{-1})^2 + 4j^{-2}]^{1/2}} = \frac{e^{\beta J}\sinh\beta H}{(e^{2\beta J}\sinh^2\beta H + e^{-2\beta J})^{1/2}} \qquad (11.97)$$

Here, we look at simple situations.

(1) When $J = 0$, we recover the simple result:

$$\langle\sigma\rangle \cong \tanh\beta H \qquad (11.98)$$

(2) When $H = 0$, $\langle\sigma\rangle = 0$ whatever value J has. The problem we pointed out in Section 11.2 does not exist in this exact solution.

(3) When $\beta J \ll 1$ (but βH is not too large),

$$\langle\sigma\rangle \equiv \tanh\beta H\left(1 + \frac{2\beta J}{\cosh^2\beta H}\right) \qquad (11.99)$$

A positive J lifts $\langle\sigma\rangle$, but its effect decreases with an increasing βH.

(4) When $\beta H \ll 1$ (but βJ is not too large),

$$\langle \sigma \rangle \cong e^{2\beta J} \beta H \tag{11.100}$$

A positive J enhances $\langle \sigma \rangle$, and a negative J suppresses $\langle \sigma \rangle$.

(5) When $e^{2\beta J} \sinh \beta H \gg 1$ (either $\beta J \gg 1$ or $\beta H \gg 1$),

$$\langle \sigma \rangle \cong 1 - \frac{1}{2e^{4\beta J} \sinh^2 \beta H} \tag{11.101}$$

Figure 11.16 shows the plot of $\langle \sigma \rangle$ as a function of βJ for several different values of βH. Note that, at $\beta H = 0$, $\langle \sigma \rangle = 0$ for all value of βJ. When $\beta H > 0$, $\langle \sigma \rangle$ increases with βJ as a sigmoidal function. With an increasing βH, $\langle \sigma \rangle$ reaches a near plateau value of 1 at a smaller βJ.

Likewise, $\langle \sigma_i \sigma_{i+1} \rangle$ is obtained using the formula:

$$\langle \sigma_i \sigma_{i+1} \rangle = \frac{1}{N\beta} \frac{\partial \ln Z_N}{\partial J} = \frac{1}{N} \frac{\partial \ln Z_N}{\partial \ln j} = \frac{j}{N} \frac{\partial \ln Z_N}{\partial j} \tag{11.102}$$

After some calculation, we obtain

$$\langle \sigma_i \sigma_{i+1} \rangle = \frac{1}{j^2 - j^{-2}} \left[j^2 + j^{-2} - \frac{2(h + h^{-1})}{\sqrt{j^4(h - h^{-1})^2 + 4}} \right] \tag{11.103}$$

In terms of H and J, it is

$$\langle \sigma_i \sigma_{i+1} \rangle = \frac{1}{\tanh 2\beta J} - \frac{\cosh \beta H}{\sinh 2\beta J (e^{4\beta J} \sinh^2 \beta H + 1)^{1/2}} \tag{11.104}$$

Here, we look at simple situations.

(1) When $H = 0$,

$$\langle \sigma_i \sigma_{i+1} \rangle = \frac{1}{\tanh 2\beta J} - \frac{1}{\sinh 2\beta J} = \tanh \beta J \tag{11.105}$$

A large βJ promotes aligned pairs, a reasonable result.

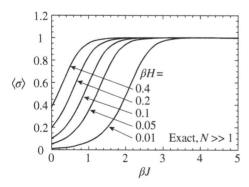

Figure 11.16 Exact $\langle \sigma \rangle$ in a 1D Ising model, plotted as a function of βJ for different values of βH. The value of βH is indicated adjacent to each curve. When $\beta H = 0$, $\langle \sigma \rangle = 0$ for all values of βJ.

(2) When $\beta J \ll 1$,

$$(e^{4\beta J}\sinh^2\beta H + 1)^{1/2} \cong (\cosh^2\beta H + 4\beta J\sinh^2\beta H)^{1/2}$$
$$\cong \cosh\beta H(1 + 4\beta J\tanh^2\beta H)^{1/2}$$
$$\cong \cosh\beta H(1 + 2\beta J\tanh^2\beta H) \qquad (11.106)$$

Then,

$$\langle\sigma_i\sigma_{i+1}\rangle \cong \frac{1}{\sinh 2\beta J}\left(\cosh 2\beta J - \frac{1}{1 + 2\beta J\tanh^2\beta H}\right) \cong \tanh^2\beta H$$
$$(11.107)$$

This result is consistent with the fact that the direction of two spins are independent of each other, i.e. $\langle\sigma_i\sigma_{i+1}\rangle = \langle\sigma_i\rangle\langle\sigma_{i+1}\rangle$, and each mean is $\tanh\beta H$ when $\beta J \ll 1$.

(3) When $\beta J \gg 1$ and $\beta H > 0$,

$$\langle\sigma_i\sigma_{i+1}\rangle \cong 1 - 2e^{-4\beta J}\left(1 + \frac{1}{\tanh\beta H}\right) \qquad (11.108)$$

Figure 11.17 shows a plot of $\langle\sigma_i\sigma_{i+1}\rangle$ as a function of βJ for different values of βH. Unlike the plot of $\langle\sigma\rangle$ in Figure 11.16, $\langle\sigma_i\sigma_{i+1}\rangle$ increases linearly with βJ when it is small for all values of βH.

11.4.3 Exact Partition Function for Arbitrary N

To obtain the exact Z_N for an arbitrary N, we need to find **B**. The eigenvectors for λ_1 and λ_2 are

$$\mathbf{b}_1 = \begin{bmatrix} 1 \\ hj(\lambda_1 - hj) \end{bmatrix} \qquad (11.109)$$

$$\mathbf{b}_2 = \begin{bmatrix} 1 \\ hj(\lambda_2 - hj) \end{bmatrix} \qquad (11.110)$$

Figure 11.17 Exact result for $\langle\sigma_i\sigma_{i+1}\rangle$, in a 1D Ising model, plotted as a function of βJ for different values of βH. The value of βH is indicated adjacent to each curve.

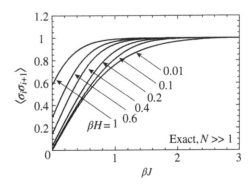

Therefore,

$$\mathbf{B} = \begin{bmatrix} 1 & 1 \\ hj(\lambda_1 - hj) & hj(\lambda_2 - hj) \end{bmatrix} \tag{11.111}$$

and

$$\mathbf{B}^{-1} = \frac{1}{hj(\lambda_2 - \lambda_1)} \begin{bmatrix} hj(\lambda_2 - hj) & -1 \\ -hj(\lambda_1 - hj) & 1 \end{bmatrix} \tag{11.112}$$

Since

$$\mathbf{\Lambda}^n = \begin{bmatrix} \lambda_1^n & 0 \\ 0 & \lambda_2^n \end{bmatrix} \tag{11.113}$$

we obtain

$$\mathbf{A}^n = \frac{1}{\lambda_2 - \lambda_1} \begin{bmatrix} (\lambda_2 - hj)\lambda_1^n - (\lambda_1 - hj)\lambda_2^n & (-\lambda_1^n + \lambda_2^n)/hj \\ hj(\lambda_1 - hj)(\lambda_2 - hj)(\lambda_1^n - \lambda_2^n) & -(\lambda_1 - hj)\lambda_1^n + (\lambda_2 - hj)\lambda_2^n \end{bmatrix} \tag{11.114}$$

With $\lambda_1 + \lambda_2 = j(h + h^{-1})$, $\lambda_1 \lambda_2 = j^2 - j^{-2}$ and $(j^2 - j^{-2})\lambda_i^{n-1} = j(h + h^{-1})\lambda_i^n - \lambda_i^{n+1}$ derived from Eq. (11.90), the matrix is rewritten to

$$\mathbf{A}^n = \frac{1}{\lambda_2 - \lambda_1} \begin{bmatrix} -(\lambda_1^{n+1} - \lambda_2^{n+1}) + \frac{j}{h}(\lambda_1^n - \lambda_2^n) & -\frac{1}{hj}(\lambda_1^n - \lambda_2^n) \\ -\frac{h}{j}(\lambda_1^n - \lambda_2^n) & -(\lambda_1^{n+1} - \lambda_2^{n+1}) + hj(\lambda_1^n - \lambda_2^n) \end{bmatrix} \tag{11.115}$$

Finally with Eq. (11.81), we arrive at

$$Z_N = \left(h + \frac{1}{h}\right) \frac{\lambda_1^N - \lambda_2^N}{\lambda_1 - \lambda_2} - 2\left(j - \frac{1}{j}\right) \frac{\lambda_1^{N-1} - \lambda_2^{N-1}}{\lambda_1 - \lambda_2} \tag{11.116}$$

This result, for $N = 2$, agrees with the one calculated in the same way as we did in Section 11.2 (Problem 11.10). We can also confirm that the result for $N = 3$ is identical to Eq. (11.14).

We discuss how to calculate the partition function of a 1D ring Ising model in Section 11.4.

Figure 11.18 shows a plot of $\ln Z_N$ as a function of j (>1; $J > 0$) for different values of h. Part (a) is for $N = 10$, and part (b) for $N = 100$. Since $\langle \sigma \rangle$ and $\langle \sigma_i \sigma_{i+1} \rangle$ are calculated by differentiating $\ln Z_N$ by h and j, respectively, these plots can tell how $\langle \sigma \rangle$ and $\langle \sigma_i \sigma_{i+1} \rangle$ change with h and j, and also with N. The main difference between the two parts is the y-axis scale, as the dominant component of $\ln Z_N$ is $N \ln \lambda_1$. Otherwise, the two parts are similar, especially when j is small. Then, $\langle \sigma \rangle$ will be similar for $N = 10$ and $N = 100$. Likewise, $\langle \sigma_i \sigma_{i+1} \rangle$ will be similar for different values of N.

When $N \gg 1$, $(\lambda_2/\lambda_1)^N$ can be neglected relative to 1. Therefore,

$$Z_N \cong \frac{\lambda_1^N}{\lambda_1 - \lambda_2} \left[\left(h + \frac{1}{h}\right) - \left(j - \frac{1}{j}\right) \frac{2}{\lambda_1} \right] \tag{11.117}$$

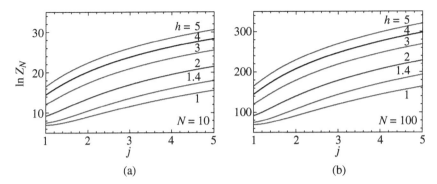

Figure 11.18 Plot of $\ln Z_N$ for a given h as a function of j. (a) $N = 10$, (b) $N = 100$. The values of h are indicated adjacent to the curves.

Figure 11.19 Ratio of two eigenvalues, plotted as a function of $j = \exp(\beta J)$ for different values of $h = \exp(\beta H)$. The value of h is indicated adjacent to each curve.

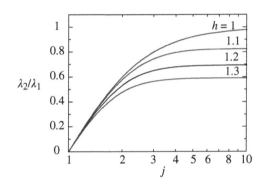

With Eq. (11.90), it is rewritten to

$$Z_N \cong \frac{\lambda_1^N}{\lambda_1 - \lambda_2}\frac{1}{j\lambda_1}\left[\lambda_1^2 - \left(j - \frac{1}{j}\right)^2\right] \tag{11.118}$$

Figure 11.19 shows a plot of λ_2/λ_1 as a function of j for $h = 1, 1.1, 1.2$, and 1.3. For $N = 100$, $0.9^N \cong 2.66 \times 10^{-5}$, indicating that $(\lambda_2/\lambda_1)^N \ll 1$ at all values of j for $h = 1, 1.1, 1.2$, and 1.3. When $h = 1$ ($H = 0$), j is large and N is small, $(\lambda_2/\lambda_1)^N$ may not be neglected, but that is a rare situation. In the first-order correction to $\ln Z_N$, we can safely neglect $(\lambda_2/\lambda_1)^N$.

The large-N asymptotes of $\ln Z$, $\langle \sigma \rangle$, and $\langle \sigma_i \sigma_{i+1} \rangle$ are calculated from Eq. (11.118). The following equations, derived from Eq. (11.90), are useful:

$$\frac{\partial \lambda_1}{\partial h} = \frac{\lambda_1}{\lambda_1 - \lambda_2}j\left(1 - \frac{1}{h^2}\right) \tag{11.119}$$

$$\frac{\partial \lambda_1}{\partial j} = \frac{1}{\lambda_1 - \lambda_2}\left[\lambda_1\left(h + \frac{1}{h}\right) - \frac{2}{j}\left(j^2 + \frac{1}{j^2}\right)\right] = \frac{\lambda_1}{j} - \frac{4}{j^3(\lambda_1 - \lambda_2)} \tag{11.120}$$

Table 11.6 Summary of results of the exact calculation of the partition function in an N-spin linear Ising model.

Function	Large-N asymptote
$\ln Z$	$N \ln \lambda_1 - \ln(\lambda_1 - \lambda_2) + \ln\left(\left(h + \dfrac{1}{h}\right) - \left(j - \dfrac{1}{j}\right)\dfrac{2}{\lambda_1}\right)$
$\langle \sigma \rangle$	$\dfrac{j(h - h^{-1})}{\lambda_1 - \lambda_2}\left\{1 - \dfrac{1}{N}\left[\dfrac{\lambda_1 + \lambda_2}{\lambda_1 - \lambda_2} - \dfrac{\lambda_1^2 + (j - j^{-1})^2}{\lambda_1^2 - (j - j^{-1})^2}\right]\right\}$
$\langle \sigma_i \sigma_{i+1} \rangle$	$1 - \dfrac{4}{j^2 \lambda_1(\lambda_1 - \lambda_2)} - \dfrac{1}{N}\left[1 - \dfrac{8}{j^2(\lambda_1 - \lambda_2)^2} + \dfrac{4}{\lambda_1(\lambda_1 - \lambda_2)}\right]$

Results are shown up to $O(N^{-1})$.

The results of the calculation are summarized in Table 11.6 together with $\ln Z$.

Depending on the situation, we can apply different approximations to λ_1, λ_2, j, and h, as we did in deriving equations in Section 11.4.2. For example, when $|\beta H| \ll 1$,

$$\langle \sigma \rangle \cong e^{2\beta J} \sinh \beta H \left(1 - \frac{1}{N} \sinh 2\beta J\right) \tag{11.121}$$

If we have data of $\langle \sigma \rangle$ as a function of N, curve fitting will allow us to estimate J, and also possibly H.

The method of transfer matrix is convenient in many systems. We use the method in Chapter 13 when we consider the helix–coil transition in polypeptides.

11.4.4 Ring Ising Model, Arbitrary N

The method of transfer matrix can also be applied to a ring arrangement. For that purpose, imagine the transfer matrix for the $(N+1)$th spin. From Eq. (11.78), we have

$$[Z_1(1) \ Z_1(-1)]\mathbf{A}^N = [Z_{N+1}(1) \ Z_{N+1}(-1)] \tag{11.122}$$

This equation is equivalent to

$$Z_1(1)(\mathbf{A}^N)_{11} + Z_1(-1)(\mathbf{A}^N)_{21} = Z_{N+1}(1) \tag{11.123}$$

$$Z_1(1)(\mathbf{A}^N)_{12} + Z_1(-1)(\mathbf{A}^N)_{22} = Z_{N+1}(-1) \tag{11.124}$$

where $(\mathbf{A}^N)_{ij}$ denotes the i–j element of matrix $\mathbf{A}^N$.

In the ring arrangement, the state of the $(N+1)$th spin must be identical to the state of the first spin. The contribution to the partition function by the state of $\sigma_1 = \sigma_{N+1} = 1$ is obtained by setting $Z_1(1) = h$ and $Z_1(-1) = 0$ in Eq. (11.123), which gives $Z_{N+1}(1) = h(\mathbf{A}^N)_{11}$. This $Z_{N+1}(1)$ counts the energy from the magnetic field twice (for the first and $(N+1)$th spins), so we need to divide

$Z_{N+1}(1)$ by h. The interaction between the Nth and $(N+1)$th spins is taken into account correctly, and so are the interactions between all the other pairs and the energy by the field. The other equation gives $Z_{N+1}(-1) = h(\mathbf{A}^N)_{12}$, but we discard it. Thus, we find that $h(\mathbf{A}^N)_{11}/h = (\mathbf{A}^N)_{11}$ is the contribution to the partition function of the N spins in the ring by all the states with $\sigma_1 = 1$. Likewise, the contribution to the partition function by the state of $\sigma_1 = \sigma_{N+1} = -1$ is obtained by setting $Z_1(1) = 0$ and $Z_1(-1) = h^{-1}$ in Eq. (11.124), which gives $Z_{N+1}(-1) = (\mathbf{A}^N)_{22}/h$. This $Z_{N+1}(-1)$ counts the energy from the magnetic field twice (for the first and $(N+1)$th spins), so we need to multiply $Z_{N+1}(-1)$ by h. We find that $(\mathbf{A}^N)_{22}$ is the contribution to the partition function of the N spins in the ring by all the states with $\sigma_1 = -1$. Combining the two contributions, we find that the partition function for the N spins in the ring arrangement is equal to the trace of matrix $\mathbf{A}^N$:

$$Z_N^{\mathrm{ring}} = (\mathbf{A}^N)_{11} + (\mathbf{A}^N)_{22} = \mathrm{tr}(\mathbf{A}^N) \tag{11.125}$$

With Eq. (11.82), the above-given equation is simplified to

$$Z_N^{\mathrm{ring}} = \mathrm{tr}(\mathbf{B}\Lambda^N\mathbf{B}^{-1}) = \mathrm{tr}(\Lambda^N) = \lambda_1^N + \lambda_2^N \tag{11.126}$$

Problem 11.11 will confirm this formula for $N = 2$ and $N = 3$.

11.4.5 Comparison of the Exact Results with Those of Mean-Field Approximations

Now that we know the exact result, we can compare the results of the two approximations methods with the exact results. We consider the limit of $N \to \infty$ only.

First, we compare the expressions of $\langle \sigma \rangle$ and $\langle \sigma_i \sigma_{i+1} \rangle$ in asymptotic situations in Tables 11.7 and 11.8, respectively.

Figure 11.20a,b compares the plots of $\langle \sigma \rangle$ as a function of βJ in a weak field ($\beta H = 0.01$) and in a strong field ($\beta H = 0.2$), respectively. Figure 11.21a,b compares the plots of $\langle \sigma_i \sigma_{i+1} \rangle$. We find that the curves of the B–W approximation

Table 11.7 Comparison of expressions of $\langle \sigma \rangle$ obtained in the transfer matrix method ($N \to \infty$) with those of approximation methods.

Method	$\|\beta J\| \ll 1$	$\|\beta H\| \ll 1, \|\beta J\|$ is not too large	$\beta J \gg 1 \ (H > 0)$
Transfer matrix	$\tanh \beta H \left(1 + \dfrac{2\beta J}{\cosh^2 \beta H}\right)$	$e^{2\beta J}\beta H$	$1 - \dfrac{e^{-4\beta J}}{2\sinh^2 \beta H}$
B–W approximation	$\tanh \beta H \left(1 + \dfrac{\beta J}{\cosh^2 \beta H}\right)$	$\dfrac{\beta H}{1 - \beta J}$	$1 - 2e^{-2(\beta H + \beta J)}$
F–H approximation	$\tanh \beta H \left(1 + \dfrac{2\beta J}{\cosh^2 \beta H}\right)$	$\dfrac{\beta H}{1 - 2\beta J}$	$1 - 2e^{-2(\beta H + 2\beta J)}$

Table 11.8 Comparison of expressions of $\langle \sigma_i \sigma_{i+1} \rangle$ obtained in the transfer matrix method ($N \to \infty$) with those of approximation methods.

Method	$|\beta J| \ll 1$	$\|\beta H\| \ll 1,\ |\beta J|$ is not too large	$\beta J \gg 1\ (H > 0)$
Transfer matrix	$t_H^2 \left[1 + \beta J \left(\dfrac{1}{t_H^2} + 2 - 3t_H^2 \right) \right]$	$t_J \left[1 + \dfrac{e^{2\beta J}}{t_J}(\beta H)^2 \right]$	$1 + 2e^{-4\beta J}\left(1 - \dfrac{1}{t_H} \right)$
B–W approximation	$t_H^2 \left(1 + \dfrac{2\beta J}{c_H^2} \right)$		$1 - 4e^{-2(\beta H + \beta J)}$
F–H approximation	$t_H^2 \left(1 + \dfrac{4\beta J}{c_H^2} \right)$		$1 - 4e^{-2(\beta H + 2\beta J)}$

c_H, coth βH; t_H, tanh βH; s_J, sinh βJ; t_J, tanh βJ.

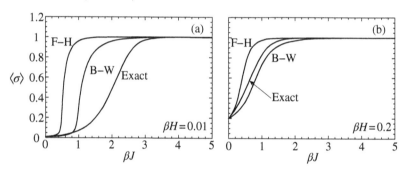

Figure 11.20 Plot of $\langle \sigma \rangle$ as a function of βJ for (a) $\beta H = 0.01$ and (b) $\beta H = 0.2$, obtained from the exact partition function ($N \to \infty$), Bragg–Williams approximation, and Flory–Huggins approximation.

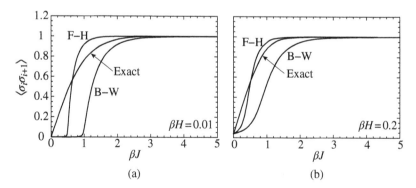

Figure 11.21 Plot of $\langle \sigma_i \sigma_{i+1} \rangle$ as a function of βJ for (a) $\beta H = 0.01$ and (b) $\beta H = 0.2$, obtained from the exact partition function ($N \to \infty$) and in B–W and F–H approximations.

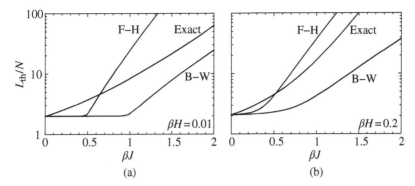

Figure 11.22 Comparison of the plots of L_{th}/N as a function of βJ for (a) $\beta H = 0.01$ and (b) $\beta H = 0.2$, obtained from the exact partition function and in B–W and F–H approximations.

runs closer to the curves of the exact result in both plots of $\langle \sigma \rangle$ compared with the curves of the F–H approximation. The opposite is true for $\langle \sigma_i \sigma_{i+1} \rangle$.

The B–W approximation is better for the overall spin orientation; the F–H approximation has an edge in the nearest-neighbor correlation. The latter approximation is designed to do well for $\langle \sigma_i \sigma_{i+1} \rangle$, as it takes into account the probabilities of the paired spin configurations. The B–W approximation lacks this consideration, and therefore underestimates $\langle \sigma_i \sigma_{i+1} \rangle$. Replacing the partition function by its peak value in the expression for the sum of states emphasizes the aligned states, and that is why the F–H approximation overestimates $\langle \sigma \rangle$.

The three curves in Figure 11.21a do not run as close to each other, as those in Figure 11.20a,b and Figure 11.22b. The two MF approximation methods fail and underestimate $\langle \sigma_i \sigma_{i+1} \rangle$ when βH and βJ are small. We need to be careful when applying the MF approaches to $\langle \sigma_i \sigma_{i+1} \rangle$.

One of the quantities calculated from $\langle \sigma_i \sigma_{i+1} \rangle$ is L_{th}, the domain size determined by thermal spin reversals. Figure 11.22a,b compares L_{th} as a function of βJ obtained from the three ways of calculating $\langle \sigma_i \sigma_{i+1} \rangle$, in two field strengths ($\beta H = 0.01$ and 0.2). Again, the three curves in Figure 11.22a do not run close to each other.

11.5 Variations of the Ising Model

11.5.1 System of Uniform Spins

We can modify the simple Ising model by changing the field and/or the interaction. In this section, we learn the modifications that Selinger and Selinger considered to describe behaviors of helical polymer chains [15]. For now, we forget about the polymer and stay with magnetic systems.

The modification we look at here is a quenched random field. By "quenching," we mean that the field is not a statistical variable such as the spin state. Although the field is random, it is fixed or static, whereas the spin states can dynamically change.

In place of a uniform H, each spin is subject to a different local field h_i. The energy of L spins ($L \gg 1$) is given as

$$E = -\sum_{i=1}^{L} h_i \sigma_i - J \sum_{i=1}^{L-1} \sigma_i \sigma_{i+1} \tag{11.127}$$

The state of the L spin system is still specified by $\sigma_1, \sigma_2, \ldots, \sigma_L$. The field at the ith spin, h_i, is not a variable, but is given. When h_i is random, the field is called a quenched random field. The interaction is the same as in the regular Ising system. The partition function for a given $h_1, h_2, \ldots, h_L$ is written as

$$Z(h_1, \cdots, h_L) = \sum_{\sigma_1 = \pm 1} \sum_{\sigma_2 = \pm 1} \cdots \sum_{\sigma_L = \pm 1} \exp\left(\beta \sum_{i=1}^{L} h_i \sigma_i + \beta J \sum_{i=1}^{L-1} \sigma_i \sigma_{i+1} \right) \tag{11.128}$$

Without prior knowledge about $h_1, h_2, \ldots$, and h_N, we cannot proceed further. Since the field is different for each spin, the MF treatment does not lead to a simple expression of Z. However, if we focus on a simple situation – a system of identical σ_i (a whole chain has the same σ_i), then we can derive simple but useful results. This situation makes the whole chain a monodomain without spin alterations, and the latter is possible when βJ is sufficiently large.

When the whole chain of spins is in a monodomain, the system's state is specified by a single variable σ, and the energy E of such a system is given as

$$E = -\sigma \sum_{i=1}^{L} h_i - (L-1)J \tag{11.129}$$

We can drop $-(L-1)J$ because it is a constant. Let us introduce the total local field h_{tot}:

$$h_{\text{tot}} \equiv \sum_{i=1}^{L} h_i \tag{11.130}$$

Then,

$$E = -\sigma h_{\text{tot}} \tag{11.131}$$

Since σ can be either 1 or -1,

$$Z = \sum_{\sigma = \pm 1} \exp(\beta \sigma h_{\text{tot}}) = 2\cosh(\beta h_{\text{tot}}) \tag{11.132}$$

The average of σ is calculated as

$$\langle \sigma \rangle = \frac{1}{Z} \sum \sigma \exp(\beta \sigma h_{\text{tot}}) = \frac{1}{h_{\text{tot}}} \frac{\partial}{\partial \beta} \ln Z = \tanh \beta h_{\text{tot}} \tag{11.133}$$

Table 11.9 Random local fields of the opposite signs.

h_i	Probability	Number of sites
h	p	l
$-h$	$1-p$	$L-l$
Total	1	L

When $h_{tot} = 0$, $\langle\sigma\rangle = 0$. With an increasing h_{tot}, the probability increases for the uniform spin to align with the majority of the fields. The details of h_i do not matter; what counts is h_{tot}. When βh_{tot} is sufficiently large, $\langle\sigma\rangle \cong 1$.

For example, if h_i's are

↑↓↓↓↑↓↑↑↑↑

it will promote σ_i's of

↑↑↑↑↑↑↑↑↑↑

rather than

↓↓↓↓↓↓↓↓↓↓

Alternatively, we can find the average from the probabilities of the two states: $\sigma = 1$ with a probability of $\exp(\beta h_{tot})$ and -1 with probability of $\exp(-\beta h_{tot})$. Since these probabilities are not normalized,

$$\langle\sigma\rangle = \frac{(1)\exp(\beta h_{tot}) + (-1)\exp(-\beta h_{tot})}{\exp(\beta h_{tot}) + \exp(-\beta h_{tot})} = \tanh \beta h_{tot} \qquad (11.134)$$

When $h_{tot} > 0$, σ is more likely to be 1 than it is -1. The greater is h_{tot}, the greater the likelihood.

Here, we consider two distributions of h_i. Then, h_{tot} also follows some distribution, and consequently $\langle\sigma\rangle$ varies, depending on h_{tot}. In Sections 11.5.2 and 11.5.3, we obtain the mean of $\langle\sigma\rangle$ for each of the two distributions of h_{tot}.

11.5.2 Random Local Fields of Opposite Directions

We consider random local fields of opposite directions. Table 11.9 lists the binary values of h_i and their probabilities.

Let l be the number of sites with $+h$ out of a total L sites. Then, the number of sites with $-h$ is $L-l$. Obviously, l is a random variable and therefore is different for each system. The l follows a binomial distribution. Its mean and variance are

$$\langle l \rangle_{rf} = Lp \qquad (11.135)$$

$$\langle \Delta l^2 \rangle_{rf} = \langle (l - \langle l \rangle_{rf})^2 \rangle_{rf} = Lp(1-p) \qquad (11.136)$$

The subscript "rf" stands for the average with respect to the distribution of h_i.

For the number of sites considered here,

$$h_{\text{tot}} = hl + (-h)(L - l) = h(2l - L) \tag{11.137}$$

Therefore, the mean and variance of h_{tot} are

$$\langle h_{\text{tot}} \rangle_{\text{rf}} = h(2\langle l \rangle_{\text{rf}} - L) = 2hL \left(p - \frac{1}{2} \right) \tag{11.138}$$

$$\langle \Delta h_{\text{tot}}^2 \rangle_{\text{rf}} = (2h)^2 \langle \Delta l^2 \rangle_{\text{rf}} = (2h)^2 Lp(1 - p) \tag{11.139}$$

When p is close to $\frac{1}{2}$, the large L makes $\langle h_{\text{tot}} \rangle_{\text{rf}}$ swing greatly around 0 for a small change in p around $\frac{1}{2}$. The situation is different for $\langle \Delta h_{\text{tot}}^2 \rangle_{\text{rf}}$. Since $p(1 - p)$ varies only slowly when p changes around $\frac{1}{2}$, $\langle \Delta h_{\text{tot}}^2 \rangle_{\text{rf}}$ remains nearly constant at around $(2h)^2 L(1/2)^2 = h^2 L$. Therefore, we can approximate the distribution of h_{tot} by a distribution with a mean of $2hL(p - \frac{1}{2})$ and variance $h^2 L$. When L is sufficiently large, the binomial distribution is close to a normal distribution with the same mean and the same variance. The distribution $P(h_{\text{tot}})$ is approximated as

$$P(h_{\text{tot}}) = \frac{1}{(2\pi h^2 L)^{1/2}} \exp \left(-\frac{[h_{\text{tot}} - 2hL(p - 1/2)]^2}{2h^2 L} \right) \tag{11.140}$$

Now, we use the distribution function $P(h_{\text{tot}})$ to calculate the mean of $\langle \sigma \rangle$, given by Eq. (11.133), under the distribution of the random fields. The double average is expressed as

$$\langle \sigma \rangle_{\text{rf}} = \int_{-\infty}^{\infty} P(h_{\text{tot}}) \langle \sigma \rangle dh_{\text{tot}} = \int_{-\infty}^{\infty} P(h_{\text{tot}}) \tanh(\beta h_{\text{tot}}) dh_{\text{tot}} \tag{11.141}$$

Figure 11.23 shows $P(h_{\text{tot}})$ and $\tanh \beta h_{\text{tot}}$ in the integrand, plotted as a function of h_{tot}. The $P(h_{\text{tot}})$ is shown for $p < \frac{1}{2}, = \frac{1}{2}$, and $> \frac{1}{2}$. The other function, $\tanh \beta h_{\text{tot}}$, is mostly -1 or 1, and the transition between them occurs in a narrow range of width β^{-1} around $h_{\text{tot}} = 0$. In contrast, $P(h_{\text{tot}})$ is distributed over a range of $L^{1/2} h$; this range is broad if $L^{1/2} h \gg \beta^{-1}$, i.e. $\beta L^{1/2} h \gg 1$. The latter will be the case when L is large.

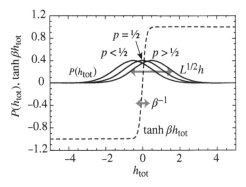

Figure 11.23 The two functions in the integrand of Eq. (11.141) are overlaid. The probability distribution $P(h_{\text{tot}})$, displayed for $p < \frac{1}{2}, = \frac{1}{2}$, and $> \frac{1}{2}$, has a width of $\sim L^{1/2} h$, whereas the other function $\tanh \beta h_{\text{tot}}$, shown as a dashed line, transitions from -1 to 1 in a narrow range of width β^{-1}.

Then, the transition in tanh βh_{tot} can be regarded as a step, and thus we can approximate it to

$$\tanh \beta h_{\text{tot}} \equiv \begin{cases} -1 & (\beta h_{\text{tot}} < 0) \\ 1 & (\beta h_{\text{tot}} > 0) \end{cases} \tag{11.142}$$

Then,

$$\langle \sigma \rangle_{\text{rf}} = -\int_{-\infty}^{0} P(h_{\text{tot}})dh_{\text{tot}} + \int_{0}^{\infty} P(h_{\text{tot}})dh_{\text{tot}} = -1 + 2\int_{0}^{\infty} P(h_{\text{tot}})dh_{\text{tot}}$$

$$= -1 + 2\int_{0}^{\infty} (2\pi h^2 L)^{-1/2} \exp\left(-\frac{[h_{\text{tot}} - 2hL(p-1/2)]^2}{2h^2 L}\right) dh_{\text{tot}} \tag{11.143}$$

Here, we change the variable of integration from h_{tot} to t defined as

$$t \equiv \frac{h_{\text{tot}} - 2hL(p-1/2)}{h(2L)^{1/2}} \tag{11.144}$$

Then,

$$\langle \sigma \rangle_{\text{rf}} = -1 + \frac{2}{\pi^{1/2}} \int_{-(2L)^{1/2}(p-1/2)}^{\infty} \exp(-t^2)dt$$

$$= -\text{erf}\left(-(2L)^{1/2}\left(p^{1/2} - \tfrac{1}{2}\right)\right) = \text{erf}\left((2L)^{1/2}\left(p - \tfrac{1}{2}\right)\right) \tag{11.145}$$

where $\text{erf}(z)$ is called an error function (Gauss error function); see Appendix A.6.

Figure 11.24 shows a plot of $\langle \sigma \rangle_{\text{rf}}$ as a function of p. At $p = \tfrac{1}{2}$ (equal probabilities for $h_i = 1$ and $h_i = -1$), $\langle h_{\text{tot}} \rangle_{\text{rf}} = 0$. With an increasing p, $\langle \sigma \rangle_{\text{rf}}$ increases as $2(2L/\pi)^{1/2}(p - \tfrac{1}{2})$, and the increase is rapid when L is large.

Now we consider how large the domain of aligned spins is. When $\langle h_{\text{tot}} \rangle_{\text{rf}} = 0$, a typical value of h_{tot} is $\langle \Delta h_{\text{tot}}^2 \rangle_{\text{rf}}^{1/2} = hL^{1/2}$. Forming a domain is a result of two competing effects – the energy due to h_{tot}, which is $\sim hL^{1/2} \times 1 = hL^{1/2}$ (note, $\sigma = 1$), and the energy of spin reversal that is J. Equating two effects gives an estimate of the domain size by the random field, L_{rf}, as

$$L_{\text{rf}} \cong (J/h)^2 \tag{11.146}$$

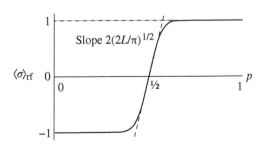

Figure 11.24 Plot of $\langle \sigma \rangle_{\text{rf}}$, the mean of σ under random local fields h_i, as a function of p, the probability of $h_i = h$.

Slope $2(2L/\pi)^{1/2}$

When $L \ll L_{\text{rf}}$, the energy required for the domain's spin reversal is too large. When $L \gg L_{\text{rf}}$, $J \ll hL^{1/2}$, and therefore the spin reversal is rampant.

We return to a 1D Ising model of N spins. Recall that the domain size within the N spins under a uniform field is limited by L_{th} and N. They are due to spin alterations and a finite chain length, respectively. Under the random field, a third mechanism to limit the domain size comes on top of the capping by L_{th} and N. All of the three mechanisms are mutually independent. When the three mechanisms of limiting the domain size are present, the overall domain size L is given by

$$\frac{1}{L} = \frac{1}{L_{\text{rf}}} + \frac{1}{L_{\text{th}}} + \frac{1}{N} \tag{11.147}$$

11.5.3 Dilute Local Fields

The other type of quenched local fields follows a distribution specified in Table 11.10. Again, each field h_i is binary, either h (>0) or 0. The probability for $h_i = h$ is r, and we assume that $r \ll 1$. The binomial distribution of h_i places a positive field once in a while, but otherwise is mostly zero.

Let l be the number of sites with h. Then, the number of sites with $h_i = 0$ is $L - l$. The random variable l follows a binomial distribution. Its mean and variance are

$$\langle l \rangle_{\text{d}} = Lr \tag{11.148}$$
$$\langle \Delta l^2 \rangle_{\text{d}} = \langle (l - \langle l \rangle_{\text{d}})^2 \rangle_{\text{d}} = Lr(1 - r) \cong Lr \tag{11.149}$$

where $r \ll 1$ was used. The subscript "d" denotes dilute fields.

Since $h_{\text{tot}} = hl$, the mean and variance of h_{tot} are

$$\langle h_{\text{tot}} \rangle_{\text{d}} = h\langle l \rangle_{\text{d}} = hLr \tag{11.150}$$
$$\langle \Delta h_{\text{tot}}^2 \rangle_{\text{d}} = h^2 \langle \Delta l^2 \rangle_{\text{d}} \cong h^2 Lr \tag{11.151}$$

To calculate the mean of $\langle \sigma \rangle$ for a given distribution of h_{tot}, we adopt a discrete version of Eq. (11.141):

$$\langle \sigma \rangle_{\text{d}} = \sum_{l=0}^{L} P_l \langle \sigma \rangle = \sum_{l=0}^{L} P_l \tanh(\beta h_{\text{tot}}) = \sum_{l=0}^{L} P_l \tanh(\beta hl) \tag{11.152}$$

Table 11.10 Dilute random local fields.

h_i	Probability	Number of sites
h	r	l
0	$1 - r$	$L - l$
Total	1	L

where $P_l = {}_LC_l\, r^l(1-r)^{L-l}$. Here, we calculate $\langle\sigma\rangle_d$ for $rL \ll 1$ and $rL \gg 1$, separately. Note that $r \ll 1$ and $L \gg 1$ can lead to any positive value of rL.

When $rL \ll 1$, on the one hand, $\langle l\rangle_d = rL \ll 1$ and $\langle\Delta l^2\rangle_d \cong rL \ll 1$. Table 11.11 lists leading terms in the series of Eq. (11.152). From the table, $\langle\sigma\rangle_d$ is calculated as

$$\langle\sigma\rangle_d = (1 - rL) \times 0 + rL \times \tanh(\beta h) = rL\tanh(\beta h) \tag{11.153}$$

When $rL \gg 1$, on the other hand, P_l is narrowly distributed; the ratio of the standard deviation to the mean is around $(rL)^{-1/2}$. Then, the series is represented by a single term with $P_l = 1$ for that the peak value of l. Then,

$$\langle\sigma\rangle_d = \tanh(\beta hLr) \tag{11.154}$$

Figure 11.25 shows a plot of $\langle\sigma\rangle_d$ as a function of rL. At a small rL, the curve runs along $rL\tanh(\beta h)$. With an increasing rL, the curve approaches $\tanh(rL\beta h)$. Overall, the plot is close to $\tanh(rL\beta h)$.

For $\langle\sigma\rangle_d$ to be close to 1, $rL > (\beta h)^{-1}$. For the spins in the monodomain of length L to be in the direction of the dilute fields, r must be sufficiently

Table 11.11 Leading terms in the series of Eq. (11.152).

l	h_{tot}	Probability P_l
0	0	$\dbinom{L}{0}(1-r)^L \cong 1 - rL$
1	h	$\dbinom{L}{1}r(1-r)^{L-1} \cong Lr$
2	$2h$	$\dbinom{L}{2}r^2(1-r)^{L-2} \cong \frac{1}{2}(Lr)^2 \ll 1$
⋮	⋮	⋮

Figure 11.25 A sketch of $\langle\sigma\rangle_d$, the average of σ under dilute local fields, plotted as a function of rL, the mean number of sites with a nonzero field.

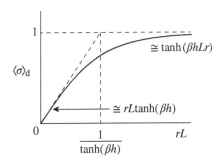

large, the temperature must be sufficiently low, or each local field must be sufficiently strong.

In the next chapter, we use the results of this chapter, especially those in the last section, to consider helical properties of a helix-forming polymer in solution. It may appear that a magnetic system of the Ising model has nothing to do with the polymer, but the common thread of the linear nature (one-dimensional) is perfect for drawing a connection. We will look at the helical polymer only, but you may be able to find other polymer systems that can be best described by the Ising model.

Problems

11.1 *Ring Ising model* ($N = 3$, 4). Calculate $\langle \sigma \rangle$ and $\langle \sigma_i \sigma_{i+1} \rangle$ for an $N = 3$ ring Ising model in Section 11.2.3.2 and an $N = 4$ ring Ising model in Section 11.2.3.3. Then, find $\langle \sigma \rangle$ and $\langle \sigma_i \sigma_{i+1} \rangle$ in the limits of $\beta J \to 0$ and $\beta J \to \infty$. You should be able to tell whether your answers are correct or not, following the same discussion as the one we had for the $N = 3$ linear Ising model in Section 11.2.3.1.

11.2 *Ring Ising model* ($N = 6$). Six spins σ_1 through σ_6 are arranged in a hexagon, as shown.

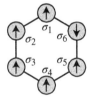

There is no external magnetic field, and the interaction exists only between an adjacent pair on the ring. The energy of spin system is given as

$$E = -J \sum_{i=1}^{6} \sigma_i \sigma_{i+1}$$

where $J > 0$ and $\sigma_7 = \sigma_1$.
(1) Complete the given table. The typical spin arrangements should show all possibilities, eliminating those that are identical by symmetry operations (rotation by 60°; reflection by a plane through the center that divides the ring into two equal parts (sans spins), inversion).

Energy level	Typical spin arrangements	Number of states
$6J$		2

(2) What is $\langle \sigma \rangle$?

(3) What is $\langle \sigma_i \sigma_{i+1} \rangle = \langle \sum \sigma_i \sigma_{i+1} \rangle / 6$?

(4) What is the low-temperature limit of $\langle \sigma_i \sigma_{i+1} \rangle$?

(5) What is the high-temperature asymptote of $\langle \sigma_i \sigma_{i+1} \rangle$?

11.3 *Coarse graining.* In the one-dimensional Ising model of N spins, we change the way the states are counted and adopt a course-grained view. In the modified counting method, a unit is a pair of adjacent spins, and the state of each unit, τ_j, is +2, 0, and −2, where $j = 1, 2, ..., \frac{1}{2}N$. Their degeneracies are 1, 2, and 1, respectively. The interaction exists between adjacent units, and is expressed as $-\gamma J \tau_j \tau_{j+1}$ where a constant γ may not be equal to one. The energy is then expressed as

$$E = -H \sum_{j=1}^{N/2} \tau_j - \gamma J \sum_{j=1}^{N/2} \tau_j \tau_{j+1}$$

Employ the Bragg–Williams (B–W) approximation to show that $\gamma = 1/2$ gives the same partition function as the one given by Eq. (11.27).

11.4 *Three-state model.* In the 1D Ising model, we change the number of states for each spin from 2 to 3:

$$\sigma_i = \begin{cases} 1 & \text{(up)} \\ -1 & \text{(down)} \\ 0 & \text{(otherwise)} \end{cases}$$

The degeneracy is 1 for each of the three states, and the energy in the presence of magnetic field H and the interaction are identical to those in Section 11.2. The expression of the energy E is identical to Eq. (11.2). Apply a B–W approximation and discuss how the mean spin changes from the one discussed in that section.

11.5 $\langle \sigma \rangle$ *in F–H approximation.* Use the expression of $\ln Z$ given by Eq. (11.58) to show that it is consistent with the formula, $\langle \sigma \rangle = (N\beta)^{-1} \partial \ln Z / \partial H$. Keep in mind that $\langle \sigma \rangle$ is a function of β, H, and J.

11.6 $\langle \sigma_i \sigma_{i+1} \rangle$ *in F–H approximation.* Apply Eq. (11.6) to Eq. (11.58) to find $\langle \sigma_i \sigma_{i+1} \rangle$ in the F–H approximation of the 1D Ising model.

11.7 *2D Ising model.* The given figure illustrates spins in a 2D Ising model on a square lattice. The interaction is between nearest neighbors. For example, the spin indicated by an outlined ellipse has four nearest neighbors indicated by gray ellipses. The system has a total of N spins ($N \gg 1$).

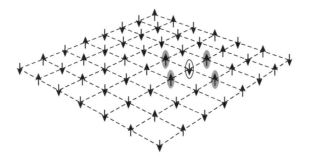

The energy of the system is given as

$$E = -H \sum_{i=1}^{N} \sigma_i - J \sum_{i,j \in NN} \sigma_i \sigma_j$$

The second series is calculated for all nearest neighbor pairs of i and j. Apply the three methods of mean-field approximation to the 2D Ising model to find the equation for $\langle \sigma \rangle$ that, when solved self-consistently, would give $\langle \sigma \rangle$.

11.8 *Polycation.* A polycation of N repeat units ($N \gg 1$) is dissolved in water. Each repeat unit has one cationic site. Small anions (acid HX) are added to the solution and some of them may pair with the cations. The state of each cation may be expressed by σ_i as

$$\sigma_i = \begin{cases} 1 & \text{(paired with } X^- \text{)} \\ 0 & \text{(not paired)} \end{cases}$$

The energy of pairing may be expressed as

$$E = \varepsilon \sum_{i=1}^{N} \sigma_i + \xi \sum_{i=1}^{N} \sigma_i \sigma_{i+1}$$

where ε (<0) represents the energy for a single cation to pair with X^-, and ξ the increase in the energy when two adjacent sites are paired. The latter may be positive or negative.

(1) What is the partition function? Use a B–W approximation.
(2) Find a self-consistent solution up to the linear order of $\beta\xi$.
(3) Draw a sketch for a plot of $\langle\sigma\rangle$ as a function of $\beta\xi$, and discuss the plot.

11.9 *Alternative listing of states, mean field.* The state of a one-dimensional Ising model of N spins, $\sigma_1, \sigma_2, \sigma_3, \ldots, \sigma_N$, can alternatively be specified by σ_1 and $\sigma_1\sigma_2, \sigma_2\sigma_3, \ldots, \sigma_{N-1}\sigma_N$. The product of an adjacent pair of spins is either +1 or −1. The energy of the N spins in magnetic field H is given as

$$E = -J\sum_{i=1}^{N-1}\sigma_i\sigma_{i+1} - H\sum_{i=1}^{N}\sigma_i \tag{11.155}$$

and the partition function is expressed as

$$Z = \sum_{\sigma_1=\pm1}\sum_{\sigma_1\sigma_2=\pm1}\sum_{\sigma_2\sigma_3=\pm1}\cdots\sum_{\sigma_{N-1}\sigma_N=\pm1}\exp\left(\beta J\sum_{i=1}^{N-1}\sigma_i\sigma_{i+1} + \beta H\sum_{i=1}^{N}\sigma_i\right) \tag{11.156}$$

We note that

$$\sigma_2 = \sigma_1 \cdot \sigma_1\sigma_2 \tag{11.157}$$

$$\sigma_3 = \sigma_1 \cdot \sigma_1\sigma_2 \cdot \sigma_2\sigma_3$$
$$\cdots \tag{11.158}$$

$$\sigma_N = \sigma_1 \cdot \sigma_1\sigma_2 \cdot \sigma_2\sigma_3 \cdot \ldots \cdot \sigma_{N-1}\sigma_N \tag{11.159}$$

We adopt a mean-field approximation for calculating the partition function. Our method is different from the Bragg–Williams approximation that replaces σ_{i+1} in $\sigma_i\sigma_{i+1}$ with $\langle\sigma\rangle$. We replace $\sigma_1\sigma_2$, etc. in Eqs. (11.157)–(11.159) with $\langle\sigma_1\sigma_2\rangle = \eta$. Thus,

$$\sigma_2 = \sigma_1\eta, \quad \sigma_3 = \sigma_1\eta^2, \quad \cdots \quad \sigma_N = \sigma_1\eta^{N-1} \tag{11.160}$$

Then,

$$\sum_{i=1}^{N}\sigma_i = \sigma_1\sum_{i=0}^{N-1}\eta^i = \sigma_1\frac{1-\eta^N}{1-\eta} \tag{11.161}$$

for $\eta < 1$. When $\eta = 1$, the series equals N. The given figure shows a plot of $(1 - \eta^N)/(1 - \eta)$ for $N = 10, 30$, and 100. Assume that N is even in this problem.

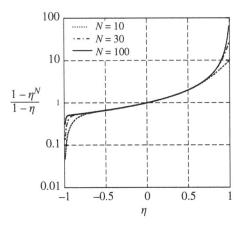

With this approximation, the partition function is expressed as

$$Z = \sum_{\sigma_1 = \pm 1} \sum_{\sigma_1 \sigma_2 = \pm 1} \sum_{\sigma_2 \sigma_3 = \pm 1} \cdots \sum_{\sigma_{N-1} \sigma_N = \pm 1} \exp\left(\beta J \sum_{i=1}^{N-1} \sigma_i \sigma_{i+1} + \beta H \sigma_1 \frac{1 - \eta^N}{1 - \eta} \right)$$

(11.162)

(1) Use Eq. (11.162) to calculate Z.
(2) Calculate $\langle \sigma_1 \sigma_2 \rangle$.
(3) Calculate $\langle \sum_{i=1}^{N} \sigma_i \rangle$.
(4) When $\eta = 1$, all the spins are aligned. Then, you can calculate the exact Z easily (without using the transfer matrix). Show that your answer of part (3) for $\eta = 1$ is identical to the one obtained from the exact Z.
(5) When $\eta = 0$, what is your answer of part (3) equal to? Is the result reasonable? Explain the result.
(2) When $\eta = -1$, what is your answer of part (3) equal to? Is the result reasonable? Explain the result.

11.10 *Exact linear Ising model.* Confirm that the partition function in the exact 1D linear Ising model agrees with the one to be obtained from discrete listing of all the states for (1) $N = 2$ and (2) $N = 3$ spins.

11.11 *Exact ring Ising model.* Confirm that the partition function in the exact 1D ring Ising model agrees with the one to be obtained from discrete listing of all the states for (1) $N = 2$ and (2) $N = 3$ spins.

11.12 *Exact Ising model, large N.* Figure 11.17 shows a plot of $\langle \sigma_i \sigma_{i+1} \rangle$ as a function of βJ for different values of βH. Draw a sketch of $\langle \sigma_i \sigma_{i+1} \rangle$ as a function of βH for different values of βJ. The range of βH should include positive and negative values of βH.

11.13 *Chiral polymer.* Consider a one-dimensional chain of spins, each capable of being in three directions labeled as -1, 0, and 1. The state of the chain consisting of N spins is specified by $\sigma_1, \sigma_2, ..., \sigma_N$, where $\sigma_i = -1$, 0, or 1 ($i = 1, ..., N$). An example is shown.

$\sigma_i = 0 \quad 0 \quad 1 \quad -1 \quad 0 \quad 1 \quad -1 \quad 0 \quad -1 \quad 0$

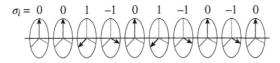

There is an interaction between an adjacent pair of spins, and the interaction ξ between σ_i and σ_{i+1} is listed in the table.

σ_i	σ_{i+1}		
	1	0	-1
1	0	$-J$	J
0	J	0	$-J$
-1	$-J$	J	0

The partition function is expressed as

$$Z = \sum_{\sigma_1} \sum_{\sigma_2} \cdots \sum_{\sigma_N} \exp\left(-\beta \sum_{i=1}^{N-1} \xi(\sigma_i, \sigma_{i+1}) \right)$$

(1) Find the transfer matrix $\mathbf{M}$ for calculating Z and the eigenvalues of $\mathbf{M}$.
(2) Assume $N \gg 1$ to obtain an approximate expression of $\ln Z$.
(3) Calculate $(N\beta)^{-1} \partial \ln Z / \partial J$. What quantity does it represent?

11.14 *Quenched local fields, $N = 3$.* We consider a 1D Ising model (linear) of 3 spins with a quenched local field in which the energy E of the system is given as

$$E = -\sum_{i=1}^{3} h_i \sigma_i - J \sum_{i=1}^{2} \sigma_i \sigma_{i+1}$$

Let us assume that $h_1 = 0$, $h_2 = h$, $h_3 = 0$ with $h > 0, J > 0$.

(1) What is the partition function Z?
(2) Calculate $\langle \sigma_1 \sigma_2 + \sigma_2 \sigma_3 \rangle$.
(3) What is $\langle \sigma_2 \rangle$?
(4) What is $\langle \sigma_1 \rangle$?
(5) Calculate the average energy $\langle E \rangle$.
(6) What is the low-temperature limit of $\langle E \rangle$? What is the state of the system in the low-temperature limit?
(7) What is the high-temperature asymptote of $\langle E \rangle$?

11.15 *Quenched random local field, short chain.* In Section 11.5, we considered a 1D Ising model of L spins in a quenched random field of $+h$ and $-h$ to obtain $\langle \sigma \rangle_{rf}$ for $\beta L^{1/2} h \gg 1$. The latter condition makes the transition of $\tanh(\beta h_{tot})$ take place in a range of h_{tot} much narrower compared with the distribution $P(h_{tot})$. This problem considers the other situation, namely, $\beta L^{1/2} h \ll 1$, i.e. the distribution of $P(h_{tot})$ is narrower compared with the transition in $\tanh(\beta h_{tot})$. $P(h_{tot})$ is distributed around $h_{tot} = h_c$, where

$$h_c \equiv 2hL(p - 1/2) \tag{11.163}$$

(1) Expand $\tanh(\beta h_{tot})$ around $h_{tot} = h_c$ into a Taylor series up to the second order of $h_{tot} - h_c$ and calculate $\langle \sigma \rangle_{rf}$.
(2) Draw a sketch for the plot of $\langle \sigma \rangle_{rf}$ as a function of p.
(3) Now combine the two situations, $\beta L^{1/2} h \gg 1$ and $\beta L^{1/2} h \ll 1$ to draw a sketch for the plot of $\langle \sigma \rangle_{rf}$ as a function of $\beta L^{1/2} h$. Assume that $p > \frac{1}{2}$.

11.16 *Quenched dilute local field, chain length.* We learned that $\langle \sigma \rangle_d \cong \tanh(\beta hrL)$ for a 1D Ising model with dilute quenched local fields. The L is the length of a monodomain, which is given as

$$\frac{1}{L} = \frac{1}{L_{th}} + \frac{1}{N}$$

Draw a sketch for a plot of $\langle \sigma \rangle_d$ vs N.

11.17 *Quenched dilute local field, normal approximation.* In Section 11.5, we used the discrete distribution (binomial distribution) to get an expression of $\langle \sigma \rangle_d$. We can get a more exact expression using a continuous approximation of the distribution. When $L \gg 1$ (but $Lr \ll 1$), the distribution of h_{tot} is approximated as

$$P(h_{tot}) = (2\pi h^2 Lr)^{-1/2} \exp\left(-\frac{(h_{tot} - hLr)^2}{2h^2 Lr}\right)$$

since the mean and variance are hLr and $h^2 Lr$, respectively. Calculate $\langle \sigma \rangle_d$.

11.18 *Adsorption.* We consider adsorption of molecules onto a planar surface that has N adsorption sites ($N \gg 1$) arranged on the grid points of a square lattice. A site is specified by two integers, i and j, for the two-dimensional lattice; (i, j) is a coordinate, but takes only integral values. The total number of (i, j) is N.

Each site can accommodate up to one molecule. The state σ_{ij} of site (i, j) is 1 when it has a molecule and 0 otherwise. When a site adsorbs a molecule of chemical potential μ, the energy of the site is lowered by ε (>0), and furthermore facilitates adsorption at the adjacent four sites. For this substrate, we can express the energy of the substrate as

$$E = -\varepsilon \sum_{ij} \sigma_{ij} - \frac{J}{2} \sum_{ij} \sigma_{ij} \sum_{k,l=\pm 1} \sigma_{i+k,j+l}$$

where J (>0) represents the interaction between occupied states on adjacent sites, and division by 2 is to cancel counting the same pair for a second time.

(1) The mean-field approximation for the interaction replaces $\sigma_{i+k,j+l}$ in the interaction term by $\langle \sigma \rangle \equiv \langle \sigma_{ij} \rangle$. What is the grand partition function Z under the mean-field approximation?

(2) Show that $\langle \sigma \rangle$ satisfies the following in the mean-field approximation

$$\langle \sigma \rangle = \frac{p}{p + q e^{-\beta(\varepsilon + 2J\langle \sigma \rangle)}}$$

where p is the pressure, and q is related to p and μ as $e^{\beta \mu} = p/q$.

(3) The given equation is a transcendental equation for $\langle \sigma \rangle$, and cannot be solved in general. However, when the temperature is sufficiently low or high, we can find an explicit result.

   (a) What is the low-temperature asymptote of $\langle \sigma \rangle$?
   (b) What is required for T to qualify as "sufficiently low"?
   (c) What is the high-temperature asymptote of $\langle \sigma \rangle$?
   (d) Draw a sketch for the plot of $\langle \sigma \rangle$ as a function of T.

(4) Calculate $\langle \sigma_{i,j} \sigma_{i,j+1} \rangle$, where $\sigma_{i,j} \sigma_{i,j+1}$ is the product of occupancy at neighboring sites. Show that $\langle \sigma_{i,j} \sigma_{i,j+1} \rangle = \langle \sigma \rangle^2$.

12

Helical Polymer

We apply what we learned in the preceding chapter to a helix-forming polymer. The helicity, right-handed or left-handed, can be mapped onto a spin, and that is why we can consider the helicity in the Ising model (Section 12.1). Our interest is in how dominant one of the handedness is. There are experimental methods to estimate the mean helicity, which we review in Section 12.2. In the section that follows, we look at helical properties of the polymer consisting of repeat units that, on the average, have nearly zero handedness.

We learned in Section 11.5 two types of quenched local fields. Both types have a bearing on the polymer, which we examine in Section 12.4.

12.1   Helix-Forming Polymer

Some polymers form a helix when dissolved in a solvent. There are two senses of the helix (see Figure 12.1). The sense is either right-handed or left-handed, named in the same way as screw threads are named. Nearly all screws we use are right-handed; when turned clockwise, the screw moves forward. In a typical helical polymer, each turn consists of three or more repeat units of the polymer.

The helical polymers have either a predetermined helical sense or a dynamically reversible sense. Proteins and polypeptides have a predetermined sense, geometrically arranged by a chiral carbon in every residue and a hydrogen bond across a few residues. A polymer that has a predetermined sense without relying on the hydrogen bonds was synthesized. A specific helical sense was incorporated into the polymer at the time of synthesis. Figure 12.2 shows the polymerization scheme. An achiral methacrylic monomer with a bulky side group (triphenyl) was polymerized by living anionic polymerization in the presence of a chiral initiator [16]. Each step of the polymerization added the next repeat unit in a specific helical sense. Once formed, the bulky triphenyl group next to the ester made it difficult to unwind the helix.

There are many synthetic polymers whose helical sense can be reversed dynamically in solutions. Examples are shown in Figure 12.3. For the polymer

Statistical Thermodynamics: Basics and Applications to Chemical Systems, First Edition. Iwao Teraoka.
© 2019 John Wiley & Sons, Inc. Published 2019 by John Wiley & Sons, Inc.
Companion website: www.wiley.com/go/Teraoka_StatsThermodynamics

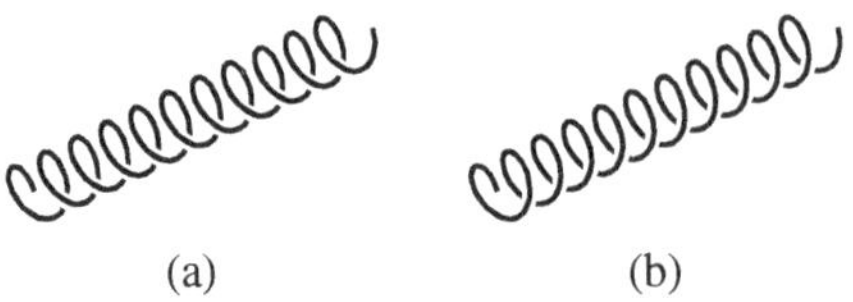

Figure 12.1 Two senses of helix.
(a) Right-handed helix, (b) left-handed helix.

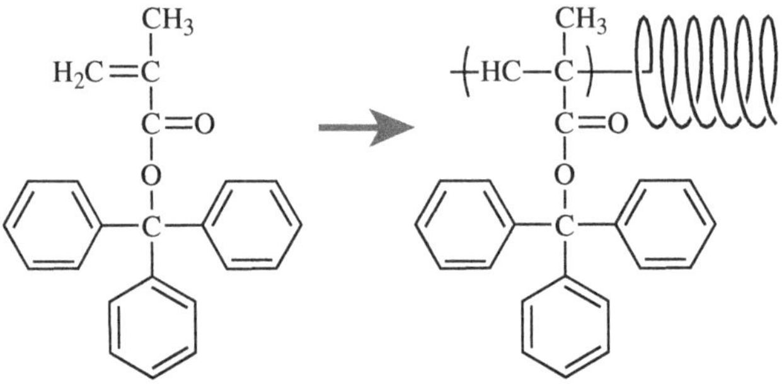

(a)   (b)

Figure 12.2 Polymerization of triphenylmethacrylate in the presence of a chiral additive leads to a helical polymer of a fixed sense. Source: Okamoto et al. 1979 [16]. Adapted with permission of American Chemical Society.

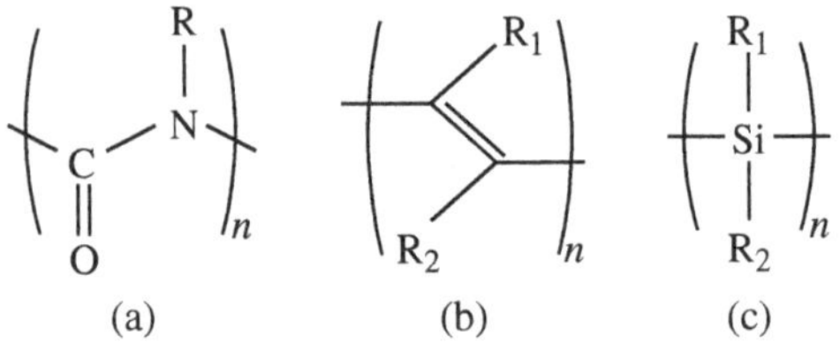

Figure 12.3 Examples of synthetic helical polymers of a reversible helical sense. (a) Polyisocyanate, (b) polyacetylene, and (c) polysilane.

(a)   (b)   (c)

to be helical and not in a random coil, twisting the polymer backbone in an identical sense must continue over several or more repeat units. Continuing the same helical sense as the preceding unit must be preferred to changing the helical sense. It means that the energy of reversing the sense is higher than the energy of continuing the same sense by more than $k_B T$, but not too large. Among these polymers, poly(n-hexyl isocyanate) (PHIC) was most extensively studied for its helical properties.[1] The n-hexyl side group (R $= C_6H_{13}$ in Figure 12.3a) makes the barrier height just right for studies at around room temperature. The polymer is soluble in different solvents, which facilitated the study. We apply the 1D Ising model to describe the helical properties of PHIC.

Figure 12.4a and b illustrates short and long polymer chains with a dynamically reversible helical sense. A short chain will be in one helical sense. On the average, there will be as many right-handed helices and as there are

---

1 Nearly all the work was conducted by Mark Green and his students at Polytechnic University, now Tandon School of Engineering of New York University, and his coworkers at other institutions.

Figure 12.4 (a) Short-chain helices at dynamic equilibrium. (b) A long-chain helix.

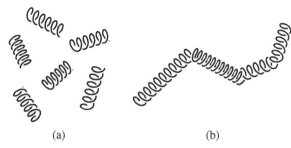

(a)   (b)

left-handed helices. There is no preference to the helical sense, unless the polymer is placed in an environment that promotes one sense over the other. For example, dissolving a chiral compound of enantiomeric purity into the solvent or employing a chiral solvent may lead to preference of one helical sense. With an increasing chain length, it becomes difficult to keep the whole chain in one helical sense. When the chain is sufficiently long, the chain will consist of domains, each being in one helical sense. A long polymer chain is a sequence of domains of alternate senses.

Since the helical sense is dynamic, the boundary of the domains can move along the polymer chain (see Figure 12.5).

If a pair of adjacent boundaries move into opposite directions to shorten the intervening domain, the movement may lead to annihilation of the domain, concatenating two adjacent domains (see Figure 12.6). Likewise, the domain can grow if its two boundaries move away from each other. A new domain with the opposite helicity can appear in the midst of an existing domain.

Figure 12.5 The boundary between domains of opposite helicities moves along the polymer chain.

Helix reversal

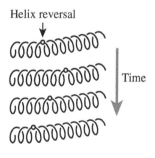

Time

Figure 12.6 Two boundaries of a domain move to shorten the domain. They may meet to annihilate the domain.

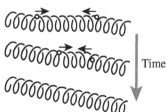

Time

The same dynamic nature will flip the helical sense in short chains. Each polymer chain will undergo a sense overturn as a whole.

12.2 Optical Rotation and Circular Dichroism

The properties of the handedness are experimentally studied by measuring the optical rotation and the circular dichroism (CD).

The optical rotation of a chiral compound (enantiomerically pure) in a solution is evaluated by passing linearly polarized light through the solution (see Figure 12.7a). Upon exiting the solution, the light remains linearly polarized, but the polarization direction is different from the one in the incident light. For this reason, a chiral compound is optically active.

The change α (in degree) of the polarization direction can be measured by optical tools. The longer the path length l (in dm) of this solution, the greater the α: $\alpha \propto l$. The α is also proportional to the concentration c (in g ml^{-1}) of the compound. Therefore,

$$\alpha = [\alpha]cl \tag{12.1}$$

where the proportionality constant $[\alpha]$ is called a **specific rotation**. It depends on the wavelength of light and is a quantity characteristic of a chiral molecule. Laevorotatory molecules such as L-glutamic acid in water have a positive $[\alpha]$, whereas dextrorotatory molecules such as D-glutamic acid and L-histidine have a negative $[\alpha]$. A 50 : 50 mixture of L-glutamic acid and D-glutamic acid (racemic mixture) in water has $\alpha = 0$. For a given mixture of enantiomers with an unknown composition, we can measure $[\alpha]$ to find the enantiomeric excess.

Another quantity characteristic of a chiral compound is a molar ellipticity $[\theta]$. It is evaluated in CD spectroscopy. Dichroism stands for two

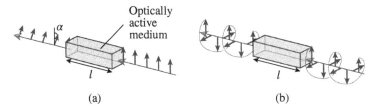

(a)   (b)

Figure 12.7 (a) Polarization direction changes when linearly polarized light passes an optically active medium such as a solution of L-amino acid. This example shows laevorotation. (b) Circularly polarized light changes its polarization direction with time and space. The period of the polarization rotation is equal to the wavelength; it is drawn out of scale with respect to the dimension of the optically active medium. Absorption of light can be different for right-circularly polarized light and left-circularly polarized light.

colors – absorption spectrum is different for two states of light. Here, the two states are right-circularly polarized light and left-circularly polarized light. Circularly polarized light changes its polarization direction as it travels. The change is 360° per wavelength or period, see Figure 12.7b. In CD spectroscopy, either right-circularly polarized light or left-circularly polarized light passes the solution under test at one time. If the solution absorbs light, exiting light will be weaker. The spectrometer measures the absorbance spectra $\mathrm{Abs_R}(\lambda)$ and $\mathrm{Abs_L}(\lambda)$ for the two polarizations, where λ is the wavelength of the light. If the solution contains an optically active compound, the two spectra can be different, and the difference is proportional to the path length l and concentration c of the compound. Therefore,

$$\mathrm{Abs_L}(\lambda) - \mathrm{Abs_R}(\lambda) = \Delta\varepsilon(\lambda)cl \tag{12.2}$$

where $\Delta\varepsilon(\lambda)$ is the difference in the molar absorptivity, and is a function of wavelength λ. Typically, the spectra are obtained in the range between ~190 and 250–300 nm. The molar ellipticity $[\theta]$ is related to $\Delta\varepsilon$ as

$$\frac{[\theta]}{\mathrm{deg}\cdot\mathrm{cm}^2\,\mathrm{dmol}^{-1}} = 3298.2\,\frac{\Delta\varepsilon}{L\cdot\mathrm{mol}^{-1}\,\mathrm{cm}^{-1}} \tag{12.3}$$

You can find how 3298.2 is derived in literature, for example, in Wiki.

CD spectroscopy is widely used for studying the conformation of a protein. A protein molecule consists of α helices, β sheets, turns, and the rest. Each constituent exhibits a specific CD spectrum, and therefore the spectroscopy allows us to find the structural composition of the protein. Here, we use the CD spectrum obtained for PHIC and its variants to quantify the helical state or the helicity of the polymers.

12.3 Pristine Poly(*n*-hexyl isocyanate)

The presence of two helical senses, and only two, and the chain structure are just right for modeling the helical polymer in the 1D Ising model. Table 12.1 shows the mapping of the sense into the spin state. A sequence of +1 domains and −1 domains can represent a sequence of right-handed helices and left-handed

Table 12.1 Designation of spin variable σ_i.

σ_i	Sense of the helix in the *i*th repeat unit
1	Right-handed
−1	Left-handed

helices. Then, the formulations and results we learned in the preceding chapter find ramifications in the properties of the helical polymer.

A pristine PHIC molecule of N repeat units dissolved in an achiral solvent has a simple expression of the energy:

$$E = -J \sum_{i=1}^{N} \sigma_i \sigma_{i+1} \tag{12.4}$$

where $J > 0$. In the ground state, the molecule does not have a helix reversal, i.e. $\sigma_1 = \sigma_2 = \cdots = \sigma_N = 1$ or $\sigma_1 = \sigma_2 = \cdots = \sigma_N = -1$, and the energy is $E = -JN$. With one helix reversal, for example, $\sigma_i = 1$ for $i = 1, \ldots, k$; -1 for $i = k+1, \ldots, N$,

$$E = -J[1 \times (N-1) + (-1) \times 1] = -J(N-2) \tag{12.5}$$

The energy is higher by $2J$ compared with the ground state. Adding a reversal increases the energy by $2J$. A chain with two reversals has an energy of $4J$ relative to the ground state.

The energy E in Eq. (12.4) does not have a component linear to σ_i, and therefore the mean of σ_i is 0. A solution of PHIC in toluene, for example, does not have an optical rotation or CD; $[\alpha] = 0$ and $[\theta] = 0$ at all wavelengths.

If the polymer has an excess right-handedness, $[\alpha]$ would be positive and $[\theta]$ would not be 0. The $[\alpha]$ and $[\theta]$ are proportional to the excess handedness. This situation parallels the one for $\langle \sigma \rangle$ in the Ising model. Recall that $N\langle \sigma \rangle$ is equal to the excess spins of $+1$ over the spins of -1. Therefore, we can use the formulas we obtained for $\langle \sigma \rangle$ in different scenarios to understand the results of $[\alpha]$ and $[\theta]$ for different helical polymers or different environments.

To estimate J, a few series of experiments were performed. One of them employed a variant of PHIC with one of the hydrogen atoms on the α carbon being deuterated [17]. The enantiomerically pure monomer was polymerized to form poly((R)-1-deuterio-n-hexyl isocyanate). Its structure is shown in Figure 12.8. Since the H–D difference is small, the deuteration imparts an extremely weak preference to one of the helical senses in the polymer. However, $[\alpha]$ and $[\theta]$ of the polymer were not small, especially when the degree of polymerization, N, exceeded 100. Figure 12.9a shows some of the results. Fractions of different molecular weights were prepared by precipitation fractionation from a broad-distribution polymer sample. Each fraction was dissolved in dichloromethane. Obviously, the polymer had more left-handed helices than

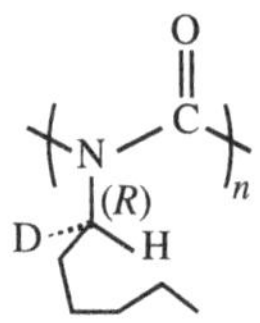

Figure 12.8 Structure of poly((R)-1-deuterio n-hexyl isocyanate).

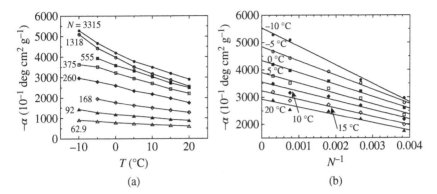

Figure 12.9 Specific optical rotation $[\alpha]$ of poly(1-deuterio-*n*-hexyl isocyanate) in dichloromethane. (a) Plot of $-[\alpha]$ as a function of temperature T for different fractions of the polymer. The values of N, degree of polymerization, are indicated adjacent to the curves. (b) Plot of $-[\alpha]$ as a function of N^{-1} at different temperatures. The temperature is indicated adjacent to each curve. Source: Panel (a) from Okamoto et al. 1996 [17]. Reproduced with permission of American Chemical Society.

it did right-handed ones. In Figure 12.9a, the lower the temperature and the greater the N, the greater the $-[\alpha]$.

We apply the Ising model (exact; large-N asymptote) to estimate J. We first note that $[\alpha]$ is related to $\langle \sigma \rangle$ by

$$\frac{[\alpha]}{[\alpha]_\infty} = \langle \sigma \rangle \tag{12.6}$$

where $[\alpha]_\infty$ is for a hypothetical polymer that has all of its repeat units in the helical sense preferred by the deuterium substitution. That could be reached in the low-temperature limit, but not at accessible temperatures.

Deuterium substitution amounts to applying a uniform, but weak field H for the spins that map onto the helical handedness. We can safely assume that βH is small, and therefore we can apply one of the formulas in Table 11.6. Equation (11.121) expresses $\langle \sigma \rangle$ with H and J. When translated into the optical rotation, we have

$$\frac{[\alpha]}{[\alpha]_\infty} \cong e^{2\beta J} \beta H \left(1 - \frac{1}{N} \sinh 2\beta J\right) \tag{12.7}$$

It may appear that $e^{2\beta J}$ makes the ratio huge. However, if βH is sufficiently small, the ratio will be a lot smaller than 1.

Equation (12.7) tells that $[\alpha]$ is a linear function of N^{-1}. We can transform Figure 12.9a into a plot of $-[\alpha]$ as a function of N^{-1} at each temperature (see Figure 12.9b). Data for the five highest molecular weight fractions are shown. It appears reasonable to fit the data by a linear equation to estimate its intercept $[\alpha]_{N\to\infty}$ and slope at each temperature. Note that $[\alpha]_{N\to\infty}$ is different from $[\alpha]_\infty$.

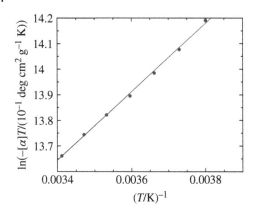

Figure 12.10 Plot of $\ln(-T[\alpha]_{N\to\infty})$ as a function of T^{-1}. Circles were obtained from $[\alpha]_{N\to\infty}$, the intercept of the linear fit in Figure 12.9b. The straight line represents the optimal fit of the data by a linear function of T^{-1}.

The latter is the low-temperature limit of $[\alpha]_{N\to\infty}$. The intercept $[\alpha]_{N\to\infty}$ should depend on T as shown:

$$\ln(-T[\alpha]_{N\to\infty}) = \frac{2J}{k_B T} + \ln\frac{-H[\alpha]_\infty}{k_B} \tag{12.8}$$

Figure 12.10 shows a plot of $\ln(-T[\alpha]_{N\to\infty})$ as a function of T^{-1}. The slope of a linear fit is $2J/k_B = 1335$ K, from which we get $2J = 11.1$ kJ mol^{-1} ($2\beta J = 4.45$ at 300 K).

Another piece of information the curve fit in Figure 12.9b offers is the slope. Equation (12.7) indicates that the ratio of the slope to the intercept should be equal to $-\sinh 2\beta J$. Here we do not do the curve fit of a plot of the ratio as a function of T^{-1} by $\sinh(2J/k_B T)$, since the slope carries greater errors than does the intercept. Rather, we compare the negative of the ratio to $\sinh(2J/k_B T)$, where $2J = 11.1$ kJ mol^{-1}. Figure 12.11 compares the two.

We now proceed to estimating the domain size L_{th} limited by thermal activation of helix reversals. For that purpose, we need to have an estimate

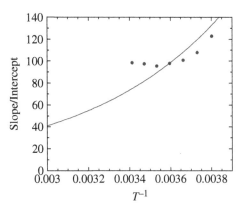

Figure 12.11 Slope to intercept ratio in the curve fit of Figure 12.10b (the sign is inverted) is plotted as a function of T^{-1}. The solid line represents $\sinh(2J/k_B T)$, where $2J = 11.1$ kJ mol^{-1}.

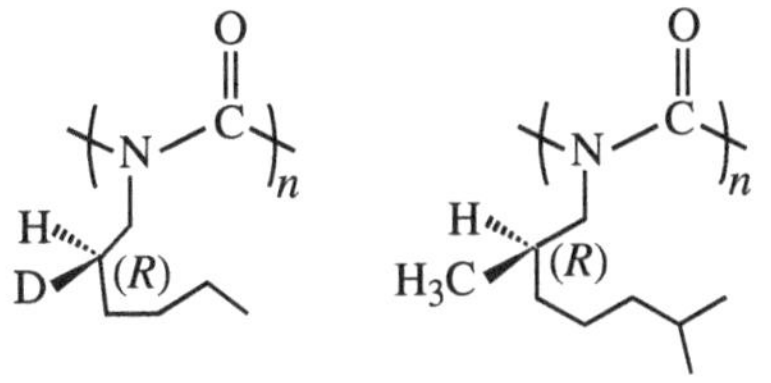

Figure 12.12 Structures of (a) poly((R)-2-deuterio n-hexyl isocyanate) and (b) poly((R)-2,6-dimethylheptyl isocyanate).

(a) (b)

of $\langle \sigma_i \sigma_{i+1} \rangle$. Equation (11.105) shows that, when $\beta H \ll 1$, $\langle \sigma_i \sigma_{i+1} \rangle \cong \tanh \beta J$. Then, with Eq. (11.11), we have

$$\langle N_{\text{alt}} \rangle = \frac{N}{2}(1 - \tanh \beta J) \cong \frac{N}{1 + e^{2\beta J}} \tag{12.9}$$

With Eq. (11.12), we have an estimate of L_{th} as

$$L_{\text{th}} = 1 + e^{2\beta J} \tag{12.10}$$

Since $2\beta J = 4.45$ at 300 K, $L_{\text{th}} = 86.6$.

Other variants to PHIC that might have a preference to a specific helical sense across the polymer chain were synthesized (see Figure 12.12). One of them had a chiral carbon at the second carbon atom in the hexyl side group through deuteration of one of its hydrogen atoms [18]. Another variation introduced methyl substitution (chiral) to the second carbon atom in addition to another methyl substitution in the sixth carbon atom (achiral) in the heptyl side group [19]. Both polymers exhibited a large value of $[\alpha]$, indicating amplification of a specific sense, made possible by a high penalty of energy against helix reversals.

It is also possible to invoke preference to a specific helical sense in plain PHIC that does not have any chiral carbon atoms. PHIC in an achiral solvent such as toluene does not induce any preference in the helical sense. However, dissolving the polymer in a chiral solvent, if the solvent is enantiomerically pure, can induce a preference in the helicity.

The optical activity of the solution is mostly from the solvent, not from PHIC. Therefore, measuring the optical rotation does not provide information on the helicity of PHIC. A CD spectrum, however, can provide the information if the solvent does not absorb in the range of wavelength where PHIC absorbs. Figure 12.13 shows a difference of the CD spectra in various chiral solvents from their respective spectrum of the pure solvent [20]. Obviously, these chiral solvents induced preference of one of the helical senses over the other in achiral PHIC.

12.4 Variations to the Helical Polymer

Now, we look at helical properties of polyisocyanates with a built-in random chirality. One is a copolymer of monomers with a chiral carbon and unmodified

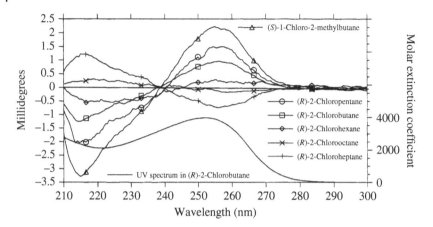

Figure 12.13 Circular dichroism spectrum of PHIC in six enantiomerically pure chiral solvents. The name of the solvent is indicated in the figure. Also shown is the absorption spectrum of PHIC. Source: Green et al. 1993 [20]. Reproduced with permission of American Chemical Society.

n-hexyl isocyanate. The other is a copolymer consisting of monomers of R isomer and those of S isomer. We use the formulas obtained in Section 11.5 to examine the helical properties.

12.4.1 Copolymer of Chiral and Achiral Isocyanate Monomers

Copolymerizing a monomer with a chiral carbon (enantiomerically pure) is a natural way to incorporate a quenched local field. First, a random copolymer of (R)-2,6-dimethyl heptyl isocyanate and n-hexyl isocyanate was synthesized (Figure 12.14a) [21]. The copolymer had ~4000 repeat units, and the mole fraction of (R)-2,6-dimethyl heptylisocyanate was just 10^{-4}. If the chiral monomer is enantiomerically pure, it will impart preference of a specific helicity to the repeat unit the monomer is bonded to. However, it turned out that its CD spectrum is not sufficiently clear to indicate a helical preference (line a in Figure 12.14c). This problem was solved by having n-butyl moiety at the β carbon of the n-hexyl side chain, as the latter stiffens the polymer chain, thus increasing the energy required for helix reversal. The structure of the copolymer is shown in Figure 12.14b, and the CD spectrum is shown in line b of the Figure 12.14c.

The effect of adding chiral repeat units amounts to adding $-\Sigma h_i \sigma_i$ to the energy, where

$$h_i = \begin{cases} h & \text{(the ith repeat unit is a chiral monomer)} \\ 0 & \text{(otherwise)} \end{cases}$$

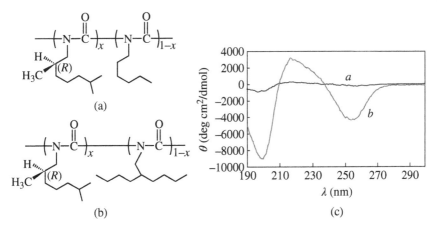

Figure 12.14 (a) Structure of a copolymer of (R)-2,6-dimethyl heptyl isocyanate and n-hexyl isocyanate. (b) Structure of a copolymer of (R)-2,6-dimethyl heptyl isocyanate and 2-butylhexyl isocyanate. (c) CD spectra of the two copolymers a and b ($N = 4000$, $x = 10^{-4}$).

Figure 12.15 Molar ellipticity $[\theta]$ of a helical copolymer containing a small fraction of repeat units with a chiral carbon, measured at 20 °C, plotted as a function of the product of molar fraction r of the chiral repeat units and the degree of polymerization, N. The solid line is an optimal fit by $\tanh(\beta h N r)$. The number labels indicate the values of r.

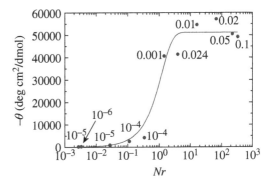

Once the copolymer is synthesized, h_i is fixed for $i = 1, \ldots, N$, where N is the number of repeat units in the polymer chain. Therefore, $-\Sigma h_i \sigma_i$ cannot be replaced with $-H\Sigma \sigma_i$; h_i is a quenched local field.

Copolymers of a structure shown in Figure 12.14b were synthesized for different monomer feeds and molecular weights, resulting in different mole fractions r of the chiral repeat units in the copolymer. Figure 12.15 shows a plot of the molar ellipticity $[\theta]$ at 254 nm as a function of Nr. Note that Nr represents the mean number of chiral repeat units per polymer chain.

The homopolymer of 2-butylhexyl isocyanate has a much more bulky side group compared with PHIC. Therefore, the energy of helix reversal is a lot greater. For this homopolymer, a ^{1}H-NMR spectrum was collected at different temperatures [22]. The protons on the α carbon exhibited splitting, but merged into one at elevated temperatures, since frequent helix reversal averaged the environment for the protons. From the temperature dependence of the split,

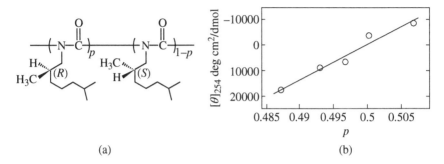

(a)   (b)

Figure 12.16 (a) Structure of a copolymer of (R)-2,6-dimethyl heptyl isocyanate and (S)-2,6-dimethyl heptyl isocyanate. (b) Molar ellipticity at 254 nm, plotted as a function of p, the molar fraction of the R isomers. Source: Panel (b) from Jha et al. 1999 [21]. Reproduced with permission of American Chemical Society.

$2J$ was estimated to be 82 kJ mol^{-1}. That would make $2\beta J = 33.7$ at 20 °C, and therefore the whole chain would be a monodomain, if N is a mere 10^4 or 10^5. We can safely assume that $L = N$ for this copolymer.

When Nr is 0.1 or less, $[\theta]$ is small, but a copolymer with r as small as 0.001 has a value of $[\theta]$ close to the saturation level – that is $[\theta]$ for copolymers of a large Nr. The curve in the figure is the optimal fit by a theoretical formula, $[\theta] = [\theta]_{sat} \tanh(\beta h Nr)$ with $\beta h = 0.631$; see Eq. (11.154). Then, $h = 1.54$ kJ mol^{-1}, a lot smaller compared with the energy for the helix reversal.

12.4.2 Copolymer of R- and S-Enantiomers of Isocyanate

The other type of quenched local fields we learned in Section 11.5 is a binomial distribution of local field, $h_i = +h$ and $-h$. The 1D Ising model with such a local field can be realized by copolymerizing a mixture of monomers that are R and S enantiomers. The structure of an example copolymer is shown in Figure 12.16a. Copolymers of different molar fractions p of the R enantiomer were synthesized, and the CD spectrum was collected [21]. Figure 12.16b is a plot of $[\theta]$ at 254 nm as a function of p. In a narrow range of p close to 0.5, the data of $[\theta]$ lie along a straight line through (0.5, 0). The slope is 1.336×10^6 deg·cm^2/dmol. It is known that $[\theta]$ at saturation is 5.2×10^4 deg·cm^2/dmol. Therefore, the slope ($\langle \sigma \rangle$ vs. p) is 25.7, which should be equal to $2(2L/\pi)^{1/2}$ (see Figure 11.24). Thus, we estimate that $L = 259$ for this copolymer.

Problems

12.1 *Estimating J.* This problem is related to the experiments partly shown in Figure 12.13. CD spectra were collected for PHIC in R-chlorobutane at three temperatures. The given figure shows a plot of ellipticity θ as a function of N^{-1} (reciprocal of the degree of polymerization). The results of a linear fit are also shown. The ellipticity θ (degree) is related to the molar ellipticity $[\theta]$ according to

$$[\theta] = \frac{100\theta}{cl}$$

where the concentration c is in $\text{mol}\,l^{-1}$, path length l is in cm, and $[\theta]$ is in $\text{degree}\cdot\text{cm}^2\,\text{dmol}^{-1}$. You can consider θ as a quantity proportional to $\langle\sigma\rangle$

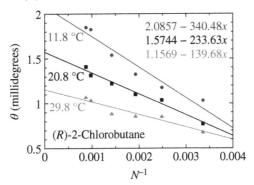

Use the values of the intercept to estimate $2J$.

12.2 *Terpolymer.* To demonstrate an extreme cooperativity of the helical sense, a random copolymer (actually a terpolymer) was synthesized from (R)-2,6-dimethyl heptyl isocyanate, (S)-2,6-dimethyl heptyl isocyanate, and achiral 2-butylhexyl isocyanate. The majority of the repeat units was the last component. The given figure depicts the structure of the terpolymer.

Let us denote by l_+ and l_- the numbers of the repeat units of (R)-2,6-dimethyl heptyl isocyanate and (S)-2,6-dimethyl heptyl isocyanate, respectively, in a total L repeat units of a given terpolymer chain. Let r be the fraction of the chiral monomers in the feed, and p the fraction of R isomers within the chiral units. The l_+ and l_- are different for each polymer chain. Their averages are Lrp and $Lr(1-p)$, respectively. The table is a list of molar fractions in the feed, the local fields in the quenched dilute random field, and the numbers of repeat units.

Repeat units	Molar fractions	Local field	Number
(R)-2,6-Dimethyl heptyl isocyanate	rp	h	l_+
(S)-2,6-Dimethyl heptyl isocyanate	$r(1-p)$	$-h$	l_-
2-Butylhexyl isocyanate	$1-r$	0	$L - l_+ - l_-$
Total	1		L

(1) Express the total magnetic field h_{tot} in a section of a uniform spin and length L using l_+ and l_-.
(2) The l_+ and l_- are distributed with a trinomial distribution. Use the formulas in Section 2.3 to find the mean and variance of h_{tot}.
(3) Assume that $r \ll 1$ but $Lr \gg 1$. Apply the Ising model with a quenched local field to draw a sketch for a plot of $\langle \sigma \rangle$ as a function of p. Indicate the slope of the tangent to the curve at $p = 1/2$.

13

Helix–Coil Transition

Transition from a helix conformation to a coil conformation of a polypeptide and vice versa is another phenomenon that a theoretical treatment similar to the one we learned in the 1D Ising model is effective. The theory was developed by Zimm and Bragg several decades ago [23]. The intuition they instilled in the model appears out-of-worldly, but it works and explains experimental results beautifully. We learn their model in this chapter.

13.1 Historical Background

A protein molecule is a sequence of amino acid residues (–NH–CHR–CO–, where R varies from residue to residue), and its sequence is given. Interactions between atoms or groups of atoms within the molecule force it to adopt a specific structure in three dimensions when the large molecule folds. We recognize a helix form (called α-helix) and a sheet form (β-sheet) in some parts of the folded structure. The α-helix is made possible through a twist favored by bulky side groups R bonded to the chiral carbon atom and supported by hydrogen bonds between the proton in NH within the residue and the oxygen atom in C=O a few residues away along the linear-chain backbone (see Figure 13.1).

When the protein is heated or its pH environment is changed, these structures may be partially damaged. Subsequent cooling may not restore the structure the protein had before the heating. Since the structure after the damage resembles a random-coil conformation of a synthetic polymer chain, this irreversible change in the denaturing is called a helix–coil transition. The interest in the transition in proteins led to studies of a similar transition in model protein molecules – typically, homopolymers of amino acid residues such as poly(L-alanine) and poly(L-lysine). We call them polypeptides, distinguished from proteins.

The optimal environment to study the helix–coil transition is a solution where polypeptide chains are isolated from each other. However, the studies are not easy, since the polypeptide has a poor solubility in water unless the

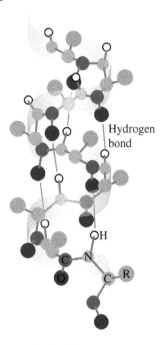

Figure 13.1 α-Helix in a protein molecule. A hydrogen bond forms between the proton in NH and the oxygen atom in C = O across a few residues. Hydrogen atoms on the carbon atoms bonded to side groups R are not shown.

Hydrogen bond

residue's side group (R) has ionizable groups, i.e. the residues are either anionic or cationic. The poor solubility is due to a scant degree of freedom for the polypeptide's conformation in solution: The α-helix is essentially just one conformation, and therefore there is no incentive to dissolve, as an increase in the entropy by dissolution is small. When R is ionizable in water, hydration lowers the free energy, promoting dissolution.

To turn the polypeptide soluble, not in water but in an organic solvent, the side group was chemically modified. This route opened up studies of the transition for many polypeptides. Since the chemical modification does not touch the amide groups, the hydrogen bonds are either intact or still possible.

Here, we look at four examples of the helix–coil transition. Transition can occur through different mechanisms, but all involve formation and destruction of the hydrogen bonds.

The first example is about poly(γ-benzyl-L-glutamate) (PBLG). To dissolve the polypeptide in an organic solvent, the carboxylic acid in the glutamate side group was esterified with benzyl alcohol. Figure 13.2 shows the optical rotation of PBLG in a mixture of ethylene dichloride and dichloroacetic acid (20 : 80 by mass) as a function of temperature [24]. Results are shown for two different molecular weights (degree of polymerization $N = 1100$ and 63 at the weight-average molecular weights). The specific rotation at high temperatures is characteristic of the α-helix. Therefore, the state of the polypeptide at low temperatures is not helix. From other studies such as viscosity of solutions, it

Figure 13.2 Specific optical rotation at 589 nm for a solution of poly(γ-benzyl-ʟ-glutamate) in a mixture of ethylene dichloride and dichloroacetic acid (20 : 80 by mass) at different temperatures. Two samples of the polypeptide with different degrees of polymerization N were employed. Solid lines are for an eye guide. Source: Data from Doty and Yang 1956 [24].

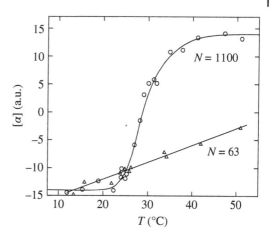

was found that the polypeptide was in a coil conformation at low temperatures. The transition between the two conformations occurred in a narrow range of temperatures for the polypeptide with $N = 1100$. The transition was diffuse for the other polypeptide with $N = 63$.

The coil state at low temperatures and the helix state at high temperatures are the opposite to the states of proteins that denature in aqueous solutions. Obviously, there is something that disrupts the hydrogen bonds within the polypeptide chain at low temperatures. That is a hydrogen bond between the NH proton of the polypeptide to one of oxygen atoms of the solvent, dichloroacetic acid. At low temperatures, this hydrogen bond dominates over the one within the polypeptide. Consequently, the polypeptide chain adopts a random-coil conformation. With an increasing temperature, this hydrogen bond weakens, and is taken over by the intrachain hydrogen bond that supports the helical conformation. Unlike the denaturation of proteins, the helix–coil transition occurs reversibly.

In the next example, the helix–coil transition was caused by a solvent composition change. The polypeptides were poly(ʟ-tyrosine) with $N = 613$ at the weight-average molecular weight and poly(ᴅ-tyrosine) with $N = 491$ [25]. Two solvents were mixed at different ratios to prepare solutions. Figure 13.3a shows the optical rotation per residue, $[R]$, for poly(ʟ-tyrosine), plotted as a function of volume percent of dimethylsulfoxide (DMSO) in propane diol. At low volume fractions of DMSO, the optical rotation was positive and large, and decreased as the volume fraction of DMSO exceeded ~70%. The transition occurred in a narrow range of the solvent composition. The value of the optical rotation per residue of poly(ʟ-tyrosine) in 100% DMSO was nearly identical to the one for tyrosine oligomer too short to form a helix. Thus, it was confirmed that poly(ʟ-tyrosine) was in coil state in 100% DMSO. The greater optical rotation in solvents with a large fraction of propane diol was ascribed to the helix

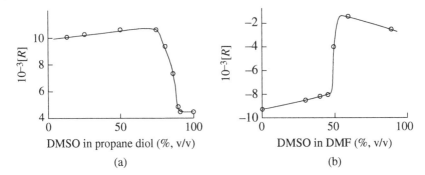

Figure 13.3 (a) Optical rotation per residue [R] at 250 nm for poly(L-tyrosine) in mixtures of propane diol and dimethylsulfoxide (DMSO). (b) Optical rotation per residue at 260 nm for poly(D-tyrosine) in mixtures of dimethylformamide (DMF) and dimethylsulfoxide (DMSO). Source: Data from Engel et al. 1971 [25].

conformation. Figure 13.3b is for the D isomer of the polypeptide, and the solvent was a mixture of DMSO and dimethylformamide (DMF). The optical rotation was negative in pure DMF, and its absolute value exhibited a sharp drop as the DMSO fraction exceeded ~50% in the mixture.

In both solution systems, increasing the volume fraction of DMSO caused a transition from helix to coil. Obviously, DMSO took over the intrachain hydrogen bonds; the oxygen atom in DMSO is a strong proton acceptor.

In the third example, the transition was triggered by a pH change, and the polymer was poly(L-lysine) [26]. Poly(L-lysine) HCl ($M_w = 70\,000\,\mathrm{g\,mol^{-1}}$; $N = 426$) was dissolved in water, and the solution's pH was adjusted by dialysis. Figure 13.4a shows circular dichroism (CD) spectra at different values of pH. At a sufficiently low pH, the spectrum was nearly identical. The spectrum changed in a narrow range of pH at around the pK_a of the conjugate acid of lysine residues when the pH increased. When the pH was sufficiently high, the spectrum was again nearly independent of pH.

At low pH values, the polymer is highly charged; and there are many water molecules around to form hydrogen bonds to NH and CO in the polymer. Therefore, the hydrogen bonds within the polymer are not preferred, forcing the polymer to adopt a coil conformation. At high pH values, the polymer is not charged, and the capacity of water molecules to hydrogen bond to the NH and CO in the polymer is reduced because the water molecules solvate OH⁻. Then, the intrachain hydrogen bonds are preferred, and the polymer takes a helix conformation. The percent helix was estimated from the CD spectrum at each pH, which is shown in Figure 13.4b.

The last example involves a neutral polypeptide in an aqueous solution at pH 7.0 [27]. A 50-mer polypeptide Ac–Y(AEAAKA)$_8$F–NH$_2$ (Y = Tyr, A = Ala, E = Glu, K = Lys, F = Phe) was dissolved in a 1 mM potassium phosphate

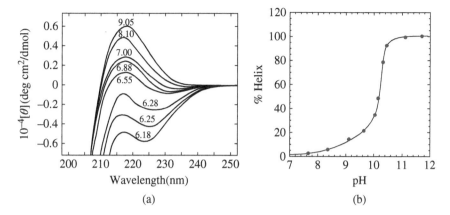

Figure 13.4 (a) CD spectra of poly(L-lysine) HCl ($M_w = 70\,000\,\text{g mol}^{-1}$; $N = 426$) at different values of pH. The number adjacent to each curve indicates the pH. (b) Percent helix estimated from the CD spectra is plotted as a function of the pH. Source: Myer 1969 [26]. Reproduced with permission of American Chemical Society.

Figure 13.5 Molar ellipticity at 222 nm of Ac–Y(AEAAKA)$_8$F–NH$_2$ at pH 7.0 in a heating scan, plotted as a function of temperature, for different concentrations of the polypeptide. Solid symbols were obtained in a cooling scan. Source: Data from Scholtz et al. 1991 [27].

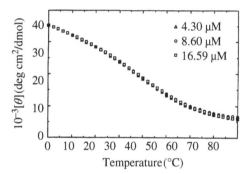

buffer at pH 7.0 with 0.1 M KF added. The zwitterionic repeat unit made the polypeptide (close to polyalanine) soluble in water. A CD spectrum was collected as the solution's temperature increased from 0 °C to 80 °C. Figure 13.5 shows the molar ellipticity at 222 nm as a function of temperature for different concentrations of the polypeptide. The ellipticity decreased gradually with an increasing temperature, indicating a diffuse transition of the short-chain polypeptide from helix to coil. The figure also shows data obtained in a cooling scan. They overlap with those in the heating scan, and thus we find that the transition is thermally reversible.

13.2 Polypeptides

Figure 13.6 is a drawing of a two-dimensional structure of a polypeptide. Each shaded rectangle represents an amino acid residue. Since the amide

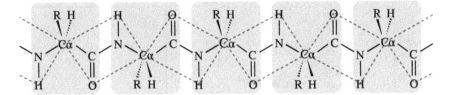

Figure 13.6 Two-dimensional structure of a polypeptide chain. A shaded rectangle represents a residue. Atoms and bonds in the dashed line are nearly coplanar.

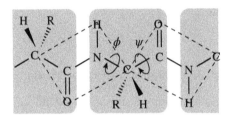

Figure 13.7 Definition of two dihedral angles ϕ and ψ.

bond is close to planar (due to resonance stabilization), the five bonds in each parallelogram (dashed line) are coplanar. Per residue, the peptide backbone has three atoms and three bonds (C(R)—C(O), C(O)—N, and N—C(R)). The N—C(R) bond can rotate around itself, and so can the C(R)—C(O) bond, but not C(O)—N.

We denote the dihedral angles of the N—C(R) bond and the C(R)—C(O) bond by ϕ and ψ, respectively (see Figure 13.7). For each combination of ϕ and ψ, there is energy per residue. If R = H (glycine), the energy would be the sum of the torsional energy similar to the one shown in Figure 5.12, plus the energy due to steric hindrance similar to the pentane effect of a polymer chain. The latter energy will be greater, if R is larger than just a hydrogen atom.

Ramachandran used a simple method based on a hard-core ball for each atom to evaluate the energy as a function of ϕ and ψ [28]. It is a topographic map with the energy in the height direction, but usually we show it as a contour plot, as most maps are drawn. Figure 13.8 shows a Ramachandran plot for poly(L-alanine) with R = CH_3 [29].

We anticipate that each plot of ϕ and ψ has local minima in the potential energy at trans (t), gauche (g), and gauche prime (g'). When plotted in the 2D contour plot, the energy will have nine local minima. The situation is different in Figure 13.8. We notice a few features in the plot.

A. The angles (ϕ, ψ) at the local minima are somewhat offset from t, g, and g'. For example, the β-sheet conformer is not at $\phi = \psi = 180°$ (tt).
B. The global energy minimum is in the region for the P_{II} conformer (gt). The next lowest energy levels are seen for the β-sheet (tt) and the right-handed α-helix (gg).

Figure 13.8 Ramachandran plot of poly(L-alanine). The darker the shade, the lower the energy. The regions of the torsional angles that lead to a right-handed α-helix, a left-handed α-helix, P_{II} conformer that does not form intrachain hydrogen bonds, a β-sheet, and a γ-turn (C_7) are indicated. Source: Jiang et al. 2013 [29]. Reproduced with permission of Royal Society of Chemistry.

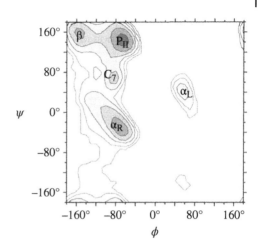

C. In contrast, gg' (C_7) and g'g (α_L) are less preferred.

D. Least favored among the nine conformers are tg', tg, g't, and g'g'.

The Ramachandran plots for the other amino acid residues except glycine are similar to the one shown in Figure 13.8.

13.3 Zimm–Bragg Model

Zimm and Bragg developed a theory based on statistical mechanics to describe a helix–coil transition in a linear chain of peptides [23]. We learn their theory in this section.

Let ϕ_i and ψ_i be the two dihedral angles of the ith residue ($i = 1, 2,..., N$) in the polypeptide chain of N residues. The conformation of the chain is specified by $\phi_1, \psi_1, \phi_2, \psi_2,..., \phi_N$, and ψ_N. We may introduce the single-residue partition function Z_1 as

$$Z_1 = \int_0^{360°} d\phi_1 \int_0^{360°} d\psi_1 \exp(-\beta E(\phi_1, \psi_1)) \tag{13.1}$$

We divide the integral into nine parts; each part is a contribution by a specific conformation. For example,

$$Z_{1gt} = \int_0^{120°} d\phi_1 \int_{120°}^{240°} d\psi_1 \exp(-\beta E(\phi_1, \psi_1)) \tag{13.2}$$

and thus

$$Z_1 = Z_{1tt} + Z_{1tg} + Z_{1tg'} + Z_{1gt} + Z_{1gg} + Z_{1gg'} + Z_{1g't} + Z_{1g'g} + Z_{1tg'g'} \tag{13.3}$$

Were it not for any interaction across several residues such as the hydrogen bonds, the partition function of the N-residue chain would be expressed as

$$Z = Z_1^N \tag{13.4}$$

We single out Z_{1gg} as the single-residue conformation responsible for the α-helix, and denote by Z_{1h} the single-residue partition function in the helix conformation. Since hydrogen bonds in the helix lower the energy,

$$Z_{1h} > Z_{1gg} \tag{13.5}$$

The remaining parts in Eq. (13.3) are for the coil conformation:

$$Z_{1c} = Z_{1tt} + Z_{1tg} + Z_{1tg'} + Z_{1gt} + Z_{1gg} + Z_{1g't} + Z_{1g'g} + Z_{g'g'} \tag{13.6}$$

We do not distinguish these eight conformers.

In the Zimm–Bragg model, the state of each residue is binary, c (= coil) or h (= helix). The model does not consider β-sheet or other conformers separately. The state of the polymer chain is, for example, ccchhhhhhccccchhhhhhhhh-hccc. Let N_h be the number of residues in the helix conformation, where N_h ranges from 0 (all coil) to N (all helix). Then, the N-residue partition function can be written as

$$Z = \sum_{N_h=0}^{N} \binom{N}{N_h} Z_{1h}^{N_h} Z_{1c}^{N-N_h} = (Z_{1c} + Z_{1h})^N \tag{13.7}$$

after taking into account the lowering in the energy by the hydrogen bonds.

This expression still does not consider cooperative nature of the α-helix. Suppose we add one residue at a time and continue to do so until amassing N residues. It is not easy to form a helix; but once formed, the helical conformation tends to continue. To take into account this cooperativeness, we specify the state of the N-residue polypeptide by N_h and n, the number of helices in the chain. We write the partition function as

$$Z = \sum Z_{1h}^{N_h} Z_{1c}^{N-N_h} \sigma^n \tag{13.8}$$

Here, we have introduced a parameter σ (<1) to indicate a penalty for starting a helix. The smaller the σ, the greater the penalty. The series is calculated for all possibilities of sequence of c and h. Each sequence has its own N_h and n.

Figure 13.9 compares two conformations that have the same number of helical residues, but different numbers of helices. Part (a) has one sequence of the helix, whereas (b) has two sequences. Since $\sigma < 1$, (a)'s contribution to Z is greater compared with (b)'s. It means that (a) is more probable than (b) is. The smaller the σ, the smaller the preferred n.

This discussion applies to a fixed sequence of c and h in each of (a) and (b). For a given N_h, different lengths of the two helices are possible. When the number of different lengths is included, structures represented by (b) may be more probable.

Figure 13.9 Two conformations of a polypeptide chain. (a) and (b) have the same number of helical residues, but their numbers of helices are different.

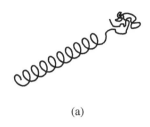

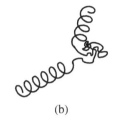

(a)   (b)

Table 13.1 Single-residue partition function of the ith residue.

	i	
$i-1$	c	h
c	Z_{1c}	$Z_{1h}\sigma$
h	Z_{1c}	Z_{1h}

The function depends on the states of the ith residue and the preceding residue.

Computation of Z by Eq. (13.8) looks formidable, since specifying N_h does not determine n; rather, both N_h and n are determined once a sequence of h and c is given. The calculation is facilitated by switching from Eq. (13.8) to the method of the transfer matrix we learned for the 1D Ising model in Section 11.4. The method lists all possibilities for the sequence of c and h. Table 13.1 shows the matrix for the ith residue that is either in c or h conformation. The contribution to the partition function by the ith residue is determined by the states of the ith residue and the preceding residue. Note that σ applies to a c—h bond, but not to an h—c bond (see Table 13.1).

Let us denote by s_i the state of the ith residue; $s_i = $ c or h. We divide Z_n, the partition function of the first n residues, into two parts:

$Z_n(s_n = c) = $ the part of Z_n with the nth reside being in coil conformation;
$Z_n(s_n = h) = $ the part of Z_n with the nth reside being in helix conformation.

Then,

$$Z_n(s_n = c) = Z_{n-1}(s_{n-1} = c)Z_{1c} + Z_{n-1}(s_{n-1} = h)Z_{1c} \tag{13.9}$$

$$Z_n(s_n = h) = Z_{n-1}(s_{n-1} = c)Z_{1h}\sigma + Z_{n-1}(s_{n-1} = h)Z_{1h} \tag{13.10}$$

In the matrix form, these two equations are rewritten to

$$[Z_n(s_n = c) \ Z_n(s_n = h)] = [Z_{n-1}(s_{n-1} = c) \ Z_{n-1}(s_{n-1} = h)] \begin{bmatrix} Z_{1c} & Z_{1h}\sigma \\ Z_{1c} & Z_{1h} \end{bmatrix}$$
$$\tag{13.11}$$

Let

$$\mathbf{U} \equiv \begin{bmatrix} Z_{1c} & Z_{1h}\sigma \\ Z_{1c} & Z_{1h} \end{bmatrix} \tag{13.12}$$

Then,

$$[Z_n(s_n = c) \; Z_n(s_n = h)] = [Z_{n-1}(s_{n-1} = c) \; Z_{n-1}(s_{n-1} = h)]\mathbf{U} \tag{13.13}$$

We can apply this recurrence relationship repeatedly to obtain

$$[Z_N(S_N = c) \; Z_N(s_N = h)] = [Z_1(s_1 = c) \; Z_1(s_1 = h)]\mathbf{U}^{N-1} \tag{13.14}$$

The partition function Z of the whole chain is

$$Z = Z_N(s_N = c) + Z_N(s_N = h) \tag{13.15}$$

In Section 11.4, we learned that, when $N \gg 1$, Z is approximated as

$$Z \equiv \lambda_1^N \tag{13.16}$$

where λ_1 is the greater of the two eigenvalues of $\mathbf{U}$. The two eigenvalues λ_1 and λ_2 are the zeros of the determinant of $\mathbf{U}$:

$$\begin{vmatrix} Z_{1c} - \lambda & Z_{1h}\sigma \\ Z_{1c} & Z_{1h} - \lambda \end{vmatrix} = (Z_{1c} - \lambda)(Z_{1h} - \lambda) - Z_{1c}Z_{1h}\sigma = 0 \tag{13.17}$$

The zeros are

$$\lambda_1 = \frac{1}{2}\{Z_{1c} + Z_{1h} + [(Z_{1c} - Z_{1h})^2 + 4Z_{1c}Z_{1h}\sigma]^{1/2}\} \tag{13.18}$$

$$\lambda_2 = \frac{1}{2}\{Z_{1c} + Z_{1h} - [(Z_{1c} - Z_{1h})^2 + 4Z_{1c}Z_{1h}\sigma]^{1/2}\} \tag{13.19}$$

Figure 13.10 shows how λ_1 and λ_2 change with Z_{1h}/Z_{1c}, the ratio of the relative magnitudes of the single-residue partition functions. The curves are shown for two values of σ. At $Z_{1h}/Z_{1c} = 1$, the two conformations are equally probable. When $Z_{1h}/Z_{1c} > 1$, the helix is preferred (before taking into account of the penalty imposed on initiating a helix); when $Z_{1h}/Z_{1c} < 1$, the coil is preferred.

The plot of λ_1/Z_{1c} is always greater than 1, and is an increasing function of Z_{1h}/Z_{1c}; the plot of λ_2/Z_{1c} is also an increasing function, but is always less than 1. Their increases are gradual when $\sigma = 10^{-2}$. For $\sigma = 10^{-4}$, there is an abrupt change in the tangential slope of the curves: The curve for λ_1/Z_{1c} runs close to a horizontal line at 1 when $Z_{1h}/Z_{1c} < 1$, and close to the diagonal line, $\lambda_1/Z_{1c} = Z_{1h}/Z_{1c}$, when $Z_{1h}/Z_{1c} > 1$. The situation is the other way around for the curve of λ_2/Z_{1c}.

The extreme situation occurs when $\sigma = 0$. Since $(x^2)^{1/2} = |x|$,

$$\lambda_1 = \begin{cases} Z_{1c} & (Z_{1h} < Z_{1c}) \\ Z_{1h} & (Z_{1h} > Z_{1c}) \end{cases} \tag{13.20}$$

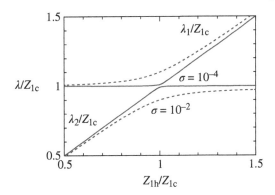

Figure 13.10 Plots of λ_1/Z_{1c} and λ_2/Z_{1c} as a function of Z_{1h}/Z_{1c} for $\sigma = 10^{-2}$ (dashed lines) and 10^{-4} (solid lines).

$$\lambda_2 = \begin{cases} Z_{1h} & (Z_{1h} < Z_{1c}) \\ Z_{1c} & (Z_{1h} > Z_{1c}) \end{cases} \qquad (13.21)$$

That is, λ_1 is equal to the greater of Z_{1h} and Z_{1c}, and λ_2 is equal to the smaller of the two. When the penalty of initiating a helix is huge, the whole chain will be either in the coil or helix, depending on which conformation is easier in a residue.

The change in Z_{1h}/Z_{1c} drives a switch from helix to coil and vice versa. In a solution, the change may be caused by a temperature change or a solvent composition change.

Now, we find what we can predict about the polypeptide chain from the statistical model. First, we consider the fraction of residues in the helix conformation. For that purpose, we calculate $\langle N_h \rangle$. With Eq. (13.8), we can derive a formula:

$$\langle N_h \rangle = \frac{1}{Z} \sum N_h Z_{1h}^{N_h} Z_{1c}^{N-N_h} \sigma^n = \frac{1}{Z} Z_{1h} \frac{\partial Z}{\partial Z_{1h}} = Z_{1h} \frac{\partial \ln Z}{\partial Z_{1h}} \qquad (13.22)$$

If $Z \cong \lambda_1{}^N$, we can proceed to calculating $\langle N_h \rangle$ to obtain

$$\frac{\langle N_h \rangle}{N} = Z_{1h} \frac{1 + [(Z_{1c} - Z_{1h})^2 + 4Z_{1c}Z_{1h}\sigma]^{-1/2}[-(Z_{1c} - Z_{1h}) + 2Z_{1c}\sigma]}{Z_{1c} + Z_{1h} + [(Z_{1c} - Z_{1h})^2 + 4Z_{1c}Z_{1h}\sigma]^{1/2}} \qquad (13.23)$$

Figure 13.11 shows a plot of $\langle N_h \rangle/N$, the fraction of helix residues, as a function of Z_{1h}/Z_{1c} for $\sigma = 10^{-2}$ and 10^{-4}. As the helix conformation becomes more favorable in the single residue, the fraction of the helix residues increases, a reasonable result. The increase is gradual for $\sigma = 10^{-2}$, but occurs in a narrow range of Z_{1h}/Z_{1c} for 10^{-4}. As initiating a helix becomes more difficult, the transition from a coil-dominated conformation to a helix-dominated conformation becomes more abrupt.

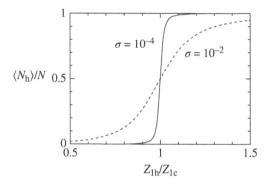

Figure 13.11 Plot of the fraction of helical residues, $\langle N_h \rangle / N$, as a function of Z_{1h}/Z_{1c} for $\sigma = 10^{-2}$ and 10^{-4}.

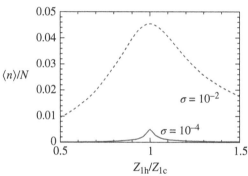

Figure 13.12 The mean number of helical sequences, divided by the number of residues in the chain, plotted as a function of Z_{1c}/Z_{1h} for $\sigma = 10^{-2}$ and 10^{-4}.

Next, we calculate $\langle n \rangle$, the mean number of helices within the chain. We can derive the following formula:

$$\langle n \rangle = \frac{1}{Z} \sum n Z_{1h}^{N_h} Z_{1c}^{N-N_h} \sigma^n = \frac{1}{Z} \sigma \frac{\partial Z}{\partial \sigma} = \sigma \frac{\partial \ln Z}{\partial \sigma} \tag{13.24}$$

With $Z \cong \lambda_1^N$, we obtain

$$\frac{\langle n \rangle}{N} = \sigma \frac{[(Z_{1c} - Z_{1h})^2 + 4Z_{1c}Z_{1h}\sigma]^{-1/2} 2Z_{1c}Z_{1h}}{Z_{1c} + Z_{1h} + [(Z_{1c} - Z_{1h})^2 + 4Z_{1c}Z_{1h}\sigma]^{1/2}} \tag{13.25}$$

Figure 13.12 shows a plot of $\langle n \rangle / N$ as a function of Z_{1h}/Z_{1c} for $\sigma = 10^{-2}$ and 10^{-4}. The number of helical sequences maximizes at $Z_{1h}/Z_{1c} = 1$. To the left of the peak, the number is less, because most of the residues are in coil. To the right of the peak, the number is fewer, since each sequence is long. Although most of the residues are in helix, they are not interrupted by intervening coil residues frequently along the chain. The smaller the σ, the fewer the number of helical sequences, a reasonable result.

The asymptotes of $\langle n \rangle / N$ are as follows. When $Z_{1h}/Z_{1c} \ll 1$, $\langle n \rangle / N$ will approach $\sigma Z_{1h}/Z_{1c}$. When $Z_{1h}/Z_{1c} \gg 1$, $\langle n \rangle / N$ will approach $\sigma Z_{1c}/Z_{1h}$.

It may be tempting to see a plot of $\langle N_h \rangle / \langle n \rangle$ that represents the mean number of residues per helical sequence. We should rather calculate $\langle N_h/n \rangle$, but it is

Figure 13.13 Ratio of the number of resides in the helix conformation to the number of helical sequences, plotted as a function of Z_{1h}/Z_{1c} for $\sigma = 10^{-2}$ and 10^{-4}.

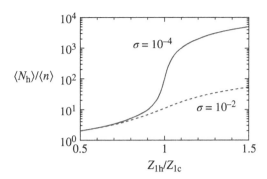

not easy, so we settle for $\langle N_h \rangle / \langle n \rangle$. Figure 13.13 shows a plot of $\langle N_h \rangle / \langle n \rangle$ as a function of Z_{1h}/Z_{1c} for $\sigma = 10^{-2}$ and 10^{-4}. The $\langle N_h \rangle / \langle n \rangle$ increases with an increasing Z_{1h}/Z_{1c}. The y axis is in a logarithmic scale, and thus we find that the number increases rapidly to $Z_{1h}/(Z_{1c}\sigma)$. The calculation shown in the figure assumes that $N \gg 1$. A finite N will limit $\langle N_h \rangle / \langle n \rangle$ to N.

Problems

13.1 *Normal transition.* We consider a helix-coil transition of a polypeptide that forms only intrachain hydrogen bonds. This type of polypeptide exhibits a normal transition, namely, helix at low temperatures. For simplicity, we write

$$Z_{1c} = \exp(-\beta E_c)$$

and

$$Z_{1h} = \exp(-\beta(E_{gg} + E_{hb,p}))$$

where $E_{hb,p}$ is the interaction due to the intrachain hydrogen bond. The latter depends on temperature T as

$$E_{hb,p} = k_{hb}(T - T_0)$$

where k_{hb} and T_0 are constants of temperature. Assume that E_c and E_{gg} are also constants of temperature.

(1) Is k_{hb} positive or negative?

(2) In the discrete model of ϕ and ψ, $\exp(-\beta E_c)$ is equal to the sum of $\exp(-\beta E_i)$, where $i, j = $ t, g, g' sans $i = j = $ g. Look at the Ramachandran plot in Figure 13.8 to compare E_c and E_{gg}.

(3) Draw a sketch for a plot of Z_{1h}/Z_{1c} as a function of temperature.

(4) Combine your sketch in (3) and Figure 13.11 to draw a sketch for a plot of $\langle N_h \rangle / N$ as a function of temperature.

13.2 *Inverted transition.* A polypeptide chain in a solvent that can form a hydrogen bond to the amide bond of the polypeptide may exhibit an inverted helix-coil transition, namely, helix at high temperatures. We include the interaction $E_{hb,s}$ in Z_{1c}:

$$Z_{1c} = \exp(-\beta(E_c + E_{hb,s}))$$

Let us express

$$E_{hb,p} - E_{hb,s} = k_{hb}(T - T_0)$$

where k_{hb} and T_0 are constants of temperature. Assume that E_c and E_{gg} are also constants of temperature.

(1) Which is required, $k_{hb} > 0$ or $k_{hb} < 0$?

(2) Draw a sketch for a plot of Z_{1h}/Z_{1c} as a function of temperature.

13.3 *Fluctuations of N_h.* Equation (13.22) can be rewritten to

$$\langle N_h \rangle = \frac{\partial \ln Z}{\partial \ln Z_{1h}}$$

(1) Show that

$$\langle \Delta N_h^2 \rangle = \frac{\partial \langle N_h \rangle}{\partial \ln Z_{1h}}$$

where $\Delta N_h = N_h - \langle N_h \rangle$.

(2) Inspect Figure 13.11 to draw a sketch for a plot of $\langle \Delta N_h^2 \rangle / N$ as a function of Z_{1h}/Z_{1c} for $\sigma = 10^{-2}$ and 10^{-4}.

14

Regular Solutions

In this chapter, we learn about thermodynamics of a binary mixture. If you remember such words as a critical temperature and a miscibility gap, you should have learned in physical chemistry the thermodynamics of mixing, at least phenomenologically. Here we apply canonical-ensemble formulation to a system of an A–B binary mixture in a liquid state. Examples include a water–acetone mixture and a water–pentanol mixture. Unlike a mixture of gas molecules, we cannot add or remove a molecule without changing the volume, that is, the mixture is incompressible. We neglect a small-volume change upon mixing due to a packing change.

First, we consider a simple mixture of A molecules and B molecules that have an equal size. Subsequently, we consider a binary mixture of different sizes. For both mixtures, we adopt a mean-field approximation for the interaction between the molecules. This model is called a Hildebrand regular solution.

14.1 Binary Mixture of Equal-Size Molecules

14.1.1 Free Energy of Mixing

Suppose we mix N_A molecules of A and N_B molecules of B. Here, N_A is not the Avogadro's number. The mixture fills a box consisting of N sites ($N = N_A + N_B$), each being occupied by either an A molecule or a B molecule (see Table 14.1). Since the two molecules have an equal size, the mole fraction $x_j = N_j/N$ ($j =$ A, B) is identical to the volume fraction.

Figure 14.1 is a two-dimensional (2D) representation of arranging A and B molecules in the mixture. In this illustration, the molecules are on lattice points of a 2D square lattice. There are $_NC_{N_A}$ ways to distribute N_A molecules of A among a total of N sites. The unoccupied sites will be automatically filled with B molecules, leaving no vacancy. Each arrangement constitutes a microstate of the mixture.

Statistical Thermodynamics: Basics and Applications to Chemical Systems, First Edition. Iwao Teraoka.
© 2019 John Wiley & Sons, Inc. Published 2019 by John Wiley & Sons, Inc.
Companion website: www.wiley.com/go/Teraoka_StatsThermodynamics

Table 14.1 Numbers of molecules and mole fractions in an A–B mixture.

Molecule	Number of molecules	Mole fraction
A	N_A	$x_A = N_A/N$
B	N_B	$x_B = N_B/N$
Total	N	1

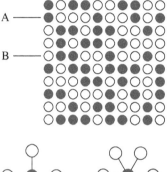

A ——
B ——

Figure 14.1 Two-dimensional representation of a mixture of A molecules and B molecules of equal size.

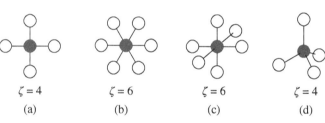

$\zeta = 4$ $\zeta = 6$ $\zeta = 6$ $\zeta = 4$

(a)   (b)   (c)   (d)

Figure 14.2 Coordination number ζ is the number of the nearest sites in a lattice model. (a) A 2D square lattice, (b) a 2D triangular lattice, (c) a 3D cubic lattice, and (d) a tetrahedral lattice.

We consider that there is an interaction between a pair of molecules on adjacent sites, and that is the only energy involved in the mixing. Each microstate has a definite arrangement of the molecules, and therefore definite numbers of A–A pairs, B–B pairs, and A–B pairs. To calculate the energy of the microstate, we need to count the number of these pairs.

In a 2D square lattice, each site has four nearest neighbors, and in a three-dimensional (3D) cubic lattice, the number is 6. We call this number a **coordination number** and denote it by ζ (see Figure 14.2). The coordination number of the triangular lattice is 6, and the one for the tetrahedral lattice is 4.

The system we consider is a mixture of A and B molecules. The partition function Z of the system is written as

$$Z = \sum_k \exp(-\beta E_k) \tag{14.1}$$

where E_k is the energy of microstate k. We denote the numbers of the A–A pairs, B–B pairs, and A–B pairs as N_{AA}, N_{AB}, and N_{BB}. We assign energy ε_{AA}, ε_{AB}, and ε_{BB} to each of the A–A, A–B, and B–B contacts. Then,

$$E_k = N_{AA}\varepsilon_{AA} + N_{AB}\varepsilon_{AB} + N_{BB}\varepsilon_{BB} \tag{14.2}$$

The three numbers are different for each microstate, but their sum is always equal to $\frac{1}{2}N\zeta$:

$$N_{AA} + N_{AB} + N_{BB} = \frac{1}{2}N\zeta \tag{14.3}$$

Each site has ζ neighbors, and $N\zeta$ would count each pair twice.

Since N_{AA}, N_{AB}, and N_{BB} are different for each k, calculating Z is all but impossible. To simplify the situation, we adopt a **mean-field approximation** and replace N_{AA}, N_{AB}, and N_{BB} with their mean values. For example, the mean number of A–A contacts is equal to the total number of contacts times the probability that both sites are A molecules:

$$\frac{1}{2}N\zeta \times \left(\frac{N_A}{N}\right)^2 = \frac{\zeta N_A^2}{2N} \tag{14.4}$$

Table 14.2 lists the mean numbers of A–B and B–B contacts as well. It is easy to confirm that the three numbers of the contacts add up to $\frac{1}{2}N\zeta$:

From Table 14.2, we find that the energy E_k is expressed as

$$E_k = \frac{\zeta}{2N}(N_A^2\varepsilon_{AA} + 2N_A N_B\varepsilon_{AB} + N_B^2\varepsilon_{BB}) \tag{14.5}$$

Then, all of $\exp(-\beta E_k)$ in the series in Eq. (14.1) are identical. Since

$$\sum_k = {}_N C_{N_A} = \frac{N!}{N_A! N_B!} \tag{14.6}$$

the partition function is

$$Z = \frac{N!}{N_A! N_B!} \exp\left(-\frac{\beta\zeta}{2N}(N_A^2\varepsilon_{AA} + 2N_A N_B\varepsilon_{AB} + N_B^2\varepsilon_{BB})\right) \tag{14.7}$$

Table 14.2 Numbers of A–A, A–B, and B–B contacts in the mean-field approximation, and their interactions.

Contact	Energy	Number of pairs
A–A	ε_{AA}	$\frac{1}{2}\zeta N_A^2/N$
B–B	ε_{BB}	$\frac{1}{2}\zeta N_B^2/N$
A–B	ε_{AB}	$\zeta N_A N_B/N$
	Total	$\frac{1}{2}N\zeta$

As you may have noticed by now, the mean-field approximation applied to a canonical ensemble converts it to a microcanonical ensemble.

The free energy of the system is

$$F = -k_B T \ln \frac{N!}{N_A! N_B!} + \frac{\zeta}{2N}(N_A^2 \varepsilon_{AA} + 2N_A N_B \varepsilon_{AB} + N_B^2 \varepsilon_{BB}) \tag{14.8}$$

Now, we compare this free energy after the mixing with the one before the mixing. In the N_A sites of A only, there are $N_A \zeta / 2$ contacts of A–A. The N_B sites of B only has $N_B \zeta / 2$ contacts of B–B. Each system of pure A and pure B has only one way to arrange the molecules. Therefore, the free energy before the mixing is

$$F_{before} = \frac{\zeta}{2}(N_A \varepsilon_{AA} + N_B \varepsilon_{BB}) \tag{14.9}$$

The change in the free energy by the mixing, $\Delta F = F - F_{before}$, is

$$\Delta F = -k_B T \ln \frac{N!}{N_A! N_B!}$$
$$+ \frac{\zeta}{2N}[N_A(N_A - N)\varepsilon_{AA} + 2N_A N_B \varepsilon_{AB} + N_B(N_B - N)\varepsilon_{BB}] \tag{14.10}$$

which is rewritten to

$$\Delta F = -k_B T \ln \frac{N!}{N_A! N_B!} + \frac{\zeta N_A N_B}{N}\left(\varepsilon_{AB} - \frac{\varepsilon_{AA} + \varepsilon_{BB}}{2}\right) \tag{14.11}$$

Let us introduce a dimensionless χ **(chi) parameter** defined as

$$\frac{\zeta}{k_B T}\left(\varepsilon_{AB} - \frac{\varepsilon_{AA} + \varepsilon_{BB}}{2}\right) \equiv \chi \tag{14.12}$$

The reason why a difference between ε_{AB} and $\frac{1}{2}(\varepsilon_{AA} + \varepsilon_{BB})$ appears in Eq. (14.12) will be obvious, if we find what happens by the mixing. In Figure 14.3, the four middle molecules change their positions, leading to creating four A–B contacts. Concomitantly, two A–A contacts and two B–B contacts are lost. As a rule of thumb, creating one A–B contact is made possible by deleting a half of A–A contact and a half of B–B contact.

The sign of χ can be positive or negative. A negative χ indicates that A–B contacts are preferred to A–A and B–B contacts, promoting the mixing. In contrast, a positive χ means that A–B contacts are disfavored, and may lead to a

Figure 14.3 In the rearrangement of A and B molecules shown here, two A–A contacts (dotted line) and two B–B contacts (dashed line) are replaced by four A–B contacts (solid line). Per newly created A–B contact, the energy changes by $\varepsilon_{AB} - \frac{1}{2}(\varepsilon_{AA} + \varepsilon_{BB})$.

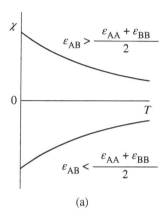
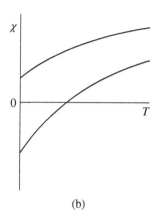

(a) (b)

Figure 14.4 If ε_{AA}, ε_{BB}, and ε_{AB} are constants of temperature T, χ will change with T as shown in (a). Depending on whether $\varepsilon_{AB} > \frac{1}{2}(\varepsilon_{AA} + \varepsilon_{BB})$ or not, the sign of χ is different, but will not change when T changes. (b) χ increases with an increasing T, if there is a hydrogen bond between A and B. χ may be negative at low temperatures.

phase separation. If ε_{AB} and $(\varepsilon_{AA} + \varepsilon_{BB})/2$ are constants of temperature, χ will decrease its magnitude as T increases, but its sign will not flip (Figure 14.4a). We see this situation in many mixtures. A different situation can occur when ε_{AB} has a component of attractive interaction that weakens with an increasing T. Hydrogen bond is one such typical interaction. Then, the negative component of ε_{AB} becomes weaker, causing χ to increase. Figure 14.4b shows two scenarios for the temperature dependence of χ.

With the χ parameter, Eq. (14.11) is rewritten to

$$\frac{\Delta F}{k_B T} = -\ln \frac{N!}{N_A! N_B!} + \chi \frac{N_A N_B}{N} \tag{14.13}$$

If $N_A \gg 1$ and $N_B \gg 1$, use of Stirling's formula converts Eq. (14.13) into

$$\frac{\Delta F}{k_B T} = -N \ln N + N_A \ln N_A + N_B \ln N_B + \chi \frac{N_A N_B}{N} \tag{14.14}$$

The free energy change per site is expressed as

$$\frac{\Delta F}{N k_B T} = x_A \ln x_A + x_B \ln x_B + \chi x_A x_B \tag{14.15}$$

The first two terms on the right-hand side are of an entropy origin (**entropy of mixing**), and is always negative, favoring the mixing. The sign of the third term is determined by the sign of χ. If $\chi < 0$, on the one hand, $\Delta F < 0$ for the whole range of x_A, indicating that the mixture is stable, and there is no miscibility gap. If $\chi > 0$ and is sufficiently large, on the other hand, ΔF may turn to positive, depending on x_A. There may be a miscibility gap. Depending on x_A, the system may separate into two phases. Here, we consider $\chi > 0$ only.

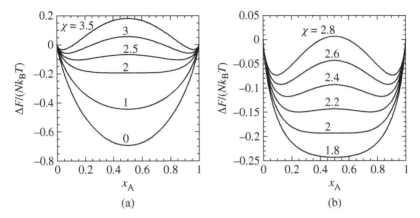

Figure 14.5 $\Delta F/(Nk_BT)$, free energy change by mixing per site, reduced by k_BT, is plotted as a function of x_A for different values of χ. (a) $\chi = 0, 1, 2, 2.5, 3$, and 3.5. (b) $\chi = 1.8, 2, 2.2, 2.4, 2.6$, and 2.8.

Figure 14.5a shows a plot of $\Delta F/(Nk_BT)$ as a function of x_A for different values of χ. Figure 14.5b is a detailed plot for χ around 2. All the curves are symmetric around a vertical line at $x_A = \frac{1}{2}$, caused by our choice of identical sizes for the two molecules. At both ends ($x_A = 0$ and 1), $\Delta F = 0$, a reasonable result since there is no mixing. The curve for $\chi = 0$ represents the negative of the entropy of mixing. The profile of the curve of $\Delta F/(Nk_BT)$ depends on χ. When $\chi < 2$, the curve minimizes at $x_A = \frac{1}{2}$, and that is the only minimum. When $\chi > 2$, the curve has two minima, one at $x_A < \frac{1}{2}$, and the other at $x_A > \frac{1}{2}$.

14.1.2 Derivatives of the Free Energy of Mixing

Ramification of a single minimum or two minima in the plot of $\Delta F/(Nk_BT)$ will be clearer when we differentiate $\Delta F/(Nk_BT)$ by x_A:

$$\frac{\partial}{\partial x_A} \frac{\Delta F}{Nk_BT} = \ln x - \ln(1-x) + \chi(1-2x_A) \tag{14.16}$$

Since $Nx_A = N_A$, it may appear that Eq. (14.16) gives the chemical potential of an A molecule reduced by k_BT, $\mu_A/(k_BT)$. It is not correct, however. Unlike a vapor system, we cannot add or remove an A molecule while holding the volume and the number of B molecules unchanged, a requirement for calculating μ_A. Recall $\mu_A = (\partial F/\partial N_A)_{T,V,N_B}$. Under the assumption of incompressibility, we cannot change N_A while holding N_B unchanged. Adding an A molecule must be accompanied by removing a B molecule to hold the volume of the system unchanged. Likewise, removing an A molecule and adding a B molecule must occur simultaneously. Therefore, $(\partial/\partial x_A)(\Delta F/Nk_BT)$ is equal to $(\mu_A - \mu_B)/(k_BT)$. We call $\mu_A - \mu_B$ a replacement chemical potential and use a

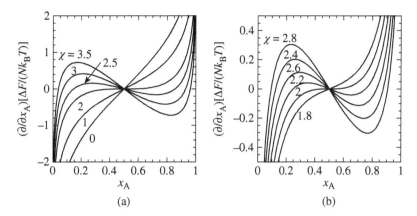

Figure 14.6 Derivative of the free energy change by mixing per site, $(\partial/\partial x_A)(\Delta F/Nk_B T) = \mu_{rep}/(k_B T)$, plotted as a function of x_A for different values of χ. (a) $\chi = 0, 1, 2, 2.5, 3$, and 3.5. (b) $\chi = 1.8, 2, 2.2, 2.4, 2.6$, and 2.8.

symbol μ_{rep}:

$$\frac{\mu_{rep}}{k_B T} \equiv \frac{\mu_A - \mu_B}{k_B T} = \ln x_A - \ln(1 - x_A) + \chi(1 - 2x_A) \tag{14.17}$$

Figure 14.6a shows plots of the derivative, $(\partial/\partial x_A)(\Delta F/(Nk_B T)) = \mu_{rep}/(k_B T)$, as a function of x_A for different values of χ. Figure 14.6b shows a plot of the same quantity for χ around 2. As you can surmise from the two-part figure, the curve is an increasing function of x_A in the whole range from 0 to 1 when $\chi < 2$. When $\chi > 2$, in contrast, the derivative curve has a local maximum at $x_A < \frac{1}{2}$ and a local minimum at $x_A > \frac{1}{2}$. The derivative curve has a horizontal tangent at both the local maximum and local minimum. At these points, $(\partial^2/\partial x_A{}^2)(\Delta F/(Nk_B T)) = 0$.

The two types of curves in Figure 14.6 are significantly different. The curve with a tangent of a positive slope in the whole range of x_A represents a normal and stable system: μ_{rep} increases with an increasing x_A. Adding an A molecule (and removing a B molecule at the same time) is more difficult when the system has already many A molecules in it than it is when the system does not have so many of them.

In contrast, the part of the curve with a negative-slope tangent indicates that the system is not stable (see Figure 14.7). The local maximum is at $x_{A,max}$, and the local minimum at $x_{A,min}$. The slope of the tangent is negative between $x_{A,max}$ and $x_{A,min}$. In this range of x_A, adding an A molecule lowers μ_{rep}, thus inviting more A molecules to get into the system. Suppose that the composition of the mixture fluctuates locally and the system happens to have a pocket of sites in which the volume fraction of A is between $x_{A,max}$ and $x_{A,min}$, and therefore the slope of the tangent is negative. That pocket will collect more A molecules from

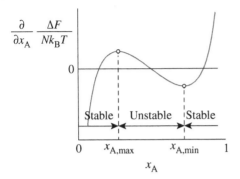

Figure 14.7 First-order derivative curve of $\Delta F/(Nk_B T)$ when it has a range of x_A with a negative-slope tangent. The curve has a local maximum at $x_{A,max}$ and a local minimum at $x_{A,min}$. Between them, the system is unstable.

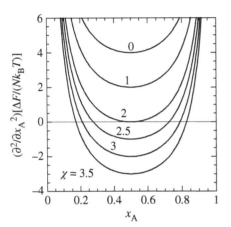

Figure 14.8 Second-order derivative of the free energy change by mixing per site, $(\partial^2/\partial x_A^2)(\Delta F/(Nk_B T))$, plotted as a function of x_A for different values of χ – 0, 1, 2, 2.5, 3, and 3.5.

its surroundings until x_A in the pocket reaches $x_{A,min}$. It means that the system is unstable. In contrast, the same system is stable if $x_A < x_{A,max}$ or $x_A > x_{A,min}$.

To locate $x_{A,max}$ and $x_{A,min}$, we differentiate $(\partial/\partial x_A)(\Delta F/(Nk_B T))$ by x_A once more:

$$\frac{\partial^2}{\partial x_A^2}\frac{\Delta F}{Nk_B T} = \frac{1}{x_A} + \frac{1}{1-x_A} - 2\chi \tag{14.18}$$

The plots of the second-order derivative are shown in Figure 14.8. All the curves are symmetric around $x_A = \frac{1}{2}$, and their differences are in the vertical position only. The curves for $\chi < 2$ are above the horizontal line at zero in the whole range of x_A; the slope of the tangent to the curve in Figure 14.6 is always positive. The mixture is stable whatever the value of x_A is.

The situation for $\chi > 2$ is illustrated in Figure 14.9, which shows the first- and second-order derivatives of $\Delta F/(Nk_B T)$ in the same chart. At $x_A = x_{A,max}$ and $x_{A,min}$, the second derivative is zero. The range of x_A that gives a negative second-order derivative is indicated by a shaded area. The latter represents an unstable mixture.

Figure 14.9 First- and second-order derivative curves of $\Delta F/(Nk_BT)$ for $\chi > 2$. At $x_{A,max}$ and $x_{A,min}$, $(\partial^2/\partial x_A^2)(\Delta F/(Nk_BT)) = 0$. The first-order derivative $(\partial/\partial x_A)(\Delta F/(Nk_BT))$ maximizes and minimizes locally at around $x_{A,max}$ and $x_{A,min}$, respectively. The slope of the tangent to the curve of $(\partial/\partial x_A)(\Delta F/(Nk_BT))$ is negative between $x_{A,max}$ and $x_{A,min}$, and the mixture is unstable.

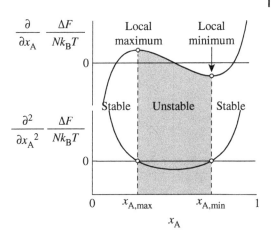

Figure 14.10 A plot of χ as a function of x_A that vanishes the second derivative of $\Delta F/(Nk_BT)$. Above the curve, the mixture is unstable; below the curve, it is stable.

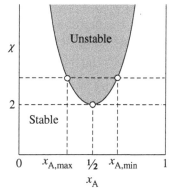

When $\chi > 2$, the second-derivative curve intersects the horizontal line at zero twice. The crossing occurs at $x_A = x_{A,max}$ and $x_{A,min}$. They satisfy

$$\frac{1}{\chi_A} + \frac{1}{1 - \chi_A} = 2\chi \qquad (14.19)$$

We can regard Eq. (14.19) as χ expressed as a function of x_A, which is shown in Figure 14.10. The curve minimizes to 2 at $x_A = \frac{1}{2}$. When $\chi > 2$, $x_A = x_{A,max}$ and $x_{A,min}$ satisfy Eq. (14.19), indicated by a dashed horizontal line above 2. We know that a mixture of x_A in the range of $x_{A,max} < x_A < x_{A,min}$ is not stable. The shaded area represents mixtures that are not stable. Below the curve, the mixture is stable, including the whole range of x_A for $\chi < 2$.

Figure 14.11 shows $\Delta F/(Nk_BT)$, $(\partial/\partial x_A)(\Delta F/(Nk_BT))$, and $(\partial^2/\partial x_A^2)(\Delta F/(Nk_BT))$ in a single chart. The curve of $\Delta F/(Nk_BT)$ has inflection points at $x_{A,max}$ and $x_{A,min}$. The two minima of $\Delta F/(Nk_BT)$ are located to the exterior of the range of $(x_{A,max}, x_{A,min})$.

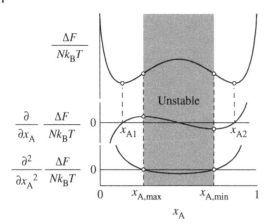

$$\frac{\Delta F}{N k_B T}$$

$$\frac{\partial}{\partial x_A} \frac{\Delta F}{N k_B T}$$

$$\frac{\partial^2}{\partial x_A^2} \frac{\Delta F}{N k_B T}$$

Figure 14.11 Free energy of mixing, $\Delta F/(N k_B T)$, its first-order derivative, and second-order derivative are shown in a single chart. The three curves are slided vertically. At $x_{A,\max}$ and $x_{A,\min}$, $(\partial^2/\partial x_A^2)(\Delta F/(N k_B T)) = 0$, $(\partial/\partial x_A)(\Delta F/(N k_B T))$ maximizes or minimizes locally, and $\Delta F/(N k_B T)$ has an inflection point. $\Delta F/(N k_B T)$ reaches local minima at x_{A1} and x_{A2}. The system is unstable between $x_{A,\max}$ and $x_{A,\min}$.

When the mixture separates into two phases (phase 1 and phase 2), their values of x_A are not $x_{A,\max}$ and $x_{A,\min}$. The two phases will be at the free energy minimum indicated by x_{A1} and x_{A2} in Figure 14.11. At the free energy minimum, $(\partial/\partial x_A)[\Delta F/(N k_B T)] = 0$. Therefore, x_{A1} and x_{A2} are the two zeros of

$$\ln x_A - \ln(1 - x_A) + \chi(1 - 2x_A) = 0 \tag{14.20}$$

excluding a trivial zero of $x_A = \frac{1}{2}$. When $\chi < 2$, the zero of Eq. (14.20) is $x_A = \frac{1}{2}$, and that is the only zero. The nontrivial zeros exist only when $\chi > 2$. We cannot express x_{A1} and x_{A2} as an explicit function of χ, but can use a numerical method to find the zeros.

14.1.3 Phase Separation

When a system with an overall composition x_A separates into two phases with x_{A1} and x_{A2}, the volumes of the two phases follow a lever rule. We denote by N_1 and N_2 the numbers of molecules (a total of A and B molecules) in the two phases (see Table 14.3).

The numbers in Table 14.3 satisfy

$$N_1 + N_2 = N, \quad x_{A1} N_1 + x_{A2} N_2 = x_A N \tag{14.21}$$

Table 14.3 Compositions and the numbers of molecules in phases 1 and 2.

Phase	Mole fraction of A	Number of molecules
1	x_{A1}	N_1
2	x_{A2}	N_2
Total	x_A	N

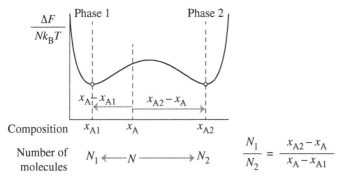

Figure 14.12 Lever rule. When a mixture of overall composition x_A separates into two phases of compositions x_{A1} and x_{A2}, the ratio of the numbers of molecules in the two phases, N_1/N_2, is equal to reciprocal of the ratio of $x_A - x_{A1}$ to $x_{A2} - x_A$.

which lead to

$$N_1 = \frac{x_{A2} - x_A}{x_{A2} - x_{A1}} N, \quad N_2 = \frac{x_A - x_{A1}}{x_{A2} - x_{A1}} N \tag{14.22}$$

Their ratio

$$\frac{N_1}{N_2} = \frac{x_{A2} - x_A}{x_A - x_{A1}} \tag{14.23}$$

follows the **lever rule**, that is, the ratio of N_1 to N_2 is equal to the reciprocal of the ratio of the distance from x_A to x_{A1}, to the distance from x_A to x_{A2}. Figure 14.12 recaps these calculations.

What about a mixture in the range between x_{A1} and $x_{A,max}$ and a mixture in the range between $x_{A,min}$ and x_{A2}? A mixture of x_A in either of the two ranges is thermodynamically stable, but is not at free energy minimum. Separating into two phases of x_{A1} and x_{A2} will lower the overall free energy. This change from a single stable phase to two phases requires agitation, and we say that the mixture in either of the two ranges is **metastable** (see Figure 14.13). If one spot within the system happens to have a composition of x_{A1} or x_{A2}, it will induce macroscopic phase separation. This mechanism of phase transition is called **nucleation and growth**.

In contrast, a mixture of x_A between $x_{A,max}$ and $x_{A,min}$ will spontaneously separate into two phases of x_{A1} and x_{A2}. This mechanism of phase transition is called **spinodal decomposition**.

The boundary between the stable region and the metastable region is given by Eq. (14.20). The boundary between the metastable region and the unstable region is given by Eq. (14.19). Each equation specifies the relationship between χ and x_A represented by a curve in the plane of χ and x_A. As shown in Figure 14.14, the two curves are nearly parabolic with an axis at $x_A = \frac{1}{2}$, and share the apex, called the **critical point**. The critical value of χ is $\chi_c = 2$, and

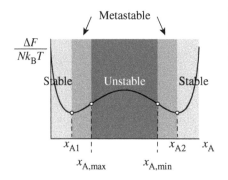

Metastable

$\frac{\Delta F}{N k_B T}$

Stable   Unstable   Stable

x_{A1} x_{A2} x_A

$x_{A,max}$ $x_{A,min}$

Figure 14.13 Three regions of composition in an A–B mixture – stable, metastable, and unstable regions – for a given value of χ (>2).

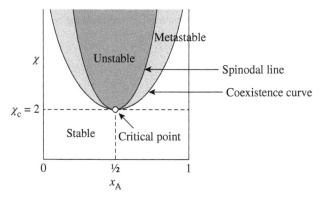

χ

Metastable

Unstable

Spinodal line

Coexistence curve

$\chi_c = 2$

Stable   Critical point

0   ½   1

x_A

Figure 14.14 Phase diagram for a binary mixture of equal-sized molecules A and B. The boundary of stable–metastable regions (coexistence curve), the boundary between the metastable and unstable regions (spinodal line), and the critical point are shown in the χ–x_A space.

the critical molar fraction is $x_A = \frac{1}{2}$. This type of plot is called a **phase diagram**. In the figure, the state of the mixture is indicated in each of the three regions. The curve for the boundary between the stable region and the metastable region is called a **coexistence curve**. The curve for the boundary between the metastable region and the unstable region is called a **spinodal line**.

As we saw in Figure 14.4, χ changes with temperature T. For most mixtures, $|\chi|$ decreases with an increasing T, as shown in Figure 14.15a. At a **critical temperature** T_c, $\chi = \chi_c = 2$. Using this χ–T relationship, we can convert the phase diagram in Figure 14.14 into another phase diagram as shown in Figure 14.15b. In the latter diagram, the ordinate is the temperature T.

The critical point is at the top of the coexistence curve, and therefore the phase diagram shown in Figure 14.15b is called an **upper critical solution temperature** (UCST)-type phase diagram. At $T > T_c$, the mixture is in a single phase regardless of x_A. There is no miscibility gap. At $T < T_c$, the mixture may separate into two phases, depending on the overall composition x_A. The two phases are a saturated solution of A in B (x_{A1}) and a saturated solution of B in

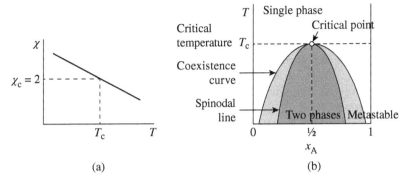

Figure 14.15 (a) Typical dependence of χ on temperature T. At the critical temperature T_c, $\chi = \chi_c = 2$. At around T_c, χ decreases with an increasing T. (b) UCST-type phase diagram in the temperature–composition plane. The coexistence curve and the spinodal line are upward convex.

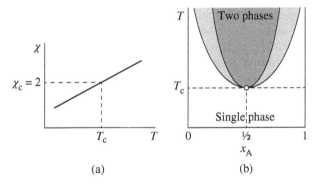

Figure 14.16 (a) Atypical dependence of χ on temperature T. At the critical temperature T_c, $\chi = \chi_c = 2$. At around T_c, χ increases with an increasing T. (b) LCST-type phase diagram in the temperature–composition plane. The coexistence curve and the spinodal line are downward convex.

A (x_{A2}). The closer the temperature to T_c, the higher the concentrations of the minority component in each solution. The miscibility gap is from x_{A1} to x_{A2}.

In some mixtures, χ increases with an increasing T. We have seen such possibilities in Figure 14.4b. In that situation, χ may reach 2 from below, and exceed 2 as T increases further (see Figure 14.16a). Then, the phase diagram will be inverted, and the critical point will be at the bottom of the parabolic curves. This type of phase diagram is called a **lower critical solution temperature (LCST)**-type phase diagram (see Figure 14.16b). At $T < T_c$, the mixture is in a single phase, regardless of x_A. There is no miscibility gap. At $T > T_c$, the mixture may separate into two phases, depending on the overall composition x_A.

Problem 14.2 considers a change in the internal energy due to demixing and the heat capacity of the two-phase solution. Problem 14.1 is a preparation for Problem 14.2.

14.2 Binary Mixture of Molecules of Different Sizes

The plots of $\Delta F/(N k_B T)$ in Figure 14.5 are symmetric around a vertical line at $x_A = \frac{1}{2}$. Therefore, a plot of ΔF as a function of $x_B = 1 - x_A$ is identical to the plot of ΔF vs. x_A. The latter is the case when molecules A and B have the same size. If their sizes are different, the curve of ΔF is not symmetric, and therefore the discussion of the phase diagram becomes a bit more complicated. In this section, we consider a binary mixture of A and B molecules in which a B molecule is twice as large as an A molecule is. To make the mixture compatible with the lattice, we consider that a B molecule consists of two atoms, each having the same size as that of an A molecule. Table 14.4 specifies the mixture. A total of N sites are occupied by N_A molecules of A and N_B molecules of B. In this section, x_A and x_B are the fractions of sites occupied by A and B molecules (i.e. volume fractions), respectively, not the molar fractions.

The model for the interactions between molecules is the same as the one we adopted in Section 14.1, that is, each pair of adjacent sites has an interaction. As each of the sites is occupied by an A molecule or a B atom, there are three types of interaction, A–A, B–B, and A–B. As we did for the mixture of equal-size molecules, we adopt a mean-field approximation to calculate the interaction. We calculate the mean numbers of A–A, B–B, and A–B contacts. There are a total $\frac{1}{2}N\zeta$ contacts between sites, as we learned in the preceding section. Out of them, N_B contacts are covalent bonds, leaving $\frac{1}{2}N\zeta - N_B$ noncovalent contacts. Table 14.5 lists the number of noncovalent contacts per molecule, and the probability for an end of a noncovalent bond to be connected to either A or B molecule. Note that there are $N\zeta - 2N_B$ ends of noncovalent contacts, if each end point is counted twice.

From the table, the number of A–A contacts is, for instance, calculated as

$$\left(\frac{N_A \zeta}{N\zeta - 2N_B} \right)^2 \left(\frac{\zeta N}{2} - N_B \right) = \frac{N_A^2 \zeta^2}{2(N\zeta - 2N_B)} \tag{14.24}$$

Table 14.4 Numbers of molecules, numbers of sites occupied, and fractions of sites in an A–B mixture of unequal sizes.

Molecule	Number of molecules	Number of sites	Fraction of sites
A	N_A	N_A	$x_A = N_A/N$
B	N_B	$2N_B$	$x_B = 2N_B/N$
	Total	$N = N_A + 2N_B$	1

Table 14.5 Number of noncovalent contacts per molecule and the probability for an end of a noncovalent bond to be connected to either A or B molecule.

Component	Number of contacts per molecule	Probability for an end of a bond to be connected to
A	ζ	$\zeta N_A/(\zeta N - 2N_B)$
B	$2(\zeta - 1)$	$2N_B(\zeta - 1)/(\zeta N - 2N_B)$
Total		1

Table 14.6 Numbers of noncovalent A–A, A–B, and B–B contacts in the mean-field approximation, and their interactions.

Contact	Energy	Number of pairs
A–A	ε_{AA}	$\tfrac{1}{2}\zeta^2 N_A^2/(\zeta N - 2N_B)$
B–B (excl. covalent bonds)	ε_{BB}	$2(\zeta - 1)^2 N_B^2/(\zeta N - 2N_B)$
A–B	ε_{AB}	$2\zeta(\zeta - 1)N_A N_B/(\zeta N - 2N_B)$
Total		$\tfrac{1}{2}\zeta N - N_B$

Table 14.6 lists the interactions and the numbers of noncovalent pairs for A–A, B–B, and A–B contacts. Confirm that the three numbers add up to $\tfrac{1}{2}N\zeta - N_B$. The symbols for the interactions are the same as those in Section 14.1.

In Section 14.1, we started with writing the partition function and used a mean-field approximation to calculate it. We learned that the resultant free energy of mixing, ΔF, consists of two parts: ΔU and $-T\Delta S$. In Eq. (14.14), $\Delta U/(k_B T) = \chi N_A N_B/N$ and $\Delta S/k_B = N \ln N - N_A \ln N_A - N_B \ln N_B$. In this section, we evaluate the internal energy change ΔU and the entropy change ΔS, separately.

First, we calculate ΔU. In pure liquid B, the number of noncovalent contacts is $(\zeta - 1)N_B$. The energy U_0 before mixing the two liquids is

$$U_0 = \varepsilon_{AA}\frac{N_A \zeta}{2} + \varepsilon_{BB}N_B(\zeta - 1) \tag{14.25}$$

We can calculate the energy U after the mixing from Table 14.6 as

$$U = \varepsilon_{AA}\frac{N_A^2\zeta^2}{2(N\zeta - 2N_B)} + \varepsilon_{BB}\frac{2N_B^2(\zeta - 1)^2}{N\zeta - 2N_B} + \varepsilon_{AB}\frac{2N_A N_B\zeta(\zeta - 1)}{N\zeta - 2N_B} \tag{14.26}$$

The change by the mixing is

$$\Delta U = \frac{2N_A N_B \zeta(\zeta - 1)}{N\zeta - 2N_B} \left(\varepsilon_{AB} - \frac{\varepsilon_{AA} + \varepsilon_{BB}}{2} \right)$$

$$= \frac{2N_A N_B \zeta(\zeta - 1)}{\zeta N_A + 2(\zeta - 1)N_B} \left(\varepsilon_{AB} - \frac{\varepsilon_{AA} + \varepsilon_{BB}}{2} \right) \tag{14.27}$$

If $\zeta \gg 1$, ΔU simplifies to

$$\Delta U \cong \frac{2\zeta N_A N_B}{N} \left(\varepsilon_{AB} - \frac{\varepsilon_{AA} + \varepsilon_{BB}}{2} \right) \tag{14.28}$$

This approximation of ΔU amounts to neglecting the covalent bond within each B molecule and calculating the numbers of A–A, B–B, and A–B contacts assuming random mixing of N_A molecules of A and $2N_B$ molecules of ½B (that is a B atom). If we use the same definition of the χ parameter as the one in Eq. (14.12), we have

$$\Delta U \equiv \frac{2N_A N_B}{N} \chi k_B T \tag{14.29}$$

Problem 14.3 considers the effect of this neglect.

Second, we consider ΔS. Before the mixing, orientation of B molecules in pure B leads to a finite number of arranging B molecules, whereas it is just one in pure A. Mixing B molecules with A molecules may increase the number of choices for their orientations. Evaluating the number of choices based on geometry is difficult. To get around this situation, we evaluate ΔS just by looking at how much the volume accessible to each molecule has increased by the mixing. Recall that, in Eq. (14.14) for a mixture of molecules of equal sizes, the entropy part was

$$\frac{\Delta S}{k_B} = N_A \ln \frac{N}{N_A} + N_B \ln \frac{N}{N_B} \tag{14.30}$$

Notice that N/N_A is equal to the ratio of the volume accessible to an A molecule after the mixing to the one before the mixing. If we apply the same principle to the mixing of molecules of different sizes, we should write

$$\frac{\Delta S}{k_B} = N_A \ln \frac{N}{N_A} + N_B \ln \frac{N}{2N_B} \tag{14.31}$$

which is further rewritten to

$$\frac{\Delta S}{N k_B} = -x_A \ln x_A - \frac{x_B}{2} \ln x_B \tag{14.32}$$

Combining Eqs. (14.29) and (14.32), we find that the free energy change per site due to the mixing is given as

$$\frac{\Delta F}{N k_B T} = x_A \ln x_A + \frac{x_B}{2} \ln x_B + \chi x_A x_B$$

$$= \frac{x_B}{2} \ln x_B + (1 - x_B) \ln(1 - x_B) + \chi x_B (1 - x_B) \tag{14.33}$$

This expression of $\Delta F/(Nk_BT)$ is different from the one in Eq. (14.15) (rewritten by replacing x_A with $1 - x_B$), and the only difference is the division by 2 in the first term. The interaction part is the same, but the entropy part is not. The contribution to ΔF by B molecules' mixing is a half of the one by monomeric B molecules, when expressed as a function of volume fraction. The size of the B molecule manifests itself only in this way.

Two panels of Figure 14.17 show $\Delta F/(Nk_BT)$ as a function of x_B for different values of χ. Later in this section, we find that $\chi_c = (3 + \sqrt{8})/4 = 1.457$ gives the critical condition for phase separation. The curves are not symmetric around a vertical line at $x_B = \frac{1}{2}$ (not shown). The asymmetricity is pronounced in the curves with χ close to χ_c, as seen in Figure 14.17b. When χ is between 1.457 and 1.801, the curve has only one local minimum although χ is greater than χ_c. When $\chi > 1.801$, the curve has two local minima, and the left minimum is closer to $x_B = 0$ than the right minimum is to $x_B = 1$.

As we did in Section 14.1, we take the derivative:

$$\frac{\partial}{\partial x_B} \frac{\Delta F}{Nk_B T} \cong \frac{1}{2}\ln x_B - \ln(1 - x_B) - \frac{1}{2} + \chi(1 - 2x_B) \tag{14.34}$$

Figure 14.18a,b shows the derivative curves. All the curves in the figure pass a common point, $x_B = \frac{1}{2}$ and $\partial[\Delta F/(Nk_BT)]/\partial x_B = \frac{1}{2}(\ln 2 - 1) = -0.153$. However, the curves are not point-symmetric around the common point, unlike those in Figure 14.6. The curves for $\chi < 1.457$ are increasing in the whole range of x_B. Each of the curves for $\chi > 1.457$ has a local maximum at $x_{B,max} < \frac{1}{2}$ and a local minimum at $x_{B,min} > \frac{1}{2}$. In the range between $x_{B,max}$ and $x_{B,min}$, the slope of the tangent to the curve is negative, indicating that the mixture is unstable.

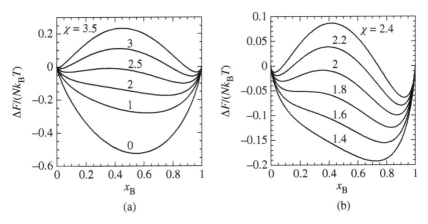

Figure 14.17 $\Delta F/(Nk_BT)$ plotted as a function of x_B in an A–B mixture with a B molecule twice as large as an A molecule, shown for (a) $\chi = 0$, 1, 2, 2.5, 3, and 3.5, and (b) $\chi = 1.4$, 1.6, 1.8, 2, 2.2, and 2.4.

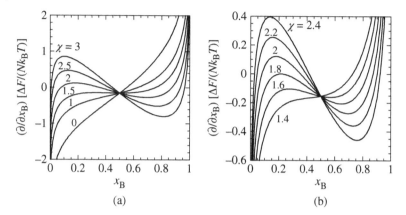

Figure 14.18 Derivative of $\Delta F/(Nk_BT)$ by x_B, shown for (a) $\chi = 0$, 1, 1.5, 2, 2.5, and 3, and (b) $\chi = 1.4$, 1.6, 1.8, 2, 2.2, and 2.4. All the curves share a point, $x_B = \frac{1}{2}$ and $(\partial/\partial x_B)[\Delta F/(Nk_BT)] = -0.153$.

The local maximum remains below 0 for χ between 1.457 and 1.801. When χ is greater than 1.801, the local maximum is above 0, and therefore the curve of $\Delta F/(Nk_BT)$ in Figure 14.17 has a local minimum.

Further differentiation of Eq. (14.34) by x_B leads to

$$\frac{\partial^2}{\partial x_B^2}\frac{\Delta F}{Nk_BT} \cong \frac{1}{2x_B} + \frac{1}{1-x_B} - 2\chi \tag{14.35}$$

The curves of the second-order derivatives are shown in Figure 14.19. When $\chi < \chi_c = 1.457$, $\partial^2[\Delta F/(Nk_BT)]/\partial x_B^2$ is always positive; there is no unstable region of x_B. When $\chi > \chi_c$, a part of the curve is negative. The intersection of the curve with the horizontal line at 0 gives $x_{B,max}$ and $x_{B,min}$, and $\partial^2[\Delta F/(Nk_BT)]/\partial x_B^2$ is negative between them. Unlike the equal-size

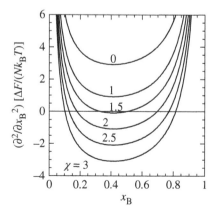

Figure 14.19 Second-order derivative of $\Delta F/(Nk_BT)$ by x_B, shown for $\chi = 0$, 1, 1.5, 2, 2.5, and 3. The curve dips below zero for $\chi > \chi_c = (3 + \sqrt{8})/4 = 1.457$.

mixture, $x_{B,min} \neq 1 - x_{B,max}$. Note that "2" in "$1/(2x_B)$" of Eq. (14.35) is due to the size of the B molecule relative to the A molecule, whereas the other "2" in "2χ" is not.

The value of χ that makes the second-derivative curve touch the horizontal line at 0 gives the critical condition. Since Eq. (14.35) being equal to 0 is rewritten to

$$4\chi x_B{}^2 - (4\chi - 1)x_B + 1 = 0 \tag{14.36}$$

the critical condition is given by the vanishing determinant of this quadratic equation. The result is

$$\chi_c = \frac{3 + \sqrt{8}}{4} \tag{14.37}$$

The critical value of x_B is

$$x_{Bc} = \sqrt{2} - 1 \tag{14.38}$$

Figure 14.20 shows $\Delta F/(Nk_B T)$ and its first and second derivatives in a single chart for $\chi > \chi_c$. The $x_{B,max}$ and $x_{B,min}$ are indicated by the vertical dashed lines. When the mixture's x_B is between $x_{B,max}$ and $x_{B,min}$, the mixture is unstable, and it will spontaneously separate into two phases. Evaluating the compositions of the two phases, x_{B1} and x_{B2}, is not as simple as finding x_B that gives the local minima in the plot of $\Delta F/(Nk_B T)$. They are rather the points of contact of the curve of $\Delta F/(Nk_B T)$ with its cotangent, as we learn subsequently. The solid horizontal line that crosses the first-derivative curve is not at $(\partial/\partial x_B)(\Delta F/(Nk_B T)) = 0$.

Now, we consider the compositions of the two phases after the phase separation. Table 14.7 lists symbols for the compositions of the two phases and their numbers of sites.

Figure 14.20 Free energy of mixing, $\Delta F/(Nk_B T)$, its first-order derivative, and second-order derivative are shown in a single chart. The three curves are slided vertically. At $x_{B,max}$ and $x_{B,min}$, $(\partial^2/\partial x_B{}^2)(\Delta F/(Nk_B T)) = 0$, $(\partial/\partial x_B)(\Delta F/(Nk_B T))$ maximizes or minimizes locally, and $\Delta F/(Nk_B T)$ has an inflection point. The two phases that the unstable mixture separates into are not specified by the local minima on the curve of $\Delta F/(Nk_B T)$, but are rather specified by a cotangent to the curve.

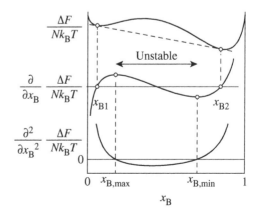

Table 14.7 Compositions and numbers of sites in two phases 1 and 2 after phase separation.

Phase	Fraction of B	Number of sites
1	x_{B1}	N_1
2	x_{B2}	N_2
Total	1	$N = N_1 + N_2$

An accounting rule gives

$$N_1 + N_2 = N, \quad x_{B1}N_1 + x_{B2}N_2 = x_B N \tag{14.39}$$

which leads to

$$N_1 = N\frac{x_{B2} - x_B}{x_{B2} - x_{B1}}, \quad N_2 = N\frac{x_B - x_{B1}}{x_{B2} - x_{B1}} \tag{14.40}$$

These expressions are identical to those for the mixture of equal-size molecules. Here, we learn how to find x_{B1} and x_{B2}.

The free energy ΔF of the demixed system (for x_B between x_{B1} and x_{B2}) consists of two parts:

$$\begin{aligned}
\Delta F(x_B) &= \frac{N_1}{N}\Delta F(x_{B1}) + \frac{N_2}{N}\Delta F(x_{B2}) \\
&= \frac{x_{B2} - x_B}{x_{B2} - x_{B1}}\Delta F(x_{B1}) + \frac{x_B - x_{B1}}{x_{B2} - x_{B1}}\Delta F(x_{B2})
\end{aligned} \tag{14.41}$$

In the plane of ΔF vs. x_B, this equation represents a straight line that passes two points on the curve of ΔF: $(x_{B1}, \Delta F(x_{B1}))$ and $(x_{B2}, \Delta F(x_{B2}))$. In contrast, the curve of ΔF in Eq. (14.33) represents ΔF before the phase separation.

The straight line will lie lowest between x_{B1} and x_{B2}, if the line is a cotangent to the curve. In Figure 14.21, the solid line is the cotangent. We can draw a straight line that intersects the curve twice between x_{B1} and x_{B2}. The dashed line is one of such lines. Obviously, the value of $\Delta F(x_B)$ on the dashed line is greater than the one on the solid line. Therefore, the cotangent provides the smallest ΔF for x_B between x_{B1} and x_{B2} after the phase separation. Note that the curve does not locally minimize at x_{B1} or x_{B2}. Recall that the curve of ΔF with χ between 1.457 and 1.801 has only one local minimum at x_B close to 1 (Figure 14.17b). We can still draw a cotangent to the curve for this range of χ.

In the mixture of equal-size molecules, x_{A1} and x_{A2} were at the minima of the curve of ΔF. They are also where the cotangent touches the curve. Thus, we find that the cotangent requirement is universal.

The two points of the tangential contact share the slope:

$$\left.\frac{\partial}{\partial x_B}\frac{\Delta F}{Nk_B T}\right|_{x_{B1}} = \left.\frac{\partial}{\partial x_B}\frac{\Delta F}{Nk_B T}\right|_{x_{B2}} \tag{14.42}$$

Figure 14.21 The cotangent to the curve of ΔF (solid line) is the lowest lying line for ΔF between x_{B1} and x_{B2} after the phase separation than any other line does that intersects the curve twice between x_{B1} and x_{B2}. The curve drawn is for $\chi = 2$.

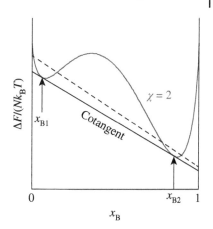

The replacement chemical potential μ_{rep} is shared by the two phases. Here, $\mu_{rep} = \mu_B - 2\mu_A$. Adding a B molecule in a fixed volume (by removing two molecules of A) increases the free energy per site by $\frac{1}{2}\mu_{rep}$. The two phases of x_{B1} and x_{B2} share μ_{rep}.

Figure 14.22 is a phase diagram in the $\chi - x_B$ plane. The cotangent specifies the coexistence curve, whereas Eq. (14.36) determines the spinodal line. The critical point is at $x_B = \sqrt{2} - 1$ and $\chi = \chi_c = (3 + \sqrt{8})/4$.

Compared with the mixture of equal-size molecules (both are size 1), χ_c is smaller; the range of χ for the single phase is less. That is because the lowering of ΔF by the entropy of mixing is less. Mixing size 1 and size 2 molecules requires a less unfavorable interaction between them compared with mixing molecules of size 1.

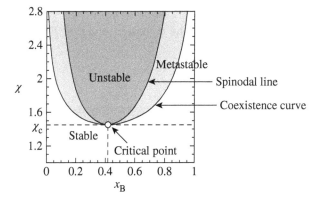

Figure 14.22 Phase diagram of the A–B mixture in which a B molecule is twice as large as an A molecule. The χ parameter is plotted as a function of the volume fraction of B. The coexistence curve, the spinodal line, and the critical point at $x_B = 0.414$ and $\chi = 1.457$ are indicated.

The formulation we adopted in this chapter can be easily extended to polymer systems. One is a solution of polymer (B) in a solvent (A) of small molecules [30]. The expression of ΔF will be different from the one in Eq. (14.33), as "2" in the denominator of the first term is replaced by the size ratio of the polymer molecule to the solvent molecule. Typically, the ratio is from 10 to 10^4. The huge size ratio leads to a heavily lopsided phase diagram, that is, the critical point will be almost next to $x_B = 0$, and χ_c is a lot smaller than 2 (see Problem 14.5).

The other polymer system of interest is a mixture of two polymers. Both terms in the entropy of mixing will be divided by the size, and therefore ΔF will be dominated by the change in the interaction. Therefore, most pairs of polymer are not miscible. Exceptions are pairs that attract each other via a specific interaction. For example, polystyrene mixes with poly(vinyl methyl ether) through the interaction between a phenylene ring and an ether oxygen. However, not all polymers that have an ether oxygen in every repeat unit mix with polystyrene.

Problems

14.1 *Near the critical point.* In Section 14.1, we considered a mixture of A and B molecules that have the same size. The critical point is $\chi = 2$ and $x_A = \frac{1}{2}$. When the system is only shallowly within the unstable region ($\chi > 2$, but $\chi - 2 \ll 1$), the points on the spinodal line are close to the critical point. The points on the coexistence curve are also close. Each of the two curves close to the critical point can be represented by x_A expressed as a function of χ.

(1) Obtain the leading term with respect to $\chi - 2$ in each of two expressions.

(2) What is the next leading term for the coexistence curve? Find its exponent with respect to $\chi - 2$.

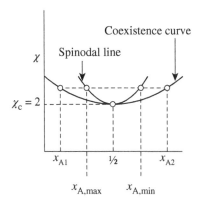

14.2 *Heat capacity.* This problem considers a change in the internal energy U when χ, initially less than 2, increases to cause a phase separation in the A–B mixture of an equal size. For simplicity, we consider $x_A = \frac{1}{2}$ for the overall composition of the mixture. Assume that the interactions, ε_{AA}, ε_{BB}, and ε_{AB} are constants.

(1) What is the internal energy in a single phase, U_s? Consider the interaction by contacts only.

(2) Use the result of Problem 14.1 to obtain the internal energy U_d in the two-phase mixture up to the leading-order correction with respect to $\chi - 2$. How much does U change by the phase separation?

(3) How does the heat capacity change in the transition?

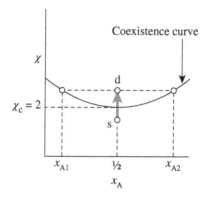

14.3 *Errors due to a finite ζ.* In Section 14.2, ΔU was evaluated assuming $\zeta \gg 1$. This problem assesses the effect of neglecting a finite value of ζ. Evaluate corrections to $O(\zeta^{-1})$.

(1) What is the correction to ΔU?

(2) What is the correction to ΔF? Also calculate its derivatives by x_B.

(3) Evaluate the correction to χ_c.

14.4 *Gas molecules on a lattice.* Section 14.1 considered a mixture of two liquids A and B. Incompressibility did not allow us to calculate the chemical potential. This problem considers a system of N_A molecules of A suspended in vacuum on a lattice of N sites, which can be configured by removing all of the B molecules. In the mean-field approximation,

$$U = \frac{\zeta}{2N} N_A^2 \varepsilon_{AA}$$

and

$$S = k_B \ln \frac{N!}{N_A!(N - N_A)!}$$

(1) What is F?

(2) Unlike the A–B mixture, we can add or remove an A molecule without changing the volume (or the number of sites). Use the formula, $\mu_A = (\partial F/\partial N_A)_{T,N}$ to calculate the chemical potential of the A molecule.

(3) The Gibbs free energy G is equal to $\mu_A N_A$. Calculate "pV."

(4) Evaluate "pV" up to $O((N_A/N)^2)$.

(5) The "pV" can be expressed as $N_A k_B T(1 + C_2 N_A/N)$, where C_2 is a constant of N_A. We can regard C_2 as the second virial coefficient. You will find that C_2 consists of two components. Discuss the origin of each of them.

14.5 *Polymer solution.* The counterpart of Eq. (14.35) in a mixture of polymer chains and small solvent molecules (their size ratio is n) is

$$\frac{\partial^2}{\partial x_B{}^2} \frac{\Delta F}{N k_B T} \equiv \frac{1}{n x_B} + \frac{1}{1 - x_B} - 2\chi$$

Find the critical value of χ.

Appendix A

Mathematics

A.1 Hyperbolic Functions

Given here is a list of definitions of hyperbolic functions.

$$\sinh x \equiv \frac{1}{2}(e^x - e^{-x}) \tag{A.1}$$

$$\cosh x \equiv \frac{1}{2}(e^x + e^{-x}) \tag{A.2}$$

$$\tanh x \equiv \frac{e^x - e^{-x}}{e^x + e^{-x}} \tag{A.3}$$

$$\coth x \equiv \frac{1}{\tanh x} \tag{A.4}$$

$$\operatorname{sech} x \equiv \frac{1}{\cosh x} \tag{A.5}$$

Figure A.1 shows the first three functions graphically.

$\sinh x$ and $\tanh x$ are odd functions of x, whereas $\cosh x$ is an even function.

The hyperbolic functions satisfy relationships similar to those of trigonometric functions. Some of the relationships are listed here.

$$\tanh x = \frac{\sinh x}{\cosh x} \tag{A.6}$$

$$\cosh^2 x - \sinh^2 x = 1 \tag{A.7}$$

Double argument formulas are also similar to those in trigonometry.

$$\sinh 2x = 2\sinh x \cosh x \tag{A.8}$$

$$\cosh 2x = 2\cosh^2 x - 1 \tag{A.9}$$

Statistical Thermodynamics: Basics and Applications to Chemical Systems, First Edition. Iwao Teraoka.
© 2019 John Wiley & Sons, Inc. Published 2019 by John Wiley & Sons, Inc.
Companion website: www.wiley.com/go/Teraoka_StatsThermodynamics

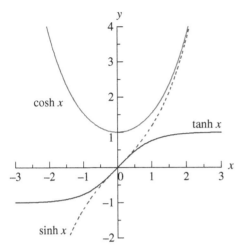

Figure A.1 Hyperbolic functions, $y = \sinh x$, $y = \cosh x$, and $y = \tanh x$.

Inverse functions are different from those of trigonometry.

$$\sinh^{-1}x = \ln(x + \sqrt{1 + x^2}) \tag{A.10}$$

$$\tanh^{-1}x = \frac{1}{2}\ln\frac{1+x}{1-x} \tag{A.11}$$

Derivatives are similar.

$$\frac{d}{dx}\sinh x = \cosh x \tag{A.12}$$

$$\frac{d}{dx}\cosh x = \sinh x \tag{A.13}$$

$$\frac{d}{dx}\tanh x = \text{sech}^2 x \tag{A.14}$$

Taylor expansion of the hyperbolic functions is listed here.

$$\sinh x = x + \frac{1}{6}x^3 + \frac{1}{120}x^5 + \cdots \tag{A.15}$$

$$\cosh x = 1 + \frac{1}{2}x^2 + \frac{1}{24}x^4 + \cdots \tag{A.16}$$

$$\tanh x = x - \frac{1}{3}x^3 + \frac{2}{15}x^5 + \cdots \tag{A.17}$$

$$\coth x = \frac{1}{x} + \frac{1}{3}x - \frac{1}{45}x^3 + \frac{2}{945}x^5 + \cdots \tag{A.18}$$

An associated function, Langevin function $L(x)$, is defined as follows:

$$L(x) \equiv \frac{1}{\tanh x} - \frac{1}{x} \tag{A.19}$$

Figure A.2 Langevin function
$y = L(x)$. The horizontal dashed line at
$y = 1$ is the large-x limit. The sloped
dashed line is the tangent to the
curve at the origin, $y = x/3$.

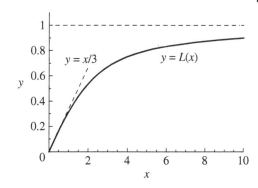

Figure A.2 shows a plot of $y = L(x)$.

The derivative and Taylor expansion of $L(x)$ are listed here.

$$\frac{dL}{dx} = \frac{1}{x^2} - \frac{1}{\sinh^2 x} \tag{A.20}$$

$$L(x) = \frac{1}{3}x - \frac{1}{45}x^3 + \cdots \tag{A.21}$$

A.2   Series

The sum of a geometric series is equal to the first term divided by (1 − common ratio):

$$\sum_{n=0}^{\infty} ar^n = \frac{a}{1 - r} \tag{A.22}$$

If the series terminates with $n = N$, the sum is equal to the first term times (1 − common ratio raised to the power of the number of terms), divided by (1 − common ratio):

$$\sum_{n=0}^{N} ar^n = \frac{a(1 - r^{N+1})}{1 - r} \tag{A.23}$$

As a corollary to Eq. (A.22), the following sum can be calculated.

$$\sum_{n=0}^{\infty} nr^{n-1} = \sum_{n=1}^{\infty} nr^{n-1} = \frac{d}{dr} \sum_{n=1}^{\infty} r^n = \frac{d}{dr} \frac{r}{1 - r} = \frac{1}{(1 - r)^2} \tag{A.24}$$

A.3   Binomial Theorem and Trinomial Theorem

Binomial and trinomial theorems are

$$\sum_{r=0}^{n} \binom{n}{r} a^r b^{n-r} = \sum_{r=0}^{n} \frac{n!}{r!(n - r)!} a^r b^{n-r} = (a + b)^n \tag{A.25}$$

$$\sum_{r=0}^{n}\sum_{q=0}^{n-r}\frac{n!}{q!r!(n-r-q)!}a^q b^r c^{n-q-r} = (a+b+c)^n \tag{A.26}$$

They are a part of a multinomial theorem:

$$\sum_{i_1,i_2,\cdots,i_k}\frac{n!}{i_1!i_2!\cdots i_k!}x_1^{i_1}x_2^{i_2}\cdots x_k^{i_k} = (x_1+x_2+\cdots+x_k)^n \tag{A.27}$$

where the sum is taken for all possible nonnegative integers i_1, i_2, ..., and i_k that satisfy $i_1 + i_2 + \cdots + i_k = n$. When $k = 2$ and 3, the multinomial theorem reduces to the binomial theorem and the trinomial theorem, respectively.

A.4   Stirling's formula

When $n \gg 1$, $n!$ can be approximated with elementary functions, A crude approximation is

$$\ln n! \cong n(\ln n - 1) \tag{A.28}$$

A more precise approximation is

$$\ln n! \equiv n(\ln n - 1) + \frac{1}{2}\ln(2\pi n) \tag{A.29}$$

A.5   Integrals

The following integral formulas are useful.

$$\int_0^\infty x^n \exp(-ax)dx = \frac{n!}{a^{n+1}} \tag{A.30}$$

$$\int_0^\infty x^{2n} \exp(-ax^2)dx = \frac{(2n-1)!!}{2(2a)^n}\sqrt{\frac{\pi}{a}} \tag{A.31}$$

$$\int_0^\infty x^{2n+1} \exp(-ax^2)dx = \frac{n!}{2a^{n+1}} \tag{A.32}$$

A.6   Error Functions

The error function, *aka* Gauss error function, is defined as

$$\text{erf}(x) \equiv \frac{2}{\pi^{1/2}}\int_0^x \exp(-t^2)dt \tag{A.33}$$

It is an odd function. Figure A.3 shows a plot of $\text{erf}(x)$.

Figure A.3 Gauss error function $y = \mathrm{erf}(x)$.

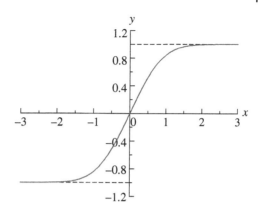

One's complement of $\mathrm{erf}(x)$ is given its own symbol:

$$\mathrm{erfc}(x) \equiv 1 - \mathrm{erf}(x) = \frac{2}{\pi^{1/2}} \int_x^\infty \exp(-t^2)\mathrm{d}t \tag{A.34}$$

For $x \gg 1$, $\mathrm{erfc}(x)$ is approximated as

$$\pi^{1/2} x \exp(x^2)\mathrm{erfc}(x) = 1 - \frac{1}{2x^2} + \frac{3}{(2x^2)^2} - \frac{15}{(2x^2)^3} + \cdots \tag{A.35}$$

A.7 Gamma Functions

The Gamma function $\Gamma(z)$ is defined as

$$\Gamma(z) = \int_0^\infty e^{-t} t^{z-1} \mathrm{d}t \tag{A.36}$$

for z with a positive real part. It satisfies

$$\Gamma(z + 1) = z\Gamma(z) \tag{A.37}$$

If the argument of the function is an integer,

$$\Gamma(n + 1) = n! \tag{A.38}$$

References

1 Einstein, A. (1906). Theorie der Strahlung und die Theorie der Spezifischen Wärme. *Ann. Phys.* 4: 180–190.

2 Debye, P. (1912). Zur Theorie der spezifischen Wärme. *Ann. Phys.* 39: 789–839.

3 Planck, M. (1900). Über eine Verbesserung der Wienschen Spektralgleichung. *Verh. Dtsch. Phys. Ges.* 2: 202–204.

4 Tolman, R.C. (1938). *The Principles of Statistical Mechanics*. Oxford, UK: Clarendon Press.

5 Frauenfelder, H. and Marden, M.C. (1989). Molecular spectroscopy. In: *A Physicist's Desk Reference* (ed. H.L. Anderson). New York, NY: American Institute of Physics.

6 Cuadros, F., Cachadiña, I., and Ahumada, W. (1996). Determination of Lennard–Jones interaction parameters using a new procedure. *Mol. Eng.* 6 (319–325).

7 Hecht, C.E. (1990). *Statistical Thermodynamics and Kinetic Theory*. New York, NY: Freeman.

8 Kaye & Laby. Tables of Physical & Chemical Constants, National Physical Laboratory. http://www.kayelaby.npl.co.uk/chemistry/#3_5 (accessed 11 July 2018).

9 McCarty, R.D. (1973). Thermodynamic properties of helium 4 from 2 to 1500 K at pressures to 10^8 Pa. *J. Phys. Chem. Ref. Data* 2: 923–1041.

10 For example, Reichel, L.E. (1980). *A Modern Course in Statistical Physics*. Austin, TX: University of Texas Press.

11 Wang, Y., Zhu, Q., and Zhang, H. (2006). Fabrication and magnetic properties of hierarchical porous hollow nickel microspheres. *J. Mater. Chem.* 16: 1212–1214.

12 Azargohar, R. and Dalai, A.K. (2006). Biochar as a precursor of activated carbon. *Appl. Biochem. Biotech.* 129–132: 762–773.

Statistical Thermodynamics: Basics and Applications to Chemical Systems, First Edition. Iwao Teraoka.
© 2019 John Wiley & Sons, Inc. Published 2019 by John Wiley & Sons, Inc.
Companion website: www.wiley.com/go/Teraoka_StatsThermodynamics

13 Garaga, M.N., Persson, M., Yaghinia, N., and Martinelli, Anna. (2016). Local coordination and dynamics of a protic ammonium based ionic liquid immobilized in nano-porous silica micro-particles probed by Raman and NMR spectroscopy. *Soft Matter* 12: 2583–2592.

14 Ho, Y.S., Porter, J.F., and McKay, G. (2002). Equilibrium isotherm studies for the sorption of divalent metal ions onto peat: copper, nickel and lead single component systems. *Water Air Soil Pollut.* 141: 1–33.

15 Selinger, J.V. and Selinger, R.L.B. (1996). Theory of chiral order in random copolymers. *Phys. Rev. Lett.* 76: 58–61.

16 Okamoto, Y., Suzuki, K., Ohta, K. et al. (1979). Optically active poly(triphenylmethyl methacrylate) with one-handed helical conformation. *J. Am. Chem. Soc.* 101: 4763–4765.

17 Okamoto, N., Mukaida, F., Gu, H. et al. (1996). Molecular weight dependence of the optical rotation of poly((R)-1-deuterio-*n*-hexyl isocyanate) in dilute solution. *Macromolecules* 29: 2878–2884.

18 Gu, H., Nakamura, Y., Sato, T. et al. (1995). Molecular-weight dependence of the optical rotation of poly((R)-2-deuterio-*n*-hexylisocyanate). *Macromolecules* 28: 1016–1024.

19 Cheon, K.S., Selinger, J.V., and Green, M.M. (2000). Designing a helical polymer that reverses its handedness at a selected, continuously variable, temperature. *Angew. Chem.* 112: 1542–1545.

20 Green, M.M., Khatri, C., and Peterson, N.C. (1993). A macromolecular conformational change driven by a minute chiral solvation energy. *J. Am. Chem. Soc.* 115: 4941–4942.

21 Jha, S.K., Cheon, K.-S., Green, M.M., and Selinger, J.V. (1999). Chiral optical properties of a helical polymer synthesized from nearly racemic chiral monomers highly diluted with achiral monomers. *J. Am. Chem. Soc.* 121: 1665–1673.

22 Ute, K., Fukunishi, Y., Jha, S.K. et al. (1999). Dynamic NMR determination of the barrier for interconversion of the left- and right-handed helical conformations in a polyisocyanate. *Macromolecules* 32: 1304–1307.

23 Zimm, B.H. and Bragg, J.K. (1959). Theory of the phase transition between helix and random coil in polypeptide chains. *J. Chem. Phys.* 31: 526–535.

24 Doty, P. and Yang, J.T. (1956). Polypeptides. VII. Poly-γ-benzyl-L-glutamate: the helix-coil transition in solution. *J. Am. Chem. Soc.* 78: 498–500.

25 Engel, J., Liehl, E., and Sorg, C. (1971). Circular dichroism, optical rotatory dispersion and helix $\rightleftarrows$ coil transition of polytyrosine and tyrosine peptides in non-aqueous solvent. *Euro. J. Biochem.* 21: 22–30.

26 Myer, Y.P. (1969). The pH-Induced helix-coil transition of poly-L-lysine and poly-L-glutamic acid and the 238-mμ dichroic band. *Macromolecules* 2: 624.

27 Scholtz, J.M., Marqusee, S., Baldwin, R.L. et al. (1991). Calorimetric determination of the enthalpy change for the α-helix to coil transition of an alanine peptide in water. *Proc. Natl. Acad. Sci. U.S.A.* 99: 2854–2858.

28 Ramachandran, G.N., Ramakrishnan, C., and Sasisekharan, V. (1963). Stereo-chemistry of polypeptide chain configurations. *J. Mol. Biol.* 7: 95–99.

29 Jiang, F., Han, W., and Wu, Y.-D. (2013). The intrinsic conformational features of amino acids from a protein coil library and their applications in force field development. *Phys. Chem. Chem. Phys.* 15: 3413–3428.

30 Teraoka, I. (2002). *Polymer Solutions: An Introduction to Physical Properties.* New York: Wiley.

Index

Statistical Thermodynamics: Basics and Applications to Chemical Systems, First Edition. Iwao Teraoka.
© 2019 John Wiley & Sons, Inc. Published 2019 by John Wiley & Sons, Inc.
Companion website: www.wiley.com/go/Teraoka_StatsThermodynamics